Distributed Security Monitoring and Resilient Cooperative Control

This book systematically addresses the security and resilience of networked agent systems, from sensor networks to general multi-agent systems, against evolving cyber threats.

Distributed Security Monitoring and Resilient Cooperative Control explores techniques for distributed security monitoring to detect sparse attacks on sensors (location-fixed/varying and sparse-fixed/varying), Byzantine node/link attacks, and vulnerability-related attacks in heterogeneous sensor networks. It provides advanced solutions for finite/fixed-time monitoring convergence under nonlinear dynamics. Regarding resilient cooperative control of multi-agent systems, the book tackles the challenge of isolation against various attacks, such as false data injection and covert, non-collusive, and collusive attacks. Practical applications range from resilient smart grids and attack-resistant drones to self-healing infrastructures. Rigorous mathematical frameworks, including graph theory and matrix analysis, establish a principled link between theoretical guarantees and the security operations of real-world networked agent systems. These frameworks offer tools to fortify critical networks against unpredictable, unstructured attacks.

This book will appeal to scholars, engineers, and practitioners involved in systems engineering, control theory, cybersecurity, and artificial intelligence.

Guanghui Wen (senior member, IEEE) is an endowed chair professor at the School of Automation, Southeast University, Nanjing, China. Prof. Wen was the recipient of the National Science Fund for Distinguished Young Scholars and the Australian Research Council Discovery Early Career Researcher Award. His current research interests include autonomous intelligent systems, complex networked systems, distributed control and optimization, resilient control, and distributed reinforcement learning.

Yuezu Lv (senior member, IEEE) is currently an associate professor at the State Key Laboratory of CNS/ATM, Beijing Institute of Technology, Beijing, China. Dr. Lv served as a program committee member for the 2019 International Workshop on Artificial Intelligence and Applications to Intelligent Manufacturing. He was a finalist for the Zhang Si-Ying Outstanding Youth Paper Award in 2015. His research interests include cooperative control of multi-agent systems and adaptive control.

Dan Zhao (member, IEEE) is currently an assistant researcher at the School of Mathematics, Southeast University, Nanjing, China. Dr. Zhao was selected for the ninth 2023 Young Elite Scientists Sponsorship by the China Association for Science and Technology. Her research interests include multi-agent systems, attack detection and isolation, and distributed resilient control.

Xuqiang Lei (member, IEEE) is currently an assistant researcher at the School of Mathematics, Southeast University, Nanjing, China. Dr. Lei was the recipient of the seventh 2023 Chinese Conference on Swarm Intelligence and Cooperative Control (CCSICC) Best Student Paper Award and the Society of Instrument and Control Engineers (SICE) Annual Conference 2023 Student Travel Grant Award. His research interests include cyber-physical systems, distributed secure state monitoring, and resilient consensus control.

Distributed Security Monitoring and Resilient Cooperative Control

Guanghui Wen, Yuezu Lv, Dan Zhao, and Xuqiang Lei

CRC Press
Taylor & Francis Group
Boca Raton London New York

CRC Press is an imprint of the
Taylor & Francis Group, an **informa** business

This book is published with financial support from the National Natural Science Foundation of China through Grant Nos. 62325304, U22B2046, U24A20279, 62306071, 62273045, and U2341213, in part by the Postdoctoral Fellowship Program of CPSF under Grant No. GZC20252266, and Jiangsu Funding Program for Excellent Postdoctoral Talent under Grant No. 2025ZB475.

MATLAB® and Simulink® are trademarks of The MathWorks, Inc. and are used with permission. The MathWorks does not warrant the accuracy of the text or exercises in this book. This book's use or discussion of MATLAB® or Simulink® software or related products does not constitute endorsement or sponsorship by The MathWorks of a particular pedagogical approach or particular use of the MATLAB® and Simulink® software.

First edition published 2026
by CRC Press
2385 NW Executive Center Drive, Suite 320, Boca Raton, FL 33431

and by CRC Press
4 Park Square, Milton Park, Abingdon, Oxon, OX14 4RN

CRC Press is an imprint of Taylor & Francis Group, LLC

ISBN: 978-1-041-03459-9 (hbk)
ISBN: 978-1-041-13463-3 (pbk)
ISBN: 978-1-003-66991-3 (ebk)

DOI: 10.1201/9781003669913

Typeset in Latin Modern font
by KnowledgeWorks Global Ltd.

Publisher's note: This book has been prepared from camera-ready copy provided by the authors.

Contents

Preface

Networked Agent Systems (NASs) comprise multiple interacting agents that cooperate through physical or cyber networks to accomplish global objectives in a distributed manner. Owing to their inherent modularity and distributed structure, NASs possess distinct advantages in robustness, scalability, and flexibility, which make them ideal modeling frameworks for a wide array of engineering systems. Representative examples include sensor networks, multi-unmanned aerial vehicle systems, intelligent transportation networks, power grids, and industrial control systems.

The past decades have witnessed significant progress in the analysis and control of cooperative behaviors for agents in NASs. Yet, with the growing interconnectivity and openness of such systems, they have become increasingly vulnerable to various attacks, including sensor attacks, jamming attacks on communication links and malicious manipulation attacks on agent behaviors. These challenges have motivated a surge of interest in developing security-aware and resilient mechanisms that enable NASs to detect, withstand, and recover from adversarial influences while maintaining their core functionalities. For instance, in a distributed sensor network deployed for environmental monitoring, a few compromised sensors injecting false data can mislead the entire system's estimation and decision-making processes, potentially causing large-scale failures or even safety hazards. Without timely detection and effective mitigation strategies, such attacks can propagate through the network and undermine global coordination goals. Therefore, ensuring the security and resilience of NASs has become a critical issue in both theory and practice.

This book is devoted to the systematic study of distributed security monitoring and resilient cooperative control in NASs under various adversarial scenarios. The focus lies in distributed algorithm design, rigorous performance analysis, and graph-theoretic conditions that ensure resilience against a broad class of attacks. Specifically, Chapter 1 provides an introduction to the key concepts of security monitoring and resilient control, accompanied by an overview of the book. Chapter 2 redefines foundational theories by synthesizing algebraic graph theory, sparse signal processing, and cyber-attack models. The proposed attack propagation dynamics model serves as a cross-domain analytical tool for subsequent chapters. Chapter 3 leverages attack suppression and isolation techniques to develop distributed security state monitoring algorithms against location-fixed and location-varying sparse sensor attacks, overcoming communication bottlenecks in traditional centralized monitoring. Chapter 4 extends to heterogeneous sensor networks, modeling sparse-varying attacks and vulnerability-correlated sparse attacks to further reveal how node heterogeneity impacts monitoring efficacy. Chapter 5 introduces finite- and fixed-time convergence mechanisms to achieve real-time security state estimation. Chapter 6 describes a dual-attack suppression mechanism based on the decoupling of high-dimensional spatial features, which is capable of detecting and responding to Byzantine attacks, covering both node-level and link-level disruptions, and based on which a theoretically guaranteed security state monitoring algorithm is designed. Chapter 7 proposes a novel concept called joint robustness to characterize topological conditions for resilience, resolving resilient consensus under directed switching topologies for both first- and second-order NASs. Chapter 8 proposes a resilient consensus framework based on attack isolation, supported by a new graph property, namely graph isolability, to ensure accurate isolation of compromised agents in general linear NASs. Chapter 9 explores more challenging scenarios involving collusive attacks. Two distributed attack isolation schemes that incorporate inclusive neighbor information and personal information to counter collusive false data injection and covert attacks, respectively, followed by the development of an imperfect isolation-based consensus protocol that balances isolation accuracy and resilience. Chapter 10 derives generalized isolation conditions to handle worse-case collusion, analyzing the interplay between node degree and graph cycle structures to guarantee successful isolation. Finally, Chapter 11 extends the framework to link attack resilience, presenting resilient consensus strategies that can tackle communication link failures or tampering, completing full-attack-surface coverage for NASs.

We gratefully acknowledge the efforts and support of our colleagues and students in advancing our research on the distributed security monitoring and resilient cooperative control. In particular, we thank Professor Guanrong (Ron) Chen of the City University of Hong Kong for his leadership in NASs and for many inspiring discussions on resilient cooperative control. We are also indebted to Professor Wei Xing Zheng at Western Sydney University, Professor Daniel W.C. Ho at the City University of Hong Kong, Professor David (Zhiwei) Gao at Northumbria University, and Professor Tingwen Huang at Shenzhen University of Advanced Technology for their constructive feedback and insightful conversations.

We further thank IEEE, Elsevier, and Taylor & Francis for granting us permission to reuse some materials from our prior publications copyrighted by these publishers. Finally, we gratefully acknowledge the financial support of the National Natural Science Foundation of China (Grants 62325304, U22B2046, U24A20279, 62306071, 62273045, and U2341213), in part by the Postdoctoral Fellowship Program of CPSF under Grant No. GZC20252266, and Jiangsu Funding Program for Excellent Postdoctoral Talent under Grant No. 2025ZB475.

Southeast University

Guanghui Wen, Dan Zhao, Xuqiang Lei
Nanjing, China
September 2025

Beijing Institute of Technology

Yuezu Lv
Beijing, China
September 2025

CHAPTER 1

Introduction

This chapter overviews some recent research progress in distributed security monitoring and resilient control of networked agent systems (NASs) against malicious cyber-attacks. Distributed security monitoring and resilient control of NASs have been a very active research topic in a wide variety of scientific communities, ranging from applied mathematics to physics, engineering, and even sociology. In Section 1.1, NASs, including sensor networks and multi-agent systems, are introduced. In Section 1.2, the research progress on security state monitoring of NASs against sparse sensor attacks is presented. In Section 1.3, the research progress of resilient consensus control of multi-agent systems against malicious attacks is presented. In Section 1.4, we conclude this chapter by presenting some future research directions from our perspective.

1.1 NETWORKED AGENT SYSTEMS

In the modern era of interconnected systems, numerous large-scale cyber-physical infrastructures ranging from industrial sensor grids to autonomous vehicular fleets can be abstracted and modeled as NASs. Broadly speaking, NASs refer to distributed systems composed of numerous agents that interact through physical or cyber communication links [1–9]. Each agent serves as a fundamental entity equipped with sensing, computation, control, and communication capabilities, their collective behavior often gives rise to global functionalities far beyond the capabilities of any individual agent. Typical examples of NASs include wireless sensor networks that monitor environmental or structural conditions [4], coordinated fleets of drones performing aerial surveillance [8], interconnected components of smart grids balancing power loads [6], and even social networks where opinions propagate through interaction [9]. Despite the diversity of applications, all these systems share a common essence: close coordination and information exchange among neighboring entities to achieve system-level objectives.

This cooperation enables NASs to accomplish a wide variety of distributed tasks, such as consensus [10], formation control [11], distributed estimation [12,13], and task allocation. These capabilities make NASs especially attractive in critical applications including autonomous transportation, energy systems, disaster response, and smart manufacturing. Their inherent scalability, modularity, and adaptability allow them

DOI: 10.1201/9781003669913-1

to operate under dynamic conditions and to respond collectively to environmental changes. Among various cooperative behaviors, consensus has emerged as one of the most fundamental problems in NASs research. It refers to the ability of agents to reach agreement on quantities of interest–such as positions, velocities, or decisions–purely through local interactions. Successful consensus not only facilitates coordination but also lays the foundation for higher-level tasks in NASs [14].

Importantly, the collective behavior of NASs is shaped not only by the internal dynamics of each agent but also by the underlying network topology through which they interact. For instance, the same consensus protocol may succeed in a connected network but fail in a sparse or dynamically switching one [15, 16]. Additionally, real-world communication networks are often subject to delays, packet losses, or even malicious attacks–all of which can profoundly influence system behavior. Understanding this interplay between agent dynamics and network structure remains a core challenge in NASs research. This book aims to systematically explore the secure cooperative mechanisms of NASs subject to uncertainties, disruptions, or adversarial interference. In the remainder of this chapter, we will review some existing results on security state monitoring and resilient consensus control against various attacks.

1.2 SECURITY STATE MONITORING OF NASS AGAINST SPARSE ATTACKS

To establish a foundation for addressing security challenges in NASs, the definitions of state monitoring in sensor networks are first formalized.

State monitoring refers to the process of estimating the global or partial state of a static/dynamic system (e.g., temperature in a power grid, traffic flow in a transportation network) through sensor measurements. In sensor networks, agents equipped with localized sensing capabilities collaboratively fuse noisy, heterogeneous, or incomplete data to reconstruct system states in real time. Mathematically, consider a discrete-time NAS which consists of one system to be monitored and m measurement sensors, and the system dynamics are represented by

$$
\begin{aligned}
x(k+1) &= f(k, x(k), u(k)), \\
y_i(k) &= g_i(k, x(k), u(k)), \ i \in [m] = \{1, 2, \ldots, m\},
\end{aligned}
\tag{1.1}
$$

where $x(k) \in \mathbb{R}^n$, $u(k) \in \mathbb{R}^l$, and $y_i(k) \in \mathbb{R}$ are the system state, the control input, and the measurement output from sensor i, $i \in [m]$, respectively. $f(\cdot, \cdot, \cdot) : [k_0, +\infty] \times \mathbb{R}^n \times \mathbb{R}^l \to \mathbb{R}^n$ represents the nonlinear dynamics of the tsystem, and $g_i(\cdot, \cdot, \cdot) : [k_0, +\infty] \times \mathbb{R}^n \times \mathbb{R}^l \to \mathbb{R}$ denotes the nonlinear measurement of i-th sensor. A particular case is the general linear time-invariant (LTI) NAS with the dynamics described by

$$
\begin{aligned}
x(k+1) &= Ax(k) + Bu(k), \\
y_i(k) &= C_i x(k) + D_i u(k), \ i \in [m],
\end{aligned}
\tag{1.2}
$$

where A, B, C_i, and D_i are system matrices with appropriate dimensions. $y(k) =$

$\mathrm{col}\{y_i(k)\} = Cx(k) + Du(k)$ denotes the global measurement output with $C = \mathrm{col}\{C_i\}$ and $D = \mathrm{col}\{D_i\}$, $i \in [m]$.

Consider that there exists an estimated variable $\hat{x}(k)$ as the monitored value of the unknown state $x(k)$. For convenience, throughout this book, we briefly refer to the NASs whose dynamics are described by (1.2) as NASs (1.2).

Definition 1.1 (State Monitoring) *For NASs (1.2), state monitoring is said to be implemented, if there exists the dynamics of $\hat{x}(k)$ such that*

$$\lim_{k \to \infty} \|\hat{x}(k) - x(k)\| = 0. \tag{1.3}$$

In addition, if the observer deployment is distributed and assumed to be one-to-one with the sensors, the estimation $\hat{x}(k)$ should be replaced as $\hat{x}_i(k)$ and $i \in [m]$. In this case, the local state estimation $\hat{x}_i(k)$ serves as the monitored value of state $x(k)$ generated by observer i at time k.

Definition 1.2 (Distributed State Monitoring) *For NASs (1.2), distributed state monitoring is said to be implemented, if for any $i, j \in [m]$,*

$$\lim_{k \to \infty} \|\hat{x}_i(k) - \hat{x}_j(k)\| = 0, \quad \lim_{k \to \infty} \|\hat{x}_i(k) - x(k)\| = 0. \tag{1.4}$$

Modern sensor networks in NASs face growing threats from sparse sensor attacks, where adversaries compromise a small subset of sensors to inject false data or suppress true measurements. These attacks are 'sparse' in that their spatial or temporal impact is localized, making them stealthy and challenging to detect. In response to sparse sensor attacks in NASs, a series of security state monitoring strategies have been developed, including static security state monitoring (based on static batch optimization) and dynamic security state monitoring (based on attack detection and isolation).

Next, a review of methodologies to security state monitoring for such attacks is highlighted, emphasizing both static and dynamic security state estimation algorithms.

A canonical sparse attack can be modeled by replacing the dynamic model in NASs (1.2) as:

$$\begin{aligned} x(k+1) &= Ax(k) + B(u(k) + w^u(k)), \\ y(k) &= Cx(k) + Du(k) + a^y(k), \end{aligned} \tag{1.5}$$

where $w^u(k)$ and $a^y(k)$ denote the r-sparse attacks injected to actuator channels and the s-sparse attacks injected to sensor channels, respectively, i.e., $|supp(w^u(k))| \leq r < l/2$ and $|supp(a^y(k))| \leq s < m/2$. The goal of the attackers is to manipulate the state estimation $\hat{x}(k)$ or $\hat{x}_i(k)$, $i \in [m]$ away from the true values, and further impact the actual system control by strategically choosing non-zero entries in $w^u(k)$ and $a^y(k)$.

To facilitate visual analysis of security state monitoring feasibility, we first formalize the notation for submatrix operations. Let $M \in \mathbb{R}^{n \times l}$ denote an arbitrary real matrix, and $\Gamma_1 \subseteq [n]$ and $\Gamma_2 \subseteq [l]$ be index subsets of rows and columns, respectively. The derived matrix $M_{(\bar{\Gamma}_1, \bar{\Gamma}_2)}$ is obtained by simultaneously removing all rows indexed by $i \in \Gamma_1$ and columns indexed by $j \in \Gamma_2$ from matrix M. Analogously, $M_{(\bar{\Gamma}_1, \cdot)}$ and $M_{(\cdot, \bar{\Gamma}_2)}$ represent the sub-matrices formed by removing solely the rows in Γ_1 or columns in Γ_2, respectively.

The objective of security state monitoring is to construct an estimation $\hat{x}(k)$ that securely and asymptotically converges to the true state $x(k)$. To rigorously characterize this property, it is crucial to introduce the following formal definitions.

Definition 1.3 (s-sparse observable [17]) *NASs (1.5) is said to be s-sparse observable if, for any subset $\Gamma_y \subset \{1, \ldots, m\}$ with $|\Gamma_y| \leq s$, the matrix pair $(C_{(\bar{\Gamma}_y, \cdot)}, A)$ is observable.*

Definition 1.4 (s-redundant strongly observable [18]) *NASs (1.5) is said to be s-redundant strongly observable if, for any subset $\Gamma_y \subset \{1, \ldots, m\}$ with $|\Gamma_y| \leq s$, the matrix quadruple $(A, B, C_{(\bar{\Gamma}_y, \cdot)}, D_{(\bar{\Gamma}_y, \cdot)})$ is strongly observable.*

Definition 1.5 ((r, s)-sparse strongly observable [19]) *NASs (1.5) is said to be (r, s)-sparse strongly observable if, for any subsets $\Gamma_u \subset \{1, \ldots, l\}$ and $\Gamma_y \subset \{1, \ldots, m\}$ with $|\Gamma_u| \leq r$ and $|\Gamma_y| \leq s$, the matrix quadruple $(A, B_{(\cdot, \bar{\Gamma}_u)}, C_{(\bar{\Gamma}_y, \cdot)}, D_{(\bar{\Gamma}_y, \bar{\Gamma}_u)})$ is strongly observable.*

1.2.1 Static security state monitoring

The methodological foundation of static security state monitoring resides in the batch processing of sensor measurements acquired over τ ($1 \leq \tau \leq n$) consecutive moments. By constructing an augmented measurement vector through temporal stacking, the monitoring problem is reformulated as a constrained batch least squares optimization framework. Then, the solution of this problem involves two critical mechanisms: (i) identification of benign communication channels and exclusion of compromised measurements (controls) through combinatorial optimization search and (ii) acceleration of convergence rate through a modified gradient descent algorithm. This framework enables secure monitoring of the system state in the presence of sparse attacks.

The augmented system obtained by accumulating τ measurements is written as follows:

$$Y(k) = \begin{bmatrix} y(k - \tau + 1) \\ \vdots \\ y(k) \end{bmatrix} = \mathcal{O}x(k - \tau + 1) + E_a^y(k) + F(U(k) + E_w^u(k)), \qquad (1.6)$$

where $U(k) = [u^T(k - \tau + 1), \ldots, u^T(k)]^T$, $E_a^y(k) = [a^{yT}(k - \tau + 1), \ldots, a^{yT}(k)]^T$, $E_w^u(k) = [w^{uT}(k - \tau + 1), \ldots, w^{uT}(k)]^T$, and

$$\mathcal{O} = \begin{bmatrix} C \\ CA \\ \vdots \\ CA^{\tau-1} \end{bmatrix}, \qquad F = \begin{bmatrix} D & 0 & \cdots & 0 & 0 \\ CB & D & \cdots & 0 & 0 \\ \vdots & \vdots & \ddots & \vdots & \vdots \\ CA^{\tau-2}B & CA^{\tau-3}B & \cdots & CB & D \end{bmatrix}.$$

The static security monitoring problem is formulated as the sparse optimization [20]:

$$\arg \min_{\hat{x} \in \mathbb{R}^n} \|Y(k) - FU(k) - \mathcal{O}\hat{x}\|_{l_0}, \tag{1.7}$$

where the l_0 norm denotes the number of non-zero elements of the vector, $\hat{x}$ is the state estimation of the static state $x(k - \tau + 1)$.

The optimization formulation (1.7) inherently constitutes a minimization non-convex problem due to the l_0-norm operator. To circumvent this computational intractability, a convex relaxation paradigm is adopted by substituting the original non-convex constraints with generalized l_1/l_r-norm surrogates ($r > 1$), as established in security state monitoring [20, 21]. Particularly, setting the regularization order $r = 2$ induces a computationally tractable quadratic program. This allows recasting the state monitoring task into a compressed sensing framework through the following regularized least-squares minimization:

$$\min_{\Gamma \in \Sigma, \; \hat{x} \in \mathbb{R}^n} \left\| Y_{(\Gamma, \cdot)}(k) - FU(k) - \mathcal{O}_{(\Gamma, \cdot)}\hat{x} \right\|_2, \tag{1.8}$$

where $\Sigma = \{\mathcal{S} \mid \mathcal{S} \subset [m], |\mathcal{S}| = s\}$ and $\mathcal{O}_{(\Gamma, \cdot)}$ denotes the sub-matrix which removes the rows from the matrix $\mathcal{O}$ whose indices do not belong to the set Γ.

The foundational requirements for attack-resilient state reconstruction under various sparsity constraints can be formally characterized as follows:

- Sparse sensor attack resilience: For s-sparse sensor attacks, the state $x(k - \tau + 1)$ is uniquely reconstructible if NASs (1.5) satisfy the $2s$-sparse observability condition, as established through geometric analysis of attack subspaces [17, 20].

- Uncertain control constraints: Under unknown exogenous inputs $u(k)$ with s-sparse sensor attacks, state monitorability demands the stronger $2s$-redundant observability condition [18], where the kernel space of observation mapping maintains dimensional sufficiency to isolate attack vectors.

- Hybrid control and sensor attack scenario: When considering concurrent r-sparse actuator intrusions and s-sparse sensor compromises, the necessary and sufficient condition evolves into a $(2r, 2s)$-sparse strong observability criterion, requiring full column rank of the augmented observability matrix $\mathcal{O}_{(\bar{\Gamma}_y, \bar{\Gamma}_u)}$ [19].

These stratified conditions fundamentally reveal that the capacity of the system to withstand combinatorial attacks is governed by its inherent redundancy degree $m \geq 2s$, quantified through the observability singular value persistence.

The static batch optimization paradigm fundamentally resides in the domain of discrete mathematics, requiring exhaustive enumeration over the power set of $\Gamma \in \Sigma$, where Γ denotes the suspect adversarial channel set. Depending on different attack exclusion algorithms, the computational complexity of the suspicious attack combination search can be C_m^s, C_{m-s}^s, or smaller. However, the essence is still not detached from the base explosion problem of combinatorial optimization, a regular NP-hard problem [22], as formally demonstrated through polynomial-time reduction from set cover problems.

In order to simplify the complexity of security state monitoring algorithms based on static batch optimization as much as possible, the current researches can be divided into three main categories: combinatorial algorithms [23–26], geometric algorithms [19–21,27], and greedy algorithms [17,18,31–33], as detailed in the following.

Combinatorial methodologies employ systematic enumeration techniques for anomaly detection in sensor networks. The work in [23] developed a combinatorial framework based on set cover theory, enabling identification of $2s$-compromised channels within m-dimensional sensor arrays under $2s$-sparse observability constraints. This approach achieves polynomial complexity reduction through search space compression. Subsequent innovations include measurement redundancy compression scheme [30], optimizing data dimensionality while preserving observability. For cyber-physical system protection, [24] introduced an attack certificate mechanism via satisfiability modulo theory, effectively pruning compromised transmission modes. The field further evolves with subspace decomposition techniques [25] and sensor classification strategies [26], collectively establishing multi-layered defense architectures against sparse adversarial attacks.

Geometric optimization paradigm reformulates the intrinsically non-convex l_0 optimization problem through norm relaxation techniques, converting it to tractable l_1/l_r ($r > 1$) convex programming. As demonstrated in previous works [19–21,27–29], the geometric perspective prioritizes state reconstruction fidelity over exact attack localization. Although introducing marginal approximation errors, this convex relaxation enables efficient computation while maintaining ϵ-suboptimal estimation guarantees. The theoretical foundation lies in geometric projection invariance under sparse attack constraints.

Greedy computational frameworks achieve near-optimal estimation through successive refinement processes. Evolutionary refinements include projected gradient descent operators [17], constrained set partitioning approach [32], and consensus voting mechanisms [33]. Further progress incorporated graph-theoretic optimization for attack graph identification with provable termination properties [34]. Building upon these foundations, investigations in [35,36] have systematically evaluated estimation resilience under limited side information availability.

1.2.2 Dynamic security state monitoring

The fundamental challenge in dynamic security state monitoring lies in integrating real-time attack detection and isolation mechanisms into Luenberger-like observers [29]. The distributed security state monitoring framework seeks to mitigate and eliminate the impact of sparse attacks by establishing collaborative mechanisms among neighboring nodes and incorporating attack suppression strategies, thereby enabling consensus-based state estimation. Unlike static batch optimization methods that produce delayed estimations, the dynamic algorithm yields asymptotic state estimates in real time, a particularly advantageous feature for industrial control systems.

Consider the LTI system (1.5), the dynamic distributed security state observer under actuator-free compromise admits the formal representation:

$$\hat{x}_i(k+1) = A\hat{x}_i(k) + Bu(k) + \alpha \sum_{j \in \mathcal{N}_i} a_{ij}(\hat{x}_j(k) - \hat{x}_i(k)) + \delta_i(k)L_i z_i(k), \qquad (1.9)$$

where $\hat{x}_i(k)$ denotes the state estimation generated by the observer i, $\mathcal{N}_i$ denotes the set of neighbors of node i, and a_{ij} is the element of the ith row and jth column of the adjacency matrix $\mathcal{A}$, defined as in subsequent Section 2.3. $z_i(k) = y_i(k) - C_i\hat{x}_i(k)$ represents the residual signal (potentially corrupted by the attack $a_i(k)$), $\delta_i(k) \in [0, 1]$ functions as the adaptive channel trust indicator for sensor i at time k and L_i is the feedback gain matrix of the state estimation to be designed.

The security guarantee hinges on constructing verifiable detection verdicts $\mathcal{P}_i(\cdot)$ that ensure proper isolation factor determination:

$$\delta_i(k) = \begin{cases} 1, & \|\mathcal{P}_i(z_i(k))\| \leq \gamma(k), \\ \frac{\gamma(k)}{\|\mathcal{P}_i(z_i(k))\|}, & \text{otherwise,} \end{cases} \qquad (1.10)$$

where $\mathcal{P}_i(z_i(k))$ defines the residual weighting operator and $\gamma(k)$ constitutes the adaptive security threshold. This saturation operator ensures $\|\delta_i(k)\mathcal{P}_i(z_i(k))\| \leq \gamma(k)$, thereby bounding malicious innovation terms as proved in [37].

Building on the aforementioned theoretical framework, dynamic security state monitoring algorithms based on attack suppression or isolation strategies have evolved rapidly. In the case of sparse attacks with fixed locations, [38] developed a security state observer with the nonlinear architecture, utilizing a switching mechanism between a supervised observer and a set of candidate observers. To address location-varying sparse attacks, [37, 39–41] proposed a distributed framework for resilient asymptotic state estimation, specifically designed to handle location-varying sparse sensor attacks. This framework employs a saturated innovation update algorithm, provided that each sensor satisfies the locally observable condition. In the domain of distributed parameter estimation, [42] introduced a recursive, distributed state estimator that uniformly updates the state using saturated adaptive estimation gains. Within the networked system structure, [43] proposed a novel attack detection threshold design to facilitate distributed security state monitoring by analyzing the convergence of global estimation errors.

In addition, security state monitoring based on dynamic observers extends beyond Luenberger-like observers to include methods involving Kalman filters, which are particularly suited for linear stochastic systems with Gaussian white noise in both dynamics and sensor measurements. By integrating Kalman filtering with resilient state monitoring techniques, this approach ensures optimal system state monitoring. [44] introduced a security state monitoring algorithm for linear dynamic systems affected by Gaussian noise, combining the satisfiability modulo theory with the Kalman filter. In [45], the optimal Kalman filter was decomposed into a weighted sum of state monitoring derived from local Kalman filters, and a convex optimization approach was proposed based on these local estimates to achieve security state monitoring. [46] addressed the distributed state monitoring problem with equality state constraints, enhancing the projection operator and covariance cross-fusion method, and developing a distributed Kalman filter to ensure the consensus and security of state monitoring.

1.3 RESILIENT CONTROL OF NASS AGAINST ADVERSARIAL ATTACKS

Distributed communication networks enable the agents dispersed across different areas to exchange information, thereby facilitating the cooperative tasks among NASs. However, this interconnectivity also introduces significant security challenges, as NASs are vulnerable to various types of attacks. To address these concerns, substantial research efforts have been devoted to enhancing the resilience of NASs, aiming to ensure acceptable performance levels even under adverse conditions.

For distributed resilient consensus against node attacks, various algorithms have been proposed to filter out misleading values, identify reliable reference points, and isolate attacked nodes.

MSR-based algorithms: To achieve resilient consensus, the Mean-Subsequence-Reduced (MSR) algorithms have been proposed to ensure that normal agents update their states using secure information [47–49]. Their core idea is to remove all potentially suspicious values received from neighboring agents during each update step. However, this removal process inevitably reduces graph connectivity, raising a critical challenge of determining the graph conditions under which resilient consensus can still be guaranteed. In [49], the concept of graph robustness was proposed to characterize network redundancy. Based on graph robustness, [49] demonstrated that resilient consensus for first-order NASs under F-total model can be achieved by the Weighted-MSR (W-MSR) algorithm if and only if the graph is $(F+1, F+1)$-robust.

Following [49], the resilient consensus of NASs has been analyzed from various perspectives. For fixed networks, [50] and [51] proposed the Double-integrator Position-based MSR (DP-MSR) algorithm for second-order NASs to reach resilient consensus. Considering the scenario where trusted nodes will not be attacked, [52] proposed the concept of graph robustness with trusted nodes, reducing the topology conditions required for resilient consensus. Resilient consensus for high-order NASs was considered in [53] and [54] by using the MSR-based algorithm separately for each dimension. For time-varying networks, [55] introduced the Sliding Weighted MSR (SW-MSR) algorithm, in which each agent stores and utilizes neighbor values within T time steps, and proved that resilient consensus can be reached for first-order NASs if the graph is $(T, 2F+1)$-robust. A further investigation for second-order NASs is given by [56]. Moreover, [57] discussed the resilient consensus problem of NASs with leaders. Note that [55–57] only provided sufficient conditions for reaching resilient consensus. Reference [58] proposed the concept of joint robustness and provided the necessary and sufficient conditions for first- and second-order NASs with time-varying communication networks to reach resilient consensus using the W-MSR and DP-MSR algorithms.

For resilient consensus under link attacks, by borrowing the concept of the F-local model from [49], a resilient consensus strategy involving trusted links and the MSR-based algorithm was proposed in [59]. And it was shown that resilient consensus could be achieved under arbitrary networks by appropriately assigning the trusted edges. Subsequently, the resilient consensus of double-integrator NASs was investigated in [60] using a similar method.

Safe-point-based algorithms: Different from the approach in [53], where the MSR-based algorithm is applied independently to each dimension of the state vector, the safe-point-based resilient vector consensus algorithms in [61,62] focus on computing a safe point within the convex hull of the vector states of normal neighbors for the normal agents to update toward it. Subsequently, [63,64] utilized the idea of Tverberg partition to compute the safe point. Based on this, an approximate distributed robust convergence algorithm is developed to achieve resilient vector consensus. Furthermore, the notion of centerpoint was introduced in [65,66] to compute the safe point. It has been demonstrated that the centerpoint-based resilient algorithm can enhance the resilience of NASs against adversarial attacks.

Attack-isolation-based algorithms: MSR-based and safe-point-based resilient algorithms can efficiently achieve consensus in integrator-type NASs under specific topology conditions. However, these algorithms cannot be directly applied for general high-order NASs. In such systems, if the compromised agents can be successfully isolated, resilient consensus can be achieved by removing the information of the isolated neighboring agents–provided that the subgraph among normal agents remains connected. Then, the resilient consensus problem for general high-order NASs transforms into an attack isolation problem. [15] proposed a complete attack isolation algorithm, which verifies whether the predefined set coincides with the attack set one by one. To alleviate the computational burden of this previous work, [67] developed a divide-and-conquer method by limiting the hop distance between any two compromised agents.

In [68–70], the attack isolation problem was investigated from the perspective of graph structure. The notion of graph isolability is introduced to ensure the accomplishment of attack isolation. Results indicate that attack isolation can be accomplished for the F-total model if the graph $\mathcal{G}$ is F-isolable [68]. Besides, in cases where multiple attacks collude to avoid detection, compromised agents can be isolated if the graph $\mathcal{G}$ is $(F,7)$-isolable [70]. Furthermore, a similar approach was used to isolate covert attacks in [69].

For resilient consensus under link attacks, [71] developed a distributed observer to detect deception attacks on each link. Based on this, a novel attack isolation algorithm was developed by constructing additional observers via sequentially removing the information associated with one of the links, such that the deception attacks can be accurately isolated. Then, an attack-isolation-based resilient algorithm was derived to achieve consensus of high-order NASs.

1.4 OVERVIEW OF THIS MONOGRAPH

This monograph systematically investigates security monitoring and resilient cooperative control in NASs, integrating foundational theory, algorithmic design, and practical validation. While both the state monitoring and cooperative control paradigms rely on distributed coordination, their focuses diverge:

i) **Objective:** State monitoring prioritizes observability, accurately inferring hidden states from measurements; whereas cooperative control emphasizes controllability, steering agent states to a common value.

ii) **Data Flow:** Monitoring systems aggregate data upward (sensors to fusion center) or distributed fusion data (sensors to distributed observers), while consensus systems propagate data laterally (peer-to-peer agent interactions).

iii) **Adversarial Impact:** Sparse sensor attacks (Section 1.2) corrupt state monitoring by altering measurements, whereas Byzantine attacks (Section 1.3) disrupt consensus by manipulating agent communication or states.

This monograph presents the framework from shallow to deep, from distributed security state monitoring against sparse attacks and resilient cooperative control against Byzantine attacks, respectively, which are organized as follows.

Chapter 1 introduces the core challenges of security monitoring and resilient control in NASs, contextualizing their importance in adversarial cyber-physical environments. It outlines the scope and methodology of the monograph.

Chapter 2 establishes the mathematical backbone, covering matrix theory, algebraic graph theory, stability analysis, and attack models (Sparse, Byzantine, Collusive). These tools underpin subsequent theoretical and algorithmic developments.

Distributed Security Monitoring Frameworks

- Chapter 3 constructs distributed algorithms for homogeneous sensor networks under location-fixed and location-varying sparse sensor attacks, emphasizing scalability and robustness.

- Chapter 4 generalizes the above framework to heterogeneous networks, addressing sparse-varying attacks and vulnerability-related sparse attacks through adaptive optimization.

- Chapter 5 enhances time-critical monitoring via finite-/fixed-time convergence algorithms, balancing speed and precision under sparse adversarial perturbations.

- Chapter 6 extends security monitoring to Byzantine threats, proposing solutions for F-local Byzantine node attacks and Byzantine link attacks.

Resilient Cooperative Control Frameworks

- Chapter 7 resolves resilient consensus for integrator-type NASs with directed switching topologies, introducing a joint robustness graph-theoretic property unifying necessary and sufficient conditions for first-/second-order systems.

- Chapter 8 tackles general linear NASs via graph isolability, enabling perfect attack isolation and zero-mistake consensus frameworks.

- Chapter 9 counters collusive attacks with two isolation schemes: one leveraging inclusive neighbor data against false injections, another using neighbor information to detect covert attacks. An improved consensus framework accommodates imperfect isolation.

- Chapter 10 generalizes graph conditions for collusive attack isolation, optimizing trade-offs between agent neighbor counts and graph cycle sizes.

- Chapter 11 addresses link attacks, completing the exploration of NAS resilience across attack vectors.

Emerging challenges include AI-driven adaptive attacks, quantum-resistant protocols, and large-scale heterogeneous NASs. This monograph advocates for cross-layer defenses, learning-augmented isolation, and rigorous resilience verification to guide next-generation security NASs design.

CHAPTER 2

Preliminaries

This chapter establishes the fundamental knowledge framework for methodological developments in subsequent discussions. Section 2.1 formalizes the mathematical symbols and operational conventions adopted throughout this book. Section 2.2 systematically examines core principles of matrix analysis, encompassing pivotal theorems such as the Kronecker product, Schur decomposition mechanism, Gershgorin's disc theorem, along with supplementary mathematical propositions. Section 2.3 develops the topological characterization of NASs through algebraic graph theory, detailing structural representations including directed/undirected graphs, connectivity robustness metrics, spanning tree configurations, and mathematical formulations through the adjacency matrix and Laplacian matrix. Notably, the robustness quantification framework introduced here constitutes a methodological cornerstone for cyber-security analysis. Section 2.4 systematically categorizes finite-time and fixed-time stability frameworks for NASs with distinct convergence rate characteristics. Section 2.5 rigorously defines heterogeneous cyber-attack paradigms in networked environments. It should be emphasized that the theoretical constructs presented in this chapter form an indispensable analytical toolkit for comprehending subsequent technical developments, with particular pedagogical value for early career researchers transitioning into this specialized domain.

2.1 NOTATIONS

$\in$	belongs to
$\subseteq (\subset)$	subset of or equal to (strict subset of)
$\supseteq (\supset)$	superset of or equal to (strict superset of)
$\nsubseteq$	not a subset of
$\rightarrow$	tends to
$\cup$	union
$\cap$	intersection
$\varnothing$	empty set
$\backslash$	excludes
$\sum$	summation operator
$\prod$	left product

DOI: 10.1201/9781003669913-2

lim	limit
max	maximum
min	minimum
$\forall$	for all
$\exists$	if there exist
∞	infinity
$\dot{f}(t)$	the derivative of f with respect to a variable t
$\mathbb{R}$	the set of real numbers
$\mathbb{R}^n$	the set of n-dimensional column real vectors
$\mathbb{R}^{m \times n}$	the set of $m \times n$-dimensional real matrices
$\mathbb{R}^+$	the set of positive real numbers
$\mathbb{Z}$	the set of integers
$\mathbb{N}$	the set of natural numbers
$\mathbb{N}^+$	the set of positive natural numbers
$\mathbb{C}$	the set of complex numbers
$[n]$	the finite natural numbers subset $\{1, \cdots, n\}$
$\mathbf{1}_n$	the n-dimensional column vector with each element being 1
$\mathbf{0}_n$	the n-dimensional column vector with each element being 0
$\mathbf{0}$	the zero vector (or matrix) with a compatible dimension
I_n	the $n \times n$-dimensional identity matrix
x^T	the transpose of a real vector x
$\|x\|$	the Euclidean (or 2-) norm of a real vector x
$\|x\|_{l_0}$	the number of nonzero elements of a vector x
$\overline{\|x\|_{l_0}}$	the number of zero elements of a vector x
$\|x\|_p$	the p-norm of a real vector x
$\mathrm{diag}\{a_i\},\ i \in [n]$	a diagonal matrix with diagonal entries a_1 to a_n
$\mathrm{col}\{A_i\},\ i \in [N]$	a block stacking matrix formed by entries A_1 to A_N
$\|A\|$	the induced 2-norm of a real matrix A
$\|A\|_{l_0}$	the number of nonzero column vectors of a real matrix A
$\overline{\|A\|_{l_0}}$	the number of zero column vectors of a real matrix A
$A^\dagger$	the Moore-Penrose pseudoinverse of the matrix A
$A > B$	the matrix $A - B$ is positive definite
$A \geq B$	the matrix $A - B$ is nonnegative definite
$\lambda_i(A)$	the i-th eigenvalue of the matrix A
$\lambda_{\max}(A)$	the largest eigenvalue of the matrix A
$\lambda_{\min}(A)$	the smallest eigenvalue of the matrix A
$rank(A)$	the rank of the matrix A
$\mathcal{R}(A)$	the range space of the matrix A
$\mathcal{N}(A)$	the null space of the matrix A
$A \otimes B$	the Kronecker product of matrices A and B
$\bar{\mathcal{S}}$	the complementary set of a subset $S \subseteq [N]$
$A_{\mathcal{S}}$	the matrix which removes the rows (or sub-matrix blocks) from the (global) matrix A whose indices do not belong to the set $\mathcal{S}$
$A_{\bar{\mathcal{S}}}$	the matrix which removes the rows (or sub-matrix blocks) from the (global) matrix A whose indices belong to the set $\mathcal{S}$

sin	the sine function
cos	the cosine function
sgn	the signum function
$\lvert \cdot \rvert$	the cardinality of a set or the module of a complex number
$\lceil \cdot \rceil$	the ceiling function (i.e., the smallest integer greater than or equal to the argument)
$\lfloor \cdot \rfloor$	the floor function (i.e., the greatest integer less than or equal to the argument)
$O(\cdot)$	the asymptotic upper bound of a function
LTI	linear time-invariant
LMIs	linear matrix inequalities
NASs	networked agent systems
SSM	security state monitoring
MSR	Mean-Subsequence-Reduced
s.t.	such that
a.s.	almost surely
$x \circ y$	the Hadamard product of the vector x with the vector y
x_i	the i-th dimensional component of the vector x
supp (x), $x \in \mathbb{R}^n$	the index corresponding to nonzero elements in the vector x, i.e., $\{i \in [n] \mid x_i \neq 0\}$
$\mathbf{1}_\Delta$	the indicator function, which equals 1 if the event Δ occurs, and 0 otherwise

2.2 MATRIX THEORY

Without going into the underlying matrix theory, this section introduces some lemmas that will be applied in subsequent chapters of the book.

Lemma 2.1 (Gershgorin's disc theorem [72]) *Let* $B = [b_{ij}] \in \mathbb{R}^{N \times N}$ *and* $R_i'(B) = \sum_{j=1, j \neq i}^{N} |b_{ij}|$, $i \in [N]$. *Then all eigenvalues of the matrix* B *are located in the union of* N *discs*

$$\bigcup_{i=1}^{N} \{z \in \mathbb{C} : |z - b_{ii}| \leq R_i'(B)\}.$$

Furthermore, if a union of k *of these* N *discs forms a connected region that is disjoint from the remaining* $N - k$ *discs, then there are exactly* k *eigenvalues of* B *in this region.*

Lemma 2.2 *[72] Suppose that matrix* $B = [b_{ij}] \in \mathbb{R}^{N \times N}$ *has* $b_{ij} \leq 0$ *for all* $i \neq j$, $i, j \in [N]$. *Then, the following statements are equivalent:*

(1) *All eigenvalues have positive real parts;*

(2) B^{-1} *exists and* B^{-1} *is nonnegative;*

(3) *There exists a diagonal matrix $\Phi = \mathrm{diag}\{\phi_1, \ldots, \phi_N\}$ with $\phi_i > 0$, $i \in [N]$, such that $B^T\Phi + \Phi B > 0$;*

(4) *B is a nonsingular M-matrix;*

where B^{-1} is said to be nonnegative if all its entries are nonnegative.

Lemma 2.3 (Lemma 3.5 **of [73])** *If the matrix $A \in \mathbb{R}^{N \times N}$ is symmetric, then all the eigenvalues of A are real.*

Lemma 2.4 (Lemma 3.9 **of [73])** *If $A \in \mathbb{R}^{N \times N}$ is positive semidefinite, then there exists a unique positive semidefinite matrix $B \in \mathbb{R}^{N \times N}$ such that $B^2 = A$. The matrix B is called the square root of A and is denoted by $A^{\frac{1}{2}}$.*

Lemma 2.5 (Theorem 4.2.2 **of [72])** *Let $A \in \mathbb{R}^{N \times N}$ be symmetric. Then,*

$$\lambda_{\min}(A) = \min_{x \neq \mathbf{0}_N} \frac{x^T A x}{x^T x} = \min_{x^T x = 1} \frac{x^T A x}{x^T x},$$

$$\lambda_{\max}(A) = \max_{x \neq \mathbf{0}_N} \frac{x^T A x}{x^T x} = \max_{x^T x = 1} \frac{x^T A x}{x^T x},$$

$$\lambda_{\min}(A)x^T x \leq x^T A x \leq \lambda_{\max}(A)x^T x, \quad \forall x \in \mathbb{R}^N.$$

Lemma 2.6 (Kronecker product [74]) *For matrices A, B, C, and D with appropriate dimensions, one has*

(1) $(A \otimes B)^T = A^T \otimes B^T$;

(2) $A \otimes (B + C) = A \otimes B + A \otimes C$;

(3) $(A \otimes B)(C \otimes D) = AC \otimes BD$;

(4) $(A \otimes B)^{-1} = A^{-1} \otimes B^{-1}$, *for any given invertible matrices A and B.*

Lemma 2.7 (Schur complement lemma [75]) *Suppose $A = A^T \in \mathbb{R}^{n \times n}$, $B = B^T \in \mathbb{R}^{m \times m}$, and $C \in \mathbb{R}^{n \times m}$. The condition*

$$\begin{bmatrix} A & C \\ C^T & B \end{bmatrix} > 0$$

is equivalent to any one of the following conditions:

(1) $B > 0$ and $A - CB^{-1}C^T > 0$;

(2) $A > 0$ and $B - C^T A^{-1} C > 0$.

Definition 2.1 *[76] A real-valued function $\alpha : [0, +\infty) \mapsto [0, +\infty)$ is said to be of class $\mathcal{K}$ if it is continuous, strictly increasing, and $\alpha(0) = 0$. If, in addition, α is unbounded, then it is said to be of class $\mathcal{K}_\infty$. A real-valued function $\beta : [0, +\infty) \times [0, +\infty) \mapsto [0, +\infty)$ is said to be of class $\mathcal{KL}$ if $\beta(\cdot, t)$ is of class $\mathcal{K}$ for each fixed $t \geq 0$, and $\beta(r, t)$ is decreasing to zero as $t \to \infty$ for each fixed $r \geq 0$.*

We shall write $\alpha \in \mathcal{K}_\infty$ and $\beta \in \mathcal{KL}$ to indicate that α is a class $\mathcal{K}_\infty$ function and β is a class $\mathcal{KL}$ function, respectively.

2.3 ALGEBRAIC GRAPH THEORY

Suppose a NAS consists of N agents that that interact with one another through a communication network, a sensing network, or a combination of both. It is natural to model the interactions among the N agents using an undirected or a directed graph $\mathcal{G} = (\mathcal{V}, \mathcal{E})$. In a directed graph $\mathcal{G} = (\mathcal{V}, \mathcal{E})$, $\mathcal{V} = [N]$ denotes the set of nodes, where the N nodes are labeled as node $1, 2, \ldots, N$ without loss generality. The edge (link) set $\mathcal{E} \subseteq \mathcal{V} \times \mathcal{V}$ represents the interactions among these nodes, where $(j, i) \in \mathcal{E}$ if and only if node i can receive information from node j. When $(j, i) \in \mathcal{E}$, node j is said to be a neighbor of node i. For a node i, $\mathcal{N}_i^1 = \{j \in \mathcal{V}|(j, i) \in \mathcal{E}\}$ denotes its one-hop neighbor set, $\mathcal{N}_i^2 = \{k \in \mathcal{V}|(j, i) \in \mathcal{E}, (k, j) \in \mathcal{E}\}$ denotes its two-hop neighbor set, $\mathcal{N}_i^{>l} = \{m \in \mathcal{V}|(j, i) \in \mathcal{E}, (k, j) \in \mathcal{E}, \ldots\}$ denotes its neighbor set more than l-hop. Note that node j is called the m-hop neighbor of node i if $j \in \mathcal{N}_i^m$, $j \in \mathcal{N}_i^l$, and $m < l$ holds. The inclusive neighbor set and the two-hop inclusive neighbor set of node i are defined as $\mathcal{J}_i = \mathcal{N}_i^1 \cup \{i\}$ and $\mathcal{J}_i^* = \mathcal{N}_i^1 \cup \mathcal{N}_i^2 \cup \{i\}$, respectively. Here, $\mathcal{G}_i(\mathcal{N}_i^l, \mathcal{E}_i)$ with $\mathcal{E}_i = \{(j, i) \in \mathcal{E}, j \in \mathcal{N}_i\}$ is called the subgraph induced by $\mathcal{N}_i^l$, $\mathcal{G}_i(\mathcal{J}_i, \mathcal{E}_i)$ with $\mathcal{E}_i = \{(j, i) \in \mathcal{E}, j \in \mathcal{J}_i\}$ is called the subgraph induced by $\mathcal{J}_i$, and $\mathcal{G}_i(\mathcal{J}_i^*, \mathcal{E}_i)$ with $\mathcal{E}_i = \{(j, i) \in \mathcal{E}, (k, j) \in \mathcal{E}, j, k \in \mathcal{J}_i^*\}$ is the one induced by $\mathcal{J}_i^*$. In this book, the one-hop neighbor set of node i is denoted by $\mathcal{N}_i$ instead of $\mathcal{N}_i^1$, and $j \in \mathcal{N}_i$ is called the neighbor of i if there is no ambiguity. For simplicity, $i^1, i^2, \ldots, i^{|\mathcal{N}_i|}$ are used to represent the one-hop neighbors of node i. The set of incoming edges of node i is denoted by $\mathcal{K}_i = \{(j, i)|j \in \mathcal{N}_i\}$. For linguistic convenience, this book does not distinguish between the terms 'system' and 'graph,' 'agent' and 'node,' or 'link' and 'edge.'

If there exists a sequence of distinct nodes $i_1, i_2, \ldots, i_m$ such that $(i, i_1), (i_1, i_2), \ldots,$ $(i_{m-1}, i_m), (i_m, k) \in \mathcal{E}$, then it is said that node i has a directed path to node k, or node k is reachable from node i. The path length between nodes i and j, denoted by $|l_{ij}|$, refers to the minimum number of nodes in a path from i to j, excluding nodes i and j themselves. Particularly, $|l_{ij}| = 0$ if nodes i and j are neighbors. Similar to the definition of a path between two nodes, edge (h, g) is said to have a directed path to edge (j, i) if there exists a finite sequence of directed edges, $(g, h_1), (h_1, h_2), \ldots, (h_k, j)$, with distinct nodes h_m, $m = 1, 2, \ldots, k$. Let $|l_{gh \rightarrow ij}|$ denote the shortest path length between edges (h, g) and (j, i), defined as the minimum number of edges on a path between (h, g) and (j, i), excluding the edges (h, g) and (j, i) themselves. In this sense, one has $|l_{gi \rightarrow ij}| = 0$ if i is the common node of two edges (g, i) and (i, j), and $|l_{gh \rightarrow ji}| = 1$ if i or j is the neighbor of g or h. In graph $\mathcal{G}$, a cycle is defined as a path that starts and ends at the same node. $\mathcal{C}_k$ denotes the cycle consisting of k nodes, and $\mathcal{C}_{>k}$, $\mathcal{C}_{\geq k}$, $\mathcal{C}_{<k}$, and $\mathcal{C}_{\leq k}$ represent the cycles containing more than, at least, less than, at most k nodes, respectively. A cycle graph is a graph containing a cycle through all nodes. The graph $\mathcal{G}$ is *strongly connected* if each node has at least one directed path to any other nodes. More generally, if there exists a node, called the root, which has at least one directed path to any other nodes, $\mathcal{G}$ is said to contain a directed *spanning tree*. Particularly, if there are more than one root node in $\mathcal{G}$, the root set of all the root nodes is denoted by $R_{\mathcal{G}}$. Denote by a_{ij} the weight of the edge (j, i), $i, j = 1, 2, \ldots, N$. It is obviously that $a_{ij} \geq 0$, where $a_{ij} > 0$ if and only

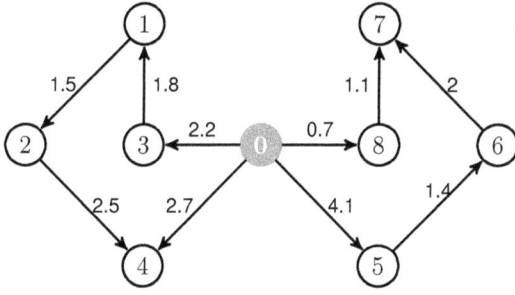

Figure 2.1 A directed graph $\mathcal{G}$ consists of 9 nodes, where the numbers around the edges represent the weights. Although $\mathcal{G}$ is not strongly connected because nodes 1–8 do not have directed paths to all other nodes, $\mathcal{G}$ contains a directed spanning tree with node 0 being the root.

if $(j,i) \in \mathcal{E}$ and $a_{ij} = 0$ otherwise. In addition, it is assumed that $a_{ii} = 0$; that is, self-loops are not considered. $\mathcal{G}$ is called an undirected graph if $(i,j) \in \mathcal{E}$ whenever $(j,i) \in \mathcal{E}$ and $a_{ij} = a_{ji}$. An undirected graph is *connected* if there exists at least one undirected path between each pair of distinct nodes. For undirected graphs, the existence of an undirected spanning tree is equivalent to being connected. However, for directed graphs, the existence of a directed spanning tree is a weaker condition than being strongly connected. Let any of the connected directed subgraph $\mathcal{G}^i \subset \mathcal{G}$ be a connected component of graph $\mathcal{G}$. Please see Fig. 2.1 for a directed graph that is not strongly connected but contains a directed spanning tree.

Let $\mathcal{A} = [a_{ij}] \in \mathbb{R}^{N \times N}$ be the adjacency matrix of the graph $\mathcal{G}$. Then the Laplacian matrix $\mathcal{L} = [l_{ij}] \in \mathbb{R}^{N \times N}$ is defined as

$$l_{ij} = \begin{cases} \sum_{j \in \mathcal{N}_i} a_{ij}, & \text{if } j = i, \\ -a_{ij}, & \text{if } j \neq i, \end{cases} \quad i,j \in \mathcal{V}. \tag{2.1}$$

If $\mathcal{G}$ is undirected, $\mathcal{L}$ is symmetric since $a_{ij} = a_{ji}$. However, when $\mathcal{G}$ is directed, $\mathcal{L}$ is not necessarily symmetric. No matter undirected or direct graphs, $\mathcal{L}$ has zero row sum. Hence, 0 is an eigenvalue of $\mathcal{L}$ with an associated eigenvector $\mathbf{1}_N$. Note that $\mathcal{L}$ is diagonally dominant and has nonnegative diagonal entries, and denote $\lambda_1(\mathcal{L}) \leq \cdots \leq \lambda_N(\mathcal{L})$ as eigenvalues of $\mathcal{L}$. According to Gershgorin's disc theorem (see Lemma 2.1), all nonzero eigenvalues of $\mathcal{L}$ have positive real parts if $\mathcal{G}$ are directed. If $\mathcal{G}$ is undirected, all nonzero eigenvalues of $\mathcal{L}$ are positive since $\mathcal{L}$ is symmetric, which indicates that all the eigenvalues are real.

Lemma 2.8 *[78] The Laplacian matrix $\mathcal{L}$ has a simple zero eigenvalue, and all other eigenvalues have positive real parts (respectively, are positive) if and only if $\mathcal{G}$ has a directed spanning tree (respectively, is connected).*

Remark 2.1 *Let $x = [x_1, \ldots, x_N]^T$. It is not difficult to obtain that $\mathcal{L}x = \left[\sum_{j=1}^N a_{1j}(x_1 - x_j), \ldots, \sum_{j=1}^N a_{Nj}(x_N - x_j) \right]^T$. If $\mathcal{G}$ is undirected, then $x^T \mathcal{L} x = \frac{1}{2} \sum_{i,j=1}^N a_{ij}(x_i - x_j)^2$. Furthermore, when $\mathcal{G}$ is connected, it follows from Lemma 2.8 that $x^T \mathcal{L} x = 0$ if and only if $x_i = x_j$ for all $i,j = 1, \ldots, N$.*

Let $\mathcal{G}$ be a directed graph with a directed spanning tree. Without loss of generality, assuming node 1 to be the root. Then $\mathcal{L}$ can be rewritten as

$$\mathcal{L} = \begin{bmatrix} 0 & \mathbf{0}_{N-1}^T \\ \mathbf{P} & \bar{\mathcal{L}} \end{bmatrix}, \quad \bar{\mathcal{L}} = \begin{bmatrix} \sum_{j \in \mathcal{N}_2} a_{2j} & -a_{23} & \cdots & -a_{2N} \\ -a_{32} & \sum_{j \in \mathcal{N}_3} a_{3j} & \cdots & -a_{3N} \\ \vdots & \vdots & \ddots & \vdots \\ -a_{N2} & -a_{N3} & \cdots & \sum_{j \in \mathcal{N}_N} a_{Nj} \end{bmatrix}, \quad (2.2)$$

where $\mathbf{P} = -[a_{21}, \ldots, a_{N1}]^T$. Then Lemma 2.8 implies that all the eigenvalues of $\bar{\mathcal{L}}$ have positive real parts. On the other hand, all the non-diagonal entries of $\bar{\mathcal{L}}$ are non-positive. Then it follows from Lemma 2.2 that $\bar{\mathcal{L}}$ is a non-singular M-matrix and there exists a diagonal matrix $\Phi = \text{diag}\{\phi_2, \ldots, \phi_N\}$ such that $\bar{\mathcal{L}}^T \Phi + \Phi \bar{\mathcal{L}} > 0$, where $\phi_i > 0$, $i = 2, \ldots, N$. Unfortunately, Lemma 2.2 has not presented any methods for selecting appropriate diagonal entries ϕ_i, which may be used for controller design.

Lemma 2.9 *[13] If $\mathcal{G}$ contains a directed spanning tree, then there exists a positive definite diagonal matrix $\Phi = \text{diag}\{\phi_2, \ldots, \phi_N\}$ such that $\bar{\mathcal{L}}^T \Phi + \Phi \bar{\mathcal{L}} > 0$. One such $\phi = [\phi_2, \ldots, \phi_N]^T$ can be obtained by solving the matrix equation $\bar{\mathcal{L}}^T \phi = \mathbf{1}_{N-1}$.*

Lemma 2.10 *[79] For an undirected and connected graph $\mathcal{G}$ with the Laplacian matrix $\mathcal{L}$, if there exists a vector $x \in \mathbb{R}^N$ satisfying $\mathbf{1}_N^T x = 0$, then one has*

$$x^T \mathcal{L} x \geq \lambda_2(\mathcal{L}) x^T x.$$

Lemma 2.11 *[80] For the Laplacian matrix $\mathcal{L}$ of the undirected and connected graph $\mathcal{G}$, there exists an orthogonal matrix $U = [U_1 \ U_2]$ with $U_2 = \frac{1}{\sqrt{N}} \mathbf{1}_N$ such that*

$$U^T \mathcal{L} U = J = \begin{bmatrix} J_1 & 0 \\ 0 & 0 \end{bmatrix}, \quad U_1^T U_2 = 0_{N-1}, \ U_2^T U_1 = 0_{N-1}^T,$$

with J_1 being a positive definite Jordan block matrix with eigenvalues as $\lambda_2(\mathcal{L}), \ldots, \lambda_N(\mathcal{L})$. Furthermore, there holds:

$$U^T \tilde{I}_N U = U^T U - U^T O_N U = \begin{bmatrix} I_{N-1} & 0 \\ 0 & 0 \end{bmatrix},$$

where $\tilde{I}_N = I_N - \frac{1}{N} \mathbf{1}_N \mathbf{1}_N^T$.

Definition 2.2 (r-reachable set [49]) *A nonempty subset $\mathcal{S} \subset \mathcal{V}$ is r-reachable ($r \in \mathbb{N}$), if there exists a node $i \in \mathcal{S}$ that contains at least r in-neighbors outside $\mathcal{S}$, i.e., $|\mathcal{N}_i \backslash \mathcal{S}| \geq r$.*

Definition 2.3 (r-robust graph [49]) *For $r \in \mathbb{N}$, a graph $\mathcal{G}$ is said to be r-robust if for all pairs of disjoint nonempty subsets $\mathcal{S}_1$, $\mathcal{S}_2 \subset \mathcal{V}$, at least one of $\mathcal{S}_1$ or $\mathcal{S}_2$ is r-reachable.*

Definition 2.4 (Strongly r-robust graph [77]) *For $r \in \mathbb{N}$, a directed graph $\mathcal{G}$ is said to be strongly r-robust if for any nonempty subset $\mathcal{S} \subset \mathcal{V}$, either $\mathcal{S}$ is r-reachable or there exists $i \in \mathcal{S}$ such that $\mathcal{V} \backslash \mathcal{S} \subseteq \mathcal{N}_i$.*

Definition 2.5 ((r, s)-robust graph [49]) *For $r \in \mathbb{N}$, a directed graph $\mathcal{G}$ is said to be (r, s)-robust if for every pair of nonempty, disjoint subsets $\mathcal{S}_k$ and $\mathcal{X}^r_{\mathcal{S}_k} = \{i \in \mathcal{S}_k : |\mathcal{N}_i \backslash \mathcal{S}_k| \geq r\}$ for $k \in \{1, 2\}$, at least one of the following conditions holds:*

(i) $|\mathcal{X}^r_{\mathcal{S}_1}| = |\mathcal{S}_1|$;

(ii) $|\mathcal{X}^r_{\mathcal{S}_2}| = |\mathcal{S}_2|$;

(iii) $|\mathcal{X}^r_{\mathcal{S}_1}| + |\mathcal{X}^r_{\mathcal{S}_2}| \geq s$.

Lemma 2.12 *Suppose a graph $\mathcal{G}$ is r-robust. Let $\mathcal{G}'$ be a graph obtained by removing $r-1$ or fewer incoming edges from each node in $\mathcal{G}$. Then $\mathcal{G}'$ contains a spanning tree rooted at some nodes.*

Lemma 2.13 *If $\mathcal{G}'$ is induced from a strongly r-robust graph $\mathcal{G}$ by arbitrarily removing at most $r-1$ incoming edges from each node, then subgraph $\mathcal{G}'$ contains a spanning tree with its root node belonging to $R_\mathcal{G}$.*

More detailed information about graph robustness can be found in [81–86], etc.

2.4 STABILITY WITH DIFFERENT CONVERGENCE RATES

This section introduces the definition and some important correlation lemmas of fast consensus convergence. For a more detailed discussion, we refer the reader to Chapter 2 in [87].

Definition 2.6 *[87–89] For the following NASs:*

$$\dot{x}_i(t) = f(t, x_i) + u_i(t), \qquad (2.3)$$

if under a suitable control protocol $u_i(t)$, $i \in \mathcal{V}$ satisfying for all $i \in \mathcal{V}$:

$$\begin{cases} \lim_{t \to T_x} \|x_i(t)\| = 0, \\ \|x_i(t)\| = 0, \quad \forall t \geq T_x. \end{cases}$$

Then, the NASs (2.3) is said to achieve

a) finite-time consensus stable if $T_x = T(\sigma) > 0$ is a settling time function with respect to the initial norm $\sigma = \max_{i \in \mathcal{V}} \|x_i(0)\|$;

b) fixed-time consensus stable if $T_x = T_{max} > 0$ is a fixed constant independent of the initial value $x_i(0)$.

Lemma 2.14 *[90] The following norm inequality property holds:*

$$\left(\sum_{i=1}^{N} y_i\right)^r \leq \sum_{i=1}^{N} y_i^r, \quad \left(\sum_{i=1}^{N} y_i\right)^h \leq N^{h-1}\sum_{i=1}^{N} y_i^h,$$

where $y_i \geq 0$ and $0 < r < 1 < h$.

Lemma 2.15 (Finite-time stable [91]) *Consider a scalar differential system as:*

$$\dot{y}(t) = -ay^p(t) - by(t), \tag{2.4}$$

where $a > 0$, $0 < p < 1$, $b \geq 0$, and $y(0) > 0$. The equilibrium $y = 0$ for (2.4) is finite-time stable with the settling time

$$T(y(0)) \leq \begin{cases} \frac{1}{b(1-p)} \ln\left(1 + \frac{b}{a}y^{1-p}(0)\right), & \text{if } b > 0, \\ \frac{1}{a(1-p)} y^{1-p}(0), & \text{if } b = 0. \end{cases}$$

Lemma 2.16 (Fixed-time stable [92]) *For the differential system of the following form*

$$\dot{y}(t) = -(ay^p(t) + by^q(t))^k,$$

where $a, b, k > 0$, $0 < pk < 1 < qk$, and $y(0) > 0$. Its equilibrium $y = 0$ is fixed-time stable with the settling time $T_{max} \leq \frac{1}{a^k(1-pk)} + \frac{1}{b^k(qk-1)}$.

2.5 DEFINITIONS OF ADVERSARIAL ATTACK MODEL

This section presents some of the definitions of attack model examined in this book.

Definition 2.7 (location-fixed s-sparse attacks [20]) *For a set of the attack sequence streams $a(k)$ that depend on time k, the attack is said to be location-fixed s-sparse attacks if it satisfies that the number of dimensions for which there exist non-zero elements does not exceed s, i.e. $|\mathcal{V}_A| \leq s$ with time-invariant index set $\mathcal{V}_A = \{i \mid a_i(k) \neq 0, \ \exists k \geq 0\}$.*

Definition 2.8 (location-varying s-sparse attacks [37, 43]) *For a set of attack sequence streams $a(k)$ that depend on time k, the attack is said to be location-varying s-sparse if it is satisfied that there are no more than s non-zero elements of $a(k)$ at any moment k, i.e., $|\mathcal{V}_{A,k}| \leq s$ with $\mathcal{V}_{A,k} = \{i \mid a_i(k) \neq 0\}$.*

Definition 2.9 (s-sparse observable [17]) *It follows from Definition 1.3, for system matrices A, C with appropriate dimensions, the matrix tuple (A, C) is said to be s-sparse observable if for any subset $\Gamma \subseteq [m]$ with $|\Gamma| = m - s$ and m is the number of columns (entries) of the (block-stacked) matrix C, there is the matrix tuple (A, C_Γ) that is observable.*

Definition 2.10 (s-sparse redundant [42]) *For matrices A, C with appropriate dimensions, the matrix tuple (A, C) is said to be s-sparse redundant if the condition $\lambda_{min}(G) > s$ holds with the matrix $G \triangleq C^T C$.*

Remark 2.2 *As reported in [42], the s-sparse redundancy is a sufficient condition for the s-sparse observability. Specifically, if $\cup_{i \in [m]}\{C_i\}$ is orthogonal, then $\lambda_{min}(G) > s$ if and only if the matrix C is one-step s-sparse observable (or equally could be called s-sparse full column-rank i.e., the matrix C_Γ is full column-rank for any subset $\Gamma \subseteq [m]$ with $|\Gamma| = m - s$ and m is the number of columns (entries) of the (block-stacked) matrix C. For further elaboration and proof, refer to [42].*

Recall that security state monitoring (SSM) works in [17, 20, 33, 37, 40], for a NAS with the following dynamics:

$$x(k + 1) = Ax(k) + Bu(k), \tag{2.5a}$$
$$y(k) = Cx(k) + a(k). \tag{2.5b}$$

It is important to note that the NAS (2.5), under location-fixed s-sparse attacks, a necessary and sufficient condition for the existence of an effective SSM algorithm is that the matrix tuple (A, C) is $2s$-sparse observable. This conclusion is mainly drawn from the analysis under a specific centralized SSM framework, which means that the defense is required to be able to exploit specific global system information. Whereas from a distributed perspective, in the absence of global measurement information, the sufficiency condition of $2s$-sparse observable has been extended in [37, 40, 42] to being $2s$-sparse redundant. In addition, this extension removes the location-fixed constraint of sparse attacks and ensures the security and feasibility of the SSM against location-varying s-sparse sensor attacks.

Definition 2.11 ((f, s)-sparse-varying attacks [93]) *For a set of attack sequence streams $a(k)$ that depend on time k, the attack is said to be (f, s)-sparse-varying if it is satisfied that the average number of non-zero elements of the attack does not exceed s at any consecutive f moments, i.e. $\sum_{j=0}^{f-1} |\mathcal{V}_{A,k-j}| \leq fs$, $\forall k \geq f$ with $\mathcal{V}_{A,k} = \{i \mid u_i(k) \neq 0\}$.*

Remark 2.3 *Different from the (location-fixed/varying) s-sparse attack, an (f, s)-sparse-varying attack does not restrict the number of attacks at each time instant to be strictly less than s. The attacker could launch a large number of attacks in a few moments, meeting the definition of sparse attacks over a given time scale. In particular, a $(1, s)$-sparse-varying attack degenerates to a location-varying s-sparse attack.*

Consider a heterogeneous sensor network model with node set $\mathcal{V}$ to construct a vulnerability metric for each heterogeneous sensor. Namely, there exists a unique mapping relation can be constructed from the sensor set $\mathcal{V}$ to the weight metric set $\Upsilon = \{\rho_1, \ldots, \rho_N\}$: $\Lambda(i) = \rho_i$ with $\rho_i \in \mathbb{R}^+$. In which, as ρ_i approaches closer to 0, it indicates that the adversary can more easily launch a successful attack on sensor i.

Definition 2.12 (Vulnerability-related s-sparse attacks) *For a set of attack sequence streams $a(k)$ that depend on time k, the attack is said to be vulnerability-related s-sparse if it is satisfied that the sum of the vulnerability metric of non-zero attacks at each time k does not exceed $\frac{s}{N}\sum_{i=1}^{N}\rho_i$, i.e.,*

$$\sum_{i\in\mathcal{V}_{A,t}}\rho_i \leq \frac{s}{N}\sum_{i=1}^{N}\rho_i.$$

Considering the stages at which adversarial attacks are launched, they can be categorized into the following two types.

Definition 2.13 (Malicious Node [49]) *A node i is said to be malicious if it transmits its true state value to all of its neighbors at every time instant but deviates from the prescribed control protocol at certain time instants.*

Definition 2.14 (Byzantine Node [49]) *A node i is said to be Byzantine if, at certain time instants, it sends different values to its neighbors and/or deviates from the prescribed control protocol.*

Consider that attacks compromise some nodes, and then the node set $\mathcal{V}$ can be divided into two subsets, including the compromised node set $\mathcal{V}_A$ and the regular one $\mathcal{V}_R = \mathcal{V}\backslash\mathcal{V}_A$. Analogously, if the attacks compromise some edges, the edge set $\mathcal{E}$ can be divided into two subsets: the compromised edge set $\mathcal{E}_A$ and the regular one $\mathcal{E}_R = \mathcal{E}\backslash\mathcal{E}_A$. To account for the upper bound on the number of compromised subsystems, the following definitions are provided.

Definition 2.15 (F-total node attack model [49]) *The NAS is said to be F-total attacked if the compromised node set $\mathcal{V}_A$ contains at most F nodes, i.e., $|\mathcal{V}_A| \leq F$.*

Definition 2.16 (F-local node attack model [49]) *The NAS is said to be F-local attacked if the compromised node set $\mathcal{V}_A$ contains at most F nodes in neighboring set $\mathcal{N}_i$ of each regular node i, i.e., $|\mathcal{N}_i\cap\mathcal{V}_A| \leq F$, $\forall i \in \mathcal{V}_R$.*

Definition 2.17 (F-total link attack model) *The NAS is said to be F-total link attacked if the compromised edge set $\mathcal{E}_A$ contains at most F edges, i.e., $|\mathcal{E}_A| \leq F$.*

Definition 2.18 (F-local link attack model [59]) *The NAS is said to be F-local link attacked if for every node $i\in\mathcal{V}$, there are at most F attacked links in $\mathcal{K}_i$, i.e., $|\{(i,j)\in\mathcal{E}_A \mid (i,j)\in\mathcal{K}_i\}|\leq F$, $\forall i\in\mathcal{V}$.*

Remark 2.4 *It is worth noting that under the F-local node attack model, there may exist more than F compromised nodes in the NAS. This implies that a NAS subject to F-local node attacks is necessarily F-total attacked, but not vice versa. Therefore, any resilient control result established for the F-local node attack model is also applicable to the F-total model, whereas the reverse is not necessarily true. A similar conclusion applies to the F-total and F-local link attack models.*

CHAPTER 3

Distributed security monitoring of homogeneous sensor networks against sparse sensor attacks

This chapter studies the problem of distributed security state monitoring (SSM) of NASs, that is, securely estimating the actual system state from a set of locally compromised measurements. The chapter begins with a review of historical work on general SSM and draws out the flaws and shortcomings of existing works. Section 3.2 studies the distributed SSM problem for NASs based on a dynamic gain adjustment mechanism against location-fixed sparse sensor attacks. This section presents a dual-layer observer framework integrating decentralized and distributed architectures by decoupling residual feedback estimation from attack suppression operations. Within this framework, a security sufficiency condition is given to guarantee SSM resilience against location-fixed s-sparse attacks, relying heavily on dynamic gain design in the information exchange between the two observer layers. Section 3.3 studies the distributed SSM problem based on an attack detection and isolation mechanism against location-varying sparse sensor attacks. Compared to location-fixed attacks, location-varying attacks are often more difficult to counter due to their unpredictability. By combining a dynamic attack detection and isolation mechanism in the construction of a Luenberger-like observer, a distributed SSM protocol against location-varying sparse attacks is developed.

3.1 INTRODUCTION

As a fusion of physical devices and the cyber world, NASs not only improve the 'intelligence' of physical systems but also inevitably expose the vulnerability of networks. Recently, numerous researchers have reported extensive findings on NASs from the perspective of security [12,35,94–97]. From the viewpoint of an attacker, various types of attack strategies have been proposed, such as stealthy attacks [94], deception attacks [98], denial-of-Service attacks [99], undetectable sparse attacks [100], etc. From the defender's point of view, the existing works have mainly concentrated on attack detection [96, 101], secure control [102], secure defense [97], and so on.

Among the numerous works, the exploration of SSM under sparse attacks has become a hot topic in the field of NASs recently, and plentiful research has been carried out primarily from the following two implementation perspectives: 1) Static batch optimization and 2) Observer-based method.

Static batch optimization: As noted in Chapter 1, an augmented system is constructed from τ-time sensor measurements, so the SSM problem is transformed into a static quadratic optimization problem. Existing algorithms addressing this problem can be broadly categorized into three classes: combination, geometry, and greedy approaches. The combination algorithms [23–25] typically rely on brute-force methods to search through all possible combinations to identify anomalous elements. Geometry-based approaches, on the other hand, relax the non-convex l_0 minimization problem into a solvable convex $l_1/l_r(r > 1)$ optimization problem [19–21, 27]. Lastly, greedy algorithms aim to achieve optimal estimation through iterative approximation methods. For instance, [17, 32, 33, 103] employed gradient-descent algorithms and projection techniques to enhance computational efficiency. However, static batch optimization is constrained by the sparsity assumption of fixed locations over the τ-time horizon. Location-varying sparse attacks, which disrupt this sparsity property, render these algorithms ineffective, thereby highlighting the need for new advancements in secure state monitoring methods.

Observer-based method: This method focuses on designing an appropriate real-time observer [15, 38, 40, 83, 104] subject to specific constraints. Observer-based real-time estimation is more convenient than the static batch approach, which can only provide delayed estimation. In [38], the SSM under sparse sensor attacks was realized by a supervisory observer along with a family of candidate nonlinear observers. Additionally, based on the Saturation Innovation Update (SIU) algorithm, the problem of distributed security state estimation against location-varying sparse sensor attacks was addressed in [37, 40–42]. However, excellent detection threshold functions are still absent for attack detection and isolation, and their construction is still quite challenging.

Motivated by the above discussion, this chapter aims to study distributed SSM problems for NASs against sparse sensor attacks. Several aspects of the current study are worth mentioning. Firstly, to overcome the parsing difficulty associated with the lack of local measurement information, a dual-layer state observation framework that integrates decentralized and distributed observers is proposed against location-fixed sparse attacks. This framework effectively separates the coupled services of residual feedback from those of attack suppression, thereby overcoming the problem of discrepancies when comparing residual information across diverse sensors. Secondly, building on the concept of dynamic average consensus, a novel dynamic gain adjustment scheme is proposed that addresses the challenges associated with the centralized design of the attack detection threshold function. Accordingly, it is unnecessary to detect or isolate malicious attack locations, and the security of the global state monitoring can be ensured through distributed dynamic gain adjustment. Finally, a new design framework and condition for the detection threshold function are proposed, which progressively reduces the stealth space of attacks. This enables precise and timely detection of malicious data, overcoming the limitations of static batch optimization algorithms in handling location-varying sparse attacks.

3.2 DISTRIBUTED SECURITY STATE MONITORING AGAINST LOCATION-FIXED SPARSE SENSOR ATTACKS

3.2.1 Model formulation

Consider a continuous-time NAS topologized as an unweighted, undirected, and connected graph $\mathcal{G}$ with N homogeneous agents and one-by-one corresponding sensors, whose dynamics are described as follows:

$$\dot{x}_i(t) = Ax_i(t) + B\left(u(t) + K\sum_{j=1}^{N} a_{ij}(x_j(t) - x_i(t))\right),$$

$$\tilde{y}_i(t) = y_i(t) + a_i(t), \quad i \in \mathcal{V}, \tag{3.1}$$

where $x_i(t) \in \mathbb{R}^n$ and $u(t) \in \mathbb{R}^p$ are the internal system state of i-th agent and the known common control input, respectively, $y_i(t) = Cx_i(t) \in \mathbb{R}^m$ denotes the nominal measurement output of i-th sensor, and $\tilde{y}_i(t)$ is the measurement output of i-th sensor that may be corrupted by a malicious attack $a_i(t)$. The attack signal $a_i(t)$ can take an arbitrary non-zero value if a successful attack is launched on the i-th sensor channel. Otherwise, it takes the value $\mathbf{0}_m$. A, B, and C are system matrices with appropriate dimensions with the pair (A, B) is controllable and (A, C) is observable, the matrix K denotes the consensus control gain such that the matrix $A - \lambda_2(\mathcal{L})BK$ is Hurwitz.

Based on the fact that (A, C) is observable, it is guaranteed that there exist a positive definite matrix $P > 0$ and an arbitrary constant $l > 0$ such that

$$PA + A^T P - 2C^T C + lP < 0. \tag{3.2}$$

This section aims to develop a distributed SSM algorithm without centralized threshold design for consensus NASs (3.1) against s-sparse sensor attacks. Referring to [24, 28, 33, 105, 106], it is assumed that the set of compromised sensors is time-invariant, and the defender cannot obtain any other information except its cardinality. Therefore, the following assumption exists for the malicious adversaries.

Assumption 3.1 *The global attack signal $a(t) = \text{col}\{a_i(t)\}_{i \in \mathcal{V}}$ is location-fixed s-sparse with $s < N/2$.*

Based on the definition of attacked set $\mathcal{V}_A = \{i \mid a_i(t) \neq \mathbf{0}_m, \ \exists t \geq 0\}$, the benign set is denoted by $\mathcal{V}_N = \{i \in \mathcal{V} | a_i(t) = \mathbf{0}_m, \ \forall t \geq 0\}$ and it satisfies $\mathcal{V}_A \cup \mathcal{V}_N = \mathcal{V}$.

Remark 3.1 *Note that the NAS (3.1) is $(N-1)$-sparse observable, i.e., the matrix pair $(A, \mathbf{1}_{\tilde{\Gamma}} \otimes C)$ is observable for any sensors set $\Gamma \subset \mathcal{V}$ with $|\Gamma| = N - 1$. It follows from [17, 19–21, 23–27, 33] that the number of compromised sensors must be strictly less than half the number of sensors, i.e., $s < N/2$. In this section, it follow this same covenant.*

3.2.2 Construction of dual-layer state observers

Due to potential unknown attack effects on sensor measurements, a single sensor cannot independently assess the reliability of its data. This necessitates intercommunication and cooperation among sensors for SSM. The challenge of distributed SSM is to mitigate the effects of malicious data through the utilization of communication interactions within networks. This approach enables the asymptotic SSM of the actual state $x_i(t)$ for each agent, even in the presence of partially manipulated sensor measurements.

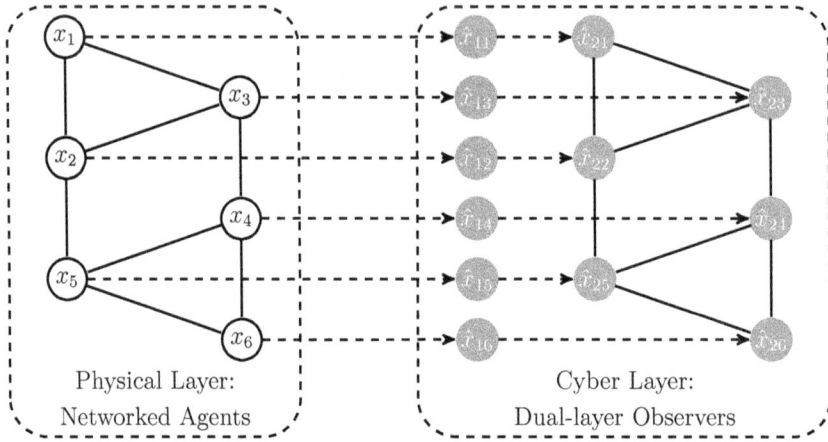

Figure 3.1 An illustration framework of SSM with decentralized and distributed observers.

Based on the above analysis and by extending the distributed observer design in the previous works [37,40], a dual-layer SSM framework that integrates decentralized and distributed observers is organized as illustrated in Fig. 3.1, where the dashed lines refer to communication channels that are vulnerable to attacks while the solid ones are channels that are immune to attack. Moreover, circles on the left of the cyber layer indicate decentralized observers $\hat{x}_{1i}$, while the right ones are distributed observers $\hat{x}_{2i}$, $i \in [6]$. Specifically, the dynamics of the i-th observer are considered to be designed as follows:

$$\dot{\hat{x}}_{1i}(t) = A\hat{x}_{1i}(t) + Bu(t) + L(\tilde{y}_i(t) - C\hat{x}_{1i}(t)), \tag{3.3}$$

$$\dot{\hat{x}}_{2i}(t) = A\hat{x}_{2i}(t) + Bu(t) + \theta \sum_{j=1}^{N} a_{ij}(\hat{x}_{2j}(t) - \hat{x}_{2i}(t))$$

$$+ \delta_i(t)l(\hat{x}_{1i}(t) - \hat{x}_{2i}(t)), \tag{3.4}$$

where $\hat{x}_{1i}(t)$ and $\hat{x}_{2i}(t)$ denote state estimations of $x_i(t)$ generated by the decentralized observer i and the distributed one, respectively. The matrix $L = P^{-1}C^T$ is the observation gain such that the matrix $A - LC$ is Hurwitz with the positive definite matrix P being a solution of the linear matrix inequality (LMI) (3.2). The parameters l and $\delta_i(t)$ denote the static and dynamic feedback gain to be designed, respectively.

Moreover, the parameter θ is the consensus control gain such that $\alpha - \lambda_2(\mathcal{L})\theta < -l$ holds with the parameter α being the solution of the following optimization problem:

$$\min \alpha \tag{3.5}$$
$$\text{s.t. } 0 < R,$$
$$\frac{1}{2}(RA + A^T R) < \alpha R.$$

To move forward, the following auxiliary decentralized monitoring system is introduced to support the analysis:

$$\dot{\tilde{x}}_{1i}(t) = (A - LC)\tilde{x}_{1i}(t) + Bu(t) + LCx_i(t), \tag{3.6}$$

where $\tilde{x}_{1i}(0) = \hat{x}_{1i}(0)$, $\forall i \in \mathcal{V}$. Further, define the error $\tilde{e}_i(t) = \tilde{x}_{1i}(t) - x_i(t)$, whose dynamics are obtained as follows:

$$\dot{\tilde{e}}_i(t) = (A - LC)\tilde{e}_i(t) - BK \sum_{j=1}^{N} a_{ij}(x_j(t) - x_i(t)).$$

Since the matrix $A - \lambda_2(\mathcal{L})BK$ is Hurwitz, one can have the conclusion that $\lim_{t\to\infty} \|x_j(t) - x_i(t)\| = 0$ holds for $\forall i, j \in \mathcal{V}$. Hence, it follows from the matrix $A - LC$ being Hurwitz that

a) $\lim_{t\to\infty} \|\tilde{e}_i(t)\| = 0$, $\forall i \in \mathcal{V}$;

b) $\tilde{x}_{1i}(t) = \hat{x}_{1i}(t)$, $\forall i \in \mathcal{V}_N$, $\forall t \geq 0$;

c) $\tilde{x}_{1i}(t) \neq \hat{x}_{1i}(t)$, $\forall i \in \mathcal{V}_A$, $\exists t \geq 0$.

The distributed state observers are secure against s-sparse sensor attacks if, for each observer $i \in \mathcal{V}$, $\lim_{t\to\infty} \|\hat{x}_{2i}(t) - x_i(t)\| = 0$ is established. To this end, the dynamic gain $\delta_i(t)$ must be carefully designed to counteract the negative impact of malicious attacks while optimally utilizing the information from benign measurements to maintain the security of the state monitoring. This implies that the design of the dynamic gain necessitates consideration of how to balance the effects of benign measurements and malicious attacks to enhance the robustness of the system against sparse attacks without compromising the monitoring performance.

3.2.3 Performance of distributed security state monitoring

In this subsection, a preliminary design of the desired form of the dynamic gain necessary for the distributed SSM is first provided. Subsequently, an adjustment method is proposed for designing distributed dynamic gains that satisfy the above requirement to resolve the distributed SSM problem; simultaneously, the feasibility of the designed dynamic gain is verified. This method enables all observers to securely estimate the state of agents utilizing benign measurement information effectively, even under the corruption of location-fixed sparse sensor attacks, through a reasonable dynamic gain adjustment strategy.

● *Sufficient conditions for security state monitoring*

Denote $e_i(t) = \hat{x}_{2i}(t) - \tilde{x}_{1i}(t)$ the local state tracking error, it follows from (3.4) and (3.6) that

$$\dot{e}_i(t) = A e_i(t) + \theta \sum_{j=1}^{N} a_{ij}(e_j(t) - e_i(t)) - \delta_i(t) l z_i(t) + \mu_i(t). \tag{3.7}$$

where $\mu_i(t) = \theta \sum_{j=1}^{N} a_{ij}\mu_{ij}(t) + LC\tilde{e}_i(t)$, $\mu_{ij}(t) = (\tilde{e}_j(t) - \tilde{e}_i(t)) + (x_j(t) - x_i(t))$, and

$$z_i(t) = \hat{x}_{2i}(t) - \hat{x}_{1i}(t) = \begin{cases} e_i(t), & \text{if } i \in \mathcal{V}_N, \\ \hat{x}_{2i}(t) - \hat{x}_{1i}(t), & \text{if } i \in \mathcal{V}_A. \end{cases}$$

For a set of data $\{\|Qe_i(t)\|, \, i \in \mathcal{V}\}$ with the matrix $Q = R^{\frac{1}{2}}$ and R defined in (3.5), sort them in descending order and denote them by $\gamma_i^e(t)$, i.e., $\gamma_1^e(t) \geq \gamma_2^e(t) \geq \cdots \geq \gamma_N^e(t)$. That is, there exists unique $j \in \mathcal{V}$ such that each $\gamma_i^e(t)$ uniquely corresponding to a $\|Qe_j(t)\|$. In particular, define $\bar{\gamma}^e(t)$ as the mean value of these data, i.e.,

$$\bar{\gamma}^e(t) = \frac{1}{N} \sum_{i=1}^{N} \gamma_i^e(t) = \frac{1}{N} \sum_{i=1}^{N} \|Qe_i(t)\|.$$

Consider the dynamic gain selected as follows:

$$\delta_i(t) = \begin{cases} \frac{\gamma_{2s+1}^e(t)}{\|Qz_i(t)\|}, & \text{if } \|z_i(t)\| \neq 0, \\ 1, & \text{if } \|z_i(t)\| = 0. \end{cases} \tag{3.8}$$

Theorem 3.1 Under the condition that Assumption 3.1 holds, if the static gain is selected to satisfy $l > \frac{\alpha N}{N-2s} + \alpha(\frac{N}{N-2s})^{\frac{3}{2}}$ and the dynamic gain is set as in (3.8), then the distributed observer (3.4) can achieve asymptotic SSM for the actual state of the agent (3.1) under any location-fixed s-sparse attack, i.e.,

$$\lim_{t \to \infty} \|\hat{x}_{2i}(t) - x_i(t)\| = \lim_{t \to \infty} \|e_i(t)\| = 0, \quad \forall i \in \mathcal{V}. \tag{3.9}$$

Proof 3.1 *Consider the Lyapunov function $V_i(t) = \|Qe_i(t))\|$, and take its differential along with (3.7) as follows:*

$$\dot{V}_i(t) = V_i^{-1}(t)\left(\frac{1}{2}e_i^T(t)(RA + A^T R)e_i(t) - \delta_i(t)l e_i^T(t)Rz_i(t)\right)$$

$$+ \theta \sum_{j \in \mathcal{N}_i} a_{ij}(V_j(t) - V_i(t)) + \|Q\mu_i(t)\|$$

$$\leq \alpha V_i(t) - \nu_i(t)l\gamma_{2s+1}^e(t) + \theta \sum_{j \in \mathcal{N}_i} a_{ij}(V_j(t) - V_i(t)) + \|Q\mu_i(t)\|, \tag{3.10}$$

where $\nu_i(t) = 1$ if $i \in \mathcal{V}_N$, otherwise $\nu_i(t) = -1$.

Denote $V(t) = \text{col}\{V_i(t)\}$ as the global form, and introduce the following two intermediate variables to demonstrate the convergence of $V(t)$:

$$\tilde{V}(t) = U_1^T V(t), \quad \bar{V}(t) = U_2^T V(t) = \sqrt{N}\bar{\gamma}^e(t),$$

where the matrices U_1 and U_2 are defined in Lemma 2.11. Hence, substituting (3.10), one has

$$\dot{\tilde{V}}(t) \leq (\alpha I_{N-1} - \theta J_1)\tilde{V}(t) - lU_1^T \nu(t)\gamma_{2s+1}^e(t) + U_1^T \mu(t), \qquad (3.11)$$

$$\dot{\bar{V}}(t) \leq \alpha\bar{V}(t) - \frac{(|\mathcal{V}_N| - |\mathcal{V}_A|)}{\sqrt{N}} l\gamma_{2s+1}^e(t) + U_2^T \mu(t)$$

$$= \alpha\bar{V}(t) - \left(\sqrt{N} - \frac{2s}{\sqrt{N}}\right) l\gamma_{2s+1}^e(t) + U_2^T \mu(t). \qquad (3.12)$$

where $\nu(t) = \mathrm{col}\{\nu_i(t)\}$, $\mu(t) = \mathrm{col}\{\|Q\mu_i(t)\|\}$, $i \in \mathcal{V}$.
Define the function $\beta(t)$ as:

$$\beta(t) = \begin{cases} \frac{\bar{\gamma}^e(t)}{\gamma_{2s+1}^e(t)}, & \text{if } 0 < \gamma_{2s+1}^e(t) \leq \bar{\gamma}^e(t), \\ 1, & \text{otherwise.} \end{cases}$$

Furthermore, the Lyapunov function $W(t) = \|\tilde{V}(t)\|$ is considered. According to Lemma 2.11 and (3.11), one has

$$\dot{W}(t) \leq (\alpha - \theta\lambda_2(\mathcal{L}))W(t) + l\|U_1^T \nu(t)\|\gamma_{2s+1}^e(t) + \|U_1^T \mu(t)\|$$

$$\overset{(a)}{\leq} -lW(t) + l\sqrt{N}\gamma_{2s+1}^e(t) + \|\mu(t)\|, \qquad (3.13)$$

where the inequality (a) is derived by $\alpha - \lambda_2(\mathcal{L})\theta < -l$ and $\|U_1\| = 1$.
It follows from (3.13) that there exists a sufficiently large $T > 0$ such that for all $t \geq T$, the inequality $0 \leq W(t) \leq \sqrt{N}\gamma_{2s+1}^e(t) + \frac{1}{l}\|\mu(t)\|$ holds.
Hence, from $V(t) = U_2\bar{V}(t) + U_1\tilde{V}(t)$, $\bar{V}(t) = \sqrt{N}\bar{\gamma}^e(t) \leq \sqrt{N}\beta(t)\gamma_{2s+1}^e(t)$ and the orthogonality of the matrix U, one has

$$\gamma_{2s+1}^e(t) > \bar{\gamma}^e(t) - \frac{1}{\sqrt{N - 2s}}W(t)$$

$$\geq \left(\beta(t) - \sqrt{\frac{N}{N - 2s}}\right)\gamma_{2s+1}^e(t) - \frac{\|\mu(t)\|}{l\sqrt{N - 2s}}.$$

Based on the previous analysis, one can conclude that $\lim_{t\to\infty} \|\mu_i(t)\| = 0$ holds due to $\lim_{t\to\infty} \|\tilde{e}_i(t)\| = 0$ and $\lim_{t\to\infty} \|x_j(t) - x_i(t)\| = 0$ for $\forall i, j \in \mathcal{V}$. Hence, one can find a $T' \geq T$ such that $\beta(t) \leq 1 + \sqrt{N/(N - 2s)}$ holds for any $t \geq T'$.
According to $\alpha - \left(1 - \frac{2s}{N}\right)\frac{l}{\beta(t)} < 0$ holding for $t \geq T'$, and from (3.12), one can obtain

$$\dot{\bar{V}}(t) \leq \left(\alpha - \left(1 - \frac{2s}{N}\right)\frac{l}{\beta(t)}\right)\bar{V}(t) + U_2^T \mu(t). \qquad (3.14)$$

It follows from $\lim_{t\to\infty} \mu(t) = \mathbf{0}_N$ that $\lim_{t\to\infty} \bar{V}(t) = 0$. Moreover, combining $\lim_{t\to\infty} \|\mu(t)\| = 0$ yields that there is $\lim_{t\to\infty} W(t) = 0$, which implies that $\lim_{t\to\infty} \tilde{V}(t) = \mathbf{0}_{N-1}$.

Finally, from $V(t) = U_2\bar{V}(t) + U_1\tilde{V}(t)$, one obtains

$$\lim_{t\to\infty} V(t) \leq \lim_{t\to\infty} U_2\bar{V}(t) + \lim_{t\to\infty} U_1\tilde{V}(t) = \mathbf{0}_N.$$

Combined with the positivity of the function $V_i(t)$, one has $\lim_{t\to\infty}\|e_i(t)\| = \lim_{t\to\infty} V_i(t) = 0$. This completes the proof.

Remark 3.2 *Note that under the dynamic gain designed in (3.8), the security of state monitoring depends on the accessibility of the value $\gamma_{2s+1}^e(t)$. Furthermore, if the average value $\bar{\gamma}^e(t)$ of $\gamma_i^e(t)$ is accessible, the static gain design can be simplified to $l > \alpha N/(N-2s)$ by replacing $\gamma_{2s+1}^e(t)$ in the dynamic gain (3.8) with $\bar{\gamma}^e(t)$. However, in practical distributed applications, the values $\gamma_{2s+1}^e(t)$ and $\bar{\gamma}^e(t)$ are both challenging to obtain. Therefore, there is an urgent need to develop an alternative design scheme, which will be discussed in the following subsection.*

● Dynamic gain adjustment strategy

To approximately obtain a valid substitute for the term $\gamma_{2s+1}^e(t)$, a distributed dynamic gain adjustment strategy is proposed, with detailed elaboration provided in Algorithm 3.1. At each time t, the i-th observer generates an approximate substitute value $\rho_i(t)$ for $\gamma_{2s+1}^e(t)$, and indicator functions $b_{1i}(t)$ and $b_{2i}(t)$ to assess the magnitude of $\rho_i(t)$.

Before moving to the main argument, an average consensus tracking problem will first be presented and analyzed. That is, for a NAS with the following dynamics:

$$\dot{w}_{1i}(t) = a(\sigma_i - w_{1i}(t)) - b\sum_{i=1}^{N} l_{ij}(w_{1j}(t) + w_{2j}(t)),$$

$$\dot{w}_{2i}(t) = b\sum_{i=1}^{N} l_{ij}w_{1j}(t),$$
(3.15)

where $a > 0$, $b > 0$, and $\sigma_i \in \mathbb{R}$ are any real numbers.

Lemma 3.1 *[30, 107] For NAS (3.15), its state $w_{1i}(t)$ will asymptotically track the average of values σ_i if the constants $a > 0$ and $t > 0$, i.e.,*

$$\lim_{t\to\infty} w_{1i}(t) = \bar{\sigma} = \frac{1}{N}\sum_{i=1}^{N} \sigma_i.$$

Proof 3.2 *Denote the overall variables $w_1(t) = \text{col}\{w_{1i}(t)\}$, $w_2(t) = \text{col}\{w_{2i}(t)\}$, and $\sigma = \text{col}\{\sigma_i\}$. Hence, one has*

$$\begin{bmatrix} \dot{\tilde{w}}_1(t) \\ \dot{\tilde{w}}_2(t) \end{bmatrix} = D_L \begin{bmatrix} \tilde{w}_1(t) \\ \tilde{w}_2(t) \end{bmatrix} + a\begin{bmatrix} U_1^T\sigma \\ 0 \end{bmatrix},$$
(3.16)

$$\dot{\bar{w}}_1(t) = a\left(\sqrt{N}\bar{\sigma} - \bar{w}_1(t)\right),$$
(3.17)

where $\tilde{w}_1(t) = U_1^T w_1(t)$, $\tilde{w}_2(t) = U_1^T w_2(t)$, $\bar{w}_1(t) = U_2^T w_1(t)$, and

---

Algorithm 3.1 Distributed dynamic gain adjustment strategy

Input: α, s, $\|z_i(t)\|$, $0 < \rho_i(0)$, $b_{1i}(0)$, $b_{2i}(0)$;

1: Pick the static gain $l > \frac{\alpha N}{N-2s} + \alpha(\frac{N}{N-2s})^{\frac{3}{2}}$;

2: Setting the consensus gain in (3.4) to be $\theta > (\alpha + l)/\lambda_2(\mathcal{L})$;

3: **if** $b_{1i}(t) \leq s$ **then**

$$\dot{\rho}_i(t) = (\alpha + l)\rho_i(t) - \theta \sum_{i=1}^{N} l_{ij}\rho_j(t); \tag{3.18}$$

4: **else**

$$\dot{\rho}_i(t) = \left(\alpha - l\left(1 - \frac{2s}{N}\right)\right)\rho_i(t) - \theta \sum_{i=1}^{N} l_{ij}\rho_j(t); \tag{3.19}$$

5: **end if**

6: **if** $\rho_i(t) \leq \|z_i(t)\|$ **then**

7:     $\dot{b}_{1i}(t) = -lb_{1i}(t) - \theta \sum_{i=1}^{N} l_{ij}(b_{1j}(t) + b_{2j}(t));$

8: **else**

9:     $\dot{b}_{1i}(t) = l(N - b_{1i}(t)) - \theta \sum_{i=1}^{N} l_{ij}(b_{1j}(t) + b_{2j}(t));$

10: **end if**

11: $\dot{b}_{2i}(t) = \theta \sum_{i=1}^{N} l_{ij}b_{1j}(t);$

12: Setting the dynamic gain as:

$$\delta_i(t) = \begin{cases} \frac{\rho_i(t)}{\|z_i(t)\|}, & \text{if } \|z_i(t)\| \neq 0, \\ 1, & \text{if } \|z_i(t)\| = 0. \end{cases} \tag{3.20}$$

Output: Dynamic gain $\delta_i(t)$.

$$D_L = -b \begin{bmatrix} 1 & 1 \\ -1 & 0 \end{bmatrix} \otimes J_1 - a \begin{bmatrix} 1 & 0 \\ 0 & 0 \end{bmatrix} \otimes I_N.$$

It is evident that, matrix D_L is Hurwitz, then for any two solutions v_1 and v_2 of (3.16), one has $\lim_{t\to\infty}(v_1(t) - v_2(t)) = 0$ since $\dot{e}_v(t) = D_L e_v(t)$ is satisfied for $e_v(t) = v_1(t) - v_2(t)$. That is, if one can prove that there exists a function $w_2^*(t)$ such that $(0, w_2^*(t))$ is a solution of (3.16), one can have the conclusion that $\lim_{t\to\infty} \tilde{w}_1(t) = 0$. By substituting $\tilde{w}_1(t) = 0$ and $\tilde{w}_2(t) = w_2^*(t)$ in (3.16), one obtains

$$\begin{aligned} bJ_1 w_2^*(t) &= aU_1^T \sigma, \\ \dot{w}_2^*(t) &= 0. \end{aligned} \tag{3.21}$$

Therefore, it makes a conjecture that $(\tilde{w}_1(t) = 0, \tilde{w}_2(t) = \frac{a}{b}J_1^{-1}U_1^T \sigma)$ constitutes a solution of (3.16).

Next, it follows from (3.17) that $\lim_{t\to\infty} \bar{w}_1(t) = \sqrt{N}\bar{\sigma}$. And according to $w_1(t) = U_1\tilde{w}_1(t) + U_2\bar{w}_1(t)$, one gets

$$\lim_{t\to\infty} w_{1i}(t) = \lim_{t\to\infty} \frac{1}{\sqrt{N}}\bar{w}_1(t) = \bar{\sigma}.$$

This completes the proof.

Back to the topic, for the data set $\{\|z_i(t)\|, \ i \in \mathcal{V}\}$, arrange its elements similarly in descending order denoted as:

$$\gamma_1^z(t) \geq \gamma_2^z(t) \geq \cdots \geq \gamma_N^z(t).$$

Since the malicious attack $a(t)$ is s-sparse, one can get that $\gamma_{2s+1}^e(t) \leq \gamma_{s+1}^z(t) \leq \gamma_1^e(t)$ holds for any $t \geq 0$.

Theorem 3.2 Under the implementation of Algorithm 3.1 with any s-sparse sensor attacks, it can be concluded that:

$$s + 1 \leq \lim_{t\to\infty} b_{1i}(t),$$

$$\lim_{t\to\infty} \gamma_{2s+1}^e(t) \leq \lim_{t\to\infty} \rho_i(t) = 0,$$

which implies that the SSM (3.9) can be reached successfully.

Proof 3.3 *According to Lemma 3.1, one concludes that $b_{1i}(t) \to \bar{b}(t)$, $\forall i \in \mathcal{V}$, where $\bar{b}(t) = |\{i \mid \rho_i(t) > \|z_i(t)\|\}|$.*

Based on (3.10), one gets that for any $i \in \mathcal{V}$,

$$\dot{V}_i(t) \leq \alpha V_i(t) + l\rho_i(t) - \theta \sum_{i=1}^{N} l_{ij}V_j(t) + \|\mu_i(t)\|. \tag{3.22}$$

Next, it will be attempted to prove that $\lim_{t\to\infty} \bar{b}(t) \geq s+1$. To this end, the cases of $b_{1i}(t) \leq s$ and $b_{1i}(t) > s$ when $i \in \mathcal{V}_N$ will be attempted to be analyzed separately.

Firstly, if there exists $b_{1i}(t) \leq s$, it follows from $\lim_{t\to\infty} \|\mu_i(t)\| = 0$, (3.18) and (3.22) that there exists a $0 < \tilde{T}_{1i} < +\infty$ such that for any $t > \tilde{T}_{1i}$, one has

$$V_i(t) \leq \rho_i(t), \ \forall i \in \mathcal{V},$$

$$\|z_i(t)\| = V_i(t) \leq \rho_i(t), \ \forall i \in \mathcal{V}_N.$$

Due to $|\mathcal{V}_N| \geq N - s \geq s + 1$, it means that $\bar{b}(t) \geq |\mathcal{V}_N| \geq s + 1$ holds when $t \geq \max\{\tilde{T}_{1i}\}$, and thus $b_{1i}(t) > s$, $\forall i \in \mathcal{V}_N$.

Secondly, if $b_{1i}(t) > s$ for all $i \in \mathcal{V}_N$, from (3.12), (3.18), and (3.19), one can similarly find a time $\tilde{T}_{2i} > 0$ such that $\dot{V}_i(t) \leq \dot{\bar{V}}(t) \leq \dot{\rho}_i(t)$ and $\|z_i(t)\| \leq \rho_i(t)$ hold for any $i \in \mathcal{V}_N$ and $t \geq \tilde{T}_{2i}$. This means that $\bar{b}(t) \geq |\mathcal{V}_N| \geq s + 1$ holds when $t \geq \max\{\tilde{T}_{2i}\}$, $i \in \mathcal{V}$.

In summary, it follows that $\lim_{t\to\infty} b_{1i}(t) = \lim_{t\to\infty} \bar{b}(t) \geq s + 1$.

Finally, from the fact that $(a + l) - \theta \lambda_2(\mathcal{L}) < 0$ *and* $\alpha - l(1 - \frac{2s}{N}) < 0$, *one can get:*

$$\lim_{t \to \infty} \rho_i(t) = \lim_{t \to \infty} \rho_j(t), \ \forall i, j \in \mathcal{V},$$

$$\lim_{t \to \infty} \gamma_{2s+1}^e(t) \le \lim_{t \to \infty} \gamma_{s+1}^z(t) \le \lim_{t \to \infty} \rho_i(t) = 0, \ \forall i \in \mathcal{V}.$$

Thus, the same effect can be achieved by using $\rho_i(t)$ *instead of* $\gamma_{2s+1}^e(t)$. *The proof is accomplished.*

3.2.4 Numerical simulations

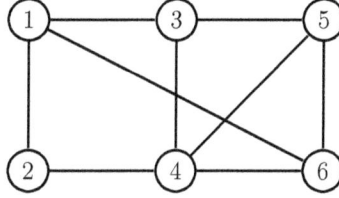

Figure 3.2   Communication topology $\mathcal{G}$ of 6 nodes.

To illustrate the effectiveness of the algorithm based on the dynamic gain adjustment strategy proposed in this section, one performed a simulation verification of the security monitoring of the IEEE 6-bus power system with three generators and six buses, benchmarked on the communication topology of Fig. 3.2. Referring to [108, 109], the small-signal method is utilized to obtain a Kron-reduced form of the power network, which is expressed as:

$$\begin{bmatrix} \dot{\zeta}(t) \\ M_g \dot{\omega}(t) \end{bmatrix} = \begin{bmatrix} 0 & I \\ L_{Ig}^T L_{II}^{-1} L_{Ig} - L_{gg} & -D_g \end{bmatrix} \begin{bmatrix} \zeta(t) \\ \omega(t) \end{bmatrix} + \begin{bmatrix} 0 & 0 \\ I & -L_{Ig}^T L_{II}^{-1} \end{bmatrix} \begin{bmatrix} p_\omega(t) \\ p_\theta(t) \end{bmatrix}, \quad (3.23)$$

where $\zeta(t) = [\zeta_1(t), \ \zeta_2(t), \ \zeta_3(t)]^T$ and $\omega(t) = [\omega_1(t), \ \omega_2(t), \ \omega_3(t)]^T$ represent the generator rotor angles and frequencies, respectively. The matrices M_g and D_g are diagonal matrices that denote the generator inertial and damping coefficients, respectively. The inputs $p_\omega(t) = [p_{1\omega}(t), \ p_{2\omega}(t), \ p_{3\omega}(t)]^T$ and $p_\theta(t) = [p_{1\theta}(t), \ p_{2\theta}(t), \ p_{3\theta}(t)]^T$ correspond to known changes in the mechanical input power and real power demand at the loads, respectively. To fit the system model given by (3.1), the system matrices can be expressed as follows:

$$A = \begin{bmatrix} 0 & I \\ M_g^{-1} \left(L_{Ig}^T L_{II}^{-1} L_{Ig} - L_{gg} \right) & -M_g^{-1} D_g \end{bmatrix},$$

$$B = \begin{bmatrix} 0 & 0 \\ M_g^{-1} & -M_g^{-1} L_{Ig}^T L_{II}^{-1} \end{bmatrix}, \quad C = I_6,$$

where the matrix parameters are set as $M_g = \text{diag}\{0.125, 0.034, 0.016\}$, $D_g = \text{diag}\{0.125, 0.068, 0.48\}$, $L_{gg} = \text{diag}\{0.058, 0.063, 0.059\}$, and

$$L_{Ig} = \begin{bmatrix} -0.058 & 0 & 0 & 0 & 0 & 0 \\ 0 & -0.063 & 0 & 0 & 0 & 0 \\ 0 & 0 & -0.059 & 0 & 0 & 0 \end{bmatrix}^T,$$

$$L_{II} = \begin{bmatrix} 0.235 & 0 & 0 & -0.085 & -0.092 & 0 \\ 0 & 0.296 & 0 & -0.161 & 0 & -0.072 \\ 0 & 0 & 0.330 & 0 & -0.170 & -0.101 \\ -0.085 & -0.161 & 0 & 0.246 & 0 & 0 \\ -0.092 & 0 & -0.170 & 0 & 0.262 & 0 \\ 0 & -0.072 & -0.101 & 0 & 0 & 0.173 \end{bmatrix}.$$

In the simulation, the initial state of the system is set as follows:

$$x_1(0) = [0.2, \ -0.5, \ 0.7, \ 0, \ 0, \ 0]^T,$$
$$x_2(0) = [0.6, \ 0.3, \ -0.2, \ 0.5, \ -0.1, \ 0.3]^T,$$
$$x_3(0) = [1.5, \ 0.1, \ -0.5, \ 1, \ 0.2, \ 0.4]^T,$$
$$x_4(0) = [-0.6, \ -0.2, \ 1, \ -0.3, \ 0.6, \ -0.4]^T,$$
$$x_5(0) = [0.4, \ -0.3, \ 0.5, \ -0.4, \ 0.7, \ 1]^T,$$
$$x_6(0) = [-0.2, \ 0.6, \ -0.3, \ 0.5, \ -0.4, \ 0.8]^T,$$

and the control input is set as $u_i(t) = 0.1\sin(\pi t) \times \mathbf{1}_9$, $i \in \mathcal{V}$. The initial monitoring values are selected randomly within the norm bound of 3. Assume that all remote transmission channels are open to attacks, but only up to two are compromised. Without loss of generality, it is assumed that $\mathcal{V}_A = \{2, 5\}$ as the attack intrusion locations set. Moreover, the attack injection $a_i(t)$ is set to

$$a_i(t) = C(\hat{x}_{2i}(t) - x_i(t)) + e^t b_i(t), \quad \text{if } i \in \mathcal{V}_A,$$

with $b_i(t)$ is a random variable obeying a standard Gaussian distribution.

In the following, an attempt will be made to verify the results derived from the Theorem 3.2. For this purpose, calculate and set the parameters as $\alpha = 7.82 * 10^{-4}$, $s = 2$, $l = 2.5064$, $\theta = 2.9041$, and the matrix K is selected as follows:

$$K = \begin{bmatrix} 15.5473 & 15.8619 & 52.1840 & 5.9653 & 5.7993 & 8.3597 \\ 12.3624 & 31.2612 & -42.5637 & 3.6241 & 11.6926 & -6.1557 \\ 16.0453 & 27.5614 & -17.1861 & 4.6591 & 9.8448 & -0.8301 \\ 2.2341 & -2.3639 & -605.8924 & 1.8565 & 1.3434 & -87.0852 \\ -10.5181 & -31.1789 & -472.4273 & -1.9783 & -8.6960 & -67.1059 \\ -19.7043 & -31.4460 & -581.7407 & -4.6643 & -7.4678 & -84.3593 \\ -6.6302 & 6.1548 & 603.5541 & -3.2463 & -0.9250 & 87.9308 \\ -18.4240 & 20.8454 & 529.1495 & -6.2331 & 3.5876 & 71.6001 \\ 13.3874 & -35.5134 & 536.0195 & 2.8497 & -14.4136 & 77.9784 \end{bmatrix}. \quad (3.24)$$

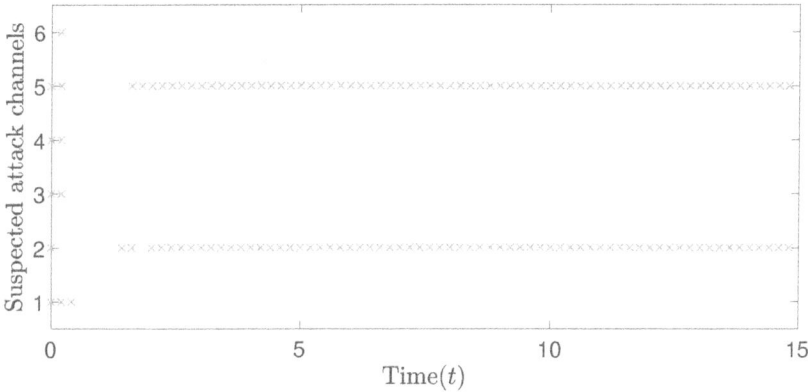

Figure 3.3 The identification result of suspicious attack channels.

To reflect the detection performance of the attack, the time-varying set $\eta(t) = \{i \mid \rho_i(t) < \|Qz_i(t)\|\}$ is defined to represent suspected compromised channels. Further, the results of suspicious attack detection are illustrated in Fig. 3.3. It can be observed that all malicious attacks were successfully identified eventually.

To demonstrate the SSM performance, Fig. 3.4 depicts the actual state trajectories (solid lines) of each power system along with their estimations (dashed lines) generated by distributed observers. And the tracking errors $e_i(t)$ between $\hat{x}_{2i}(t)$ and $\tilde{x}_{1i}(t)$ are illustrated in Fig. 3.5. It is evident that the proposed dual-layer observer structure remains unaffected by location-fixed 2-sparse sensor attacks and continues to achieve an SSM of the true state of power systems.

Fig. 3.6 shows the evolution curves of the dynamic gain parameter $\rho_i(t)$ in comparison to $\gamma^e_{2s+1}(t)$ and $\bar{\gamma}^e(t)$, from which it can be observed that the function $\rho_i(t)$ is in accordance with the conclusion in Theorem 3.2, i.e., $\lim_{t \to \infty} \gamma^e_{2s+1}(t) \leq \lim_{t \to \infty} \rho_i(t) = 0$. Furthermore, the profile of the function $b_{1i}(t)$ can be found in Fig. 3.7, and its final converged value also satisfies the conclusion of Theorem 3.2.

3.3 DISTRIBUTED SECURITY STATE MONITORING AGAINST LOCATION-VARYING SPARSE SENSOR ATTACKS

3.3.1 Model formulation

In this section, a discrete-time NAS is considered, which consists of N homogeneous agents and N one-by-one wireless sensors in a vulnerable cyber environment with the undirected topology $\mathcal{G}$. The dynamics of the i-th agent state $x_i(k) \in \mathbb{R}^n$ and corresponding sensor measurement $\tilde{y}_i(k) \in \mathbb{R}^m$ under location-varying sparse attacks are described as follows:

$$x_i(k+1) = Ax_i(k) + B\left(u(k) + K\sum_{j=1}^{N} a_{ij}(x_j(k) - x_i(k))\right), \qquad (3.25a)$$

$$\tilde{y}_i(k) = y_i(k) + a_i(k) = Cx_i(k) + a_i(k), \qquad (3.25b)$$

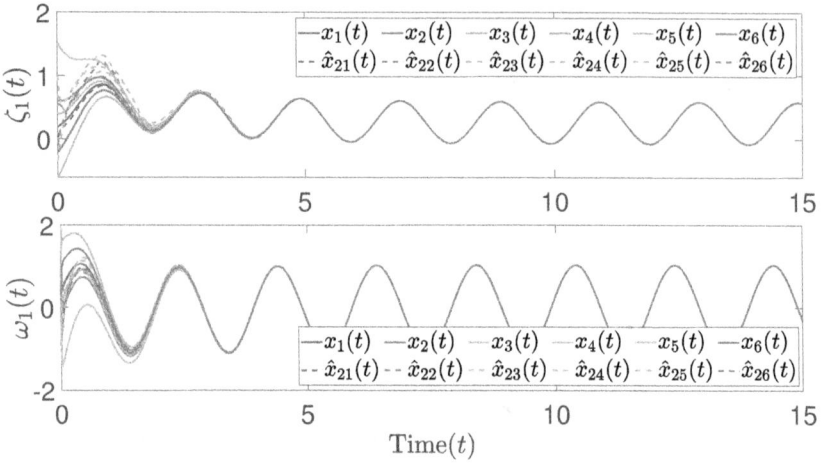

(a) $j = 1$

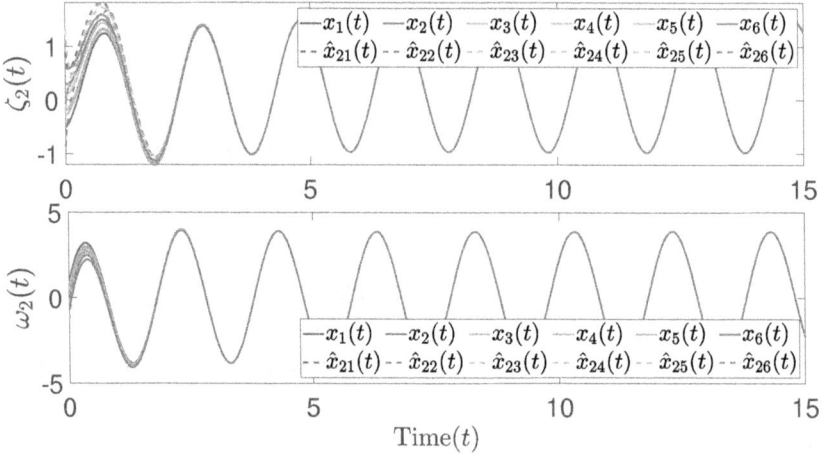

(b) $j = 2$

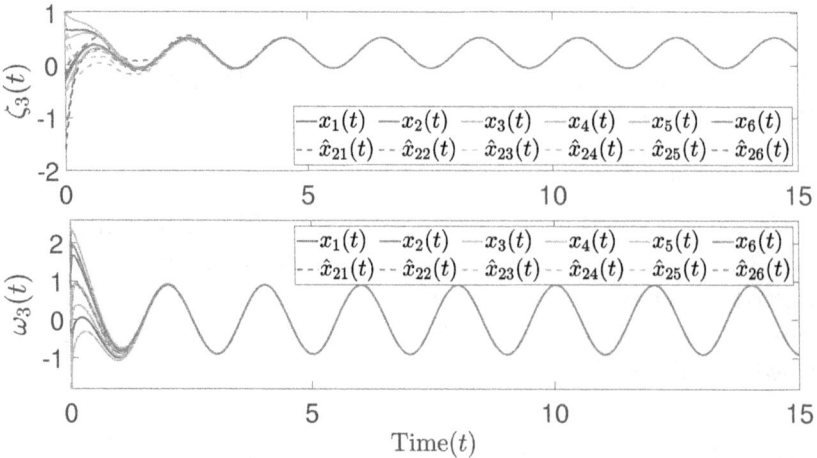

(c) $j = 3$

Figure 3.4 The system state trajectories of $\zeta_j(t)$, $\omega_j(t)$, $j \in [3]$ of power systems and their estimations $\hat{x}_{2i}(t)$.

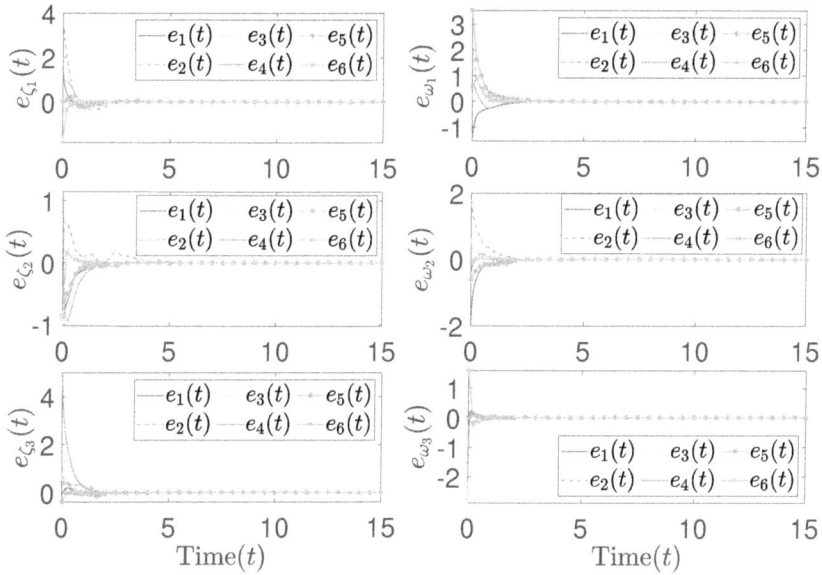

Figure 3.5 The curve of tracking errors $e_i(t) = \hat{x}_{2i}(t) - \tilde{x}_{1i}(t)$.

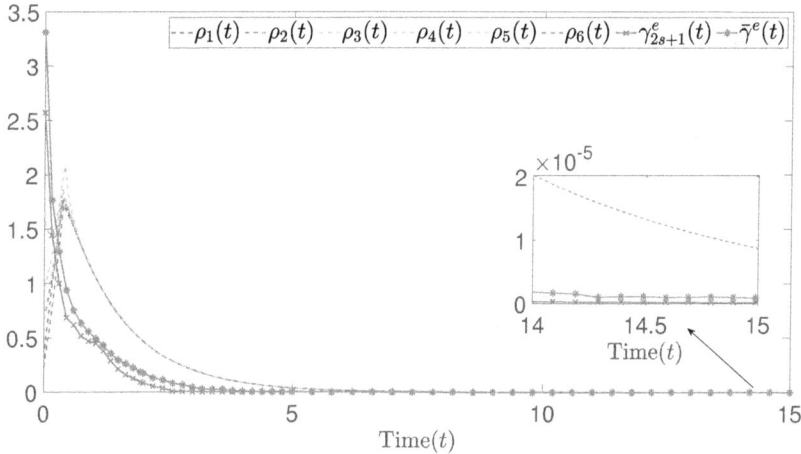

Figure 3.6 Dynamic trajectories of parameters $\rho_i(t)$, $\gamma^e_{2s+1}(t)$, and $\bar{\gamma}^e(t)$.

where $u(k) \in \mathbb{R}^p$ is the known internal control input, $y_i(k) = Cx_i(k)$ is the nominal sensor measurement, and $a_i(k)$ denotes the unknown external attack injection on the i-th sensor channel at time k. A, B, and C are internal system matrices with appropriate dimensions, and K represents the inner consensus control gain matrix.

The following assumptions and lemma about the NAS model are constructed.

Assumption 3.2 The matrix pair (A, B) is controllable and (C, A) is observable.

Assumption 3.3 The product of all unstable eigenvalues of the matrix A satisfies

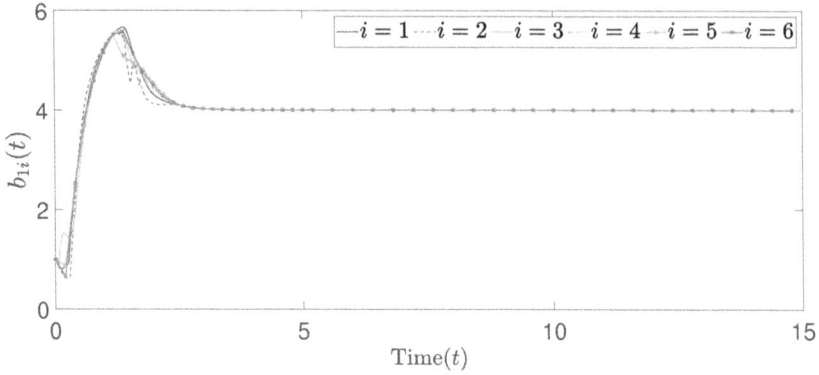

Figure 3.7 The profile of the indicator function $b_{1i}(t)$.

the following inequality:

$$\prod_{j \in \Gamma} \left| \lambda_j^u(A) \right| < \frac{\lambda_2(\mathcal{L}) + \lambda_N(\mathcal{L})}{\lambda_N(\mathcal{L}) - \lambda_2(\mathcal{L})},$$

where the set $\Gamma = \{j \mid |\lambda_j(A)| > 1\}$ denotes the indicator set of all unstable eigenvalues of the matrix A, $\lambda_2(\mathcal{L})$ and $\lambda_N(\mathcal{L})$ are the second smallest and largest eigenvalue of the Laplacian matrix $\mathcal{L}$, respectively.

Lemma 3.2 (Global consensus [80]) Under Assumptions 3.2–3.3, consider the NAS (3.25a) with the undirected and connected topology $\mathcal{G}$, if there exist matrices $R > 0$ and $\tilde{K}$ such that the following LMIs are satisfied for $j \in \{2, N\}$:

$$\begin{bmatrix} -R & RA_0^T - \lambda_j(\mathcal{L})\tilde{K}^T B^T \\ * & -R \end{bmatrix} < 0,$$

then the global consensus of (3.25a) can be reached with the control gain designed as $K = \tilde{K}R^{-1}$, i.e., $\lim_{k \to \infty} \|x_i(k) - x_j(k)\| = 0$, $\forall i, j \in \mathcal{V}$.

Out of the unpredictability of attacks, the fewer constraints imposed on an attack mean that the system is more defensive. Therefore, it is considered here that an attacker can arbitrarily select no more than a certain number of measurement transmissions to be modified to any false value at any moment, and the attack location can be different at each moment. The specific types include but are not limited to the following types: stealthy attacks [94], spoofing attacks [101], FDI attacks [110], and other data tampering attacks. Therefore, the defender knows nothing about the attack except the limit on the number of attacks. The size of this limit depends on the resources the attacker can afford and the bandwidth constraints of the network or an attack resilience index that the defender can tolerate.

Similarly, let $\mathcal{V}_{A,k} = \{i \in \mathcal{V} \mid a_i(k) \neq \mathbf{0}_m\}$ be defined as the corresponding indicator set of attacks at time k, and the set $\mathcal{V}_{N,k} = \mathcal{V} \backslash \mathcal{V}_{A,k}$ denotes the set of indicators that are non-compromised at moment k. Then, with the attack set $\mathcal{V}_{A,k}$, the following assumption about location-varying sparse attacks is given.

Assumption 3.4 *The attacks $a(k) = \text{col}\{a_i(k)\}$, $i \in \mathcal{V}$ is location-varying s-sparse, i.e., $|\mathcal{V}_{A,k}| \le s < N/2$ holds for each moment.*

3.3.2 Distributed security monitoring based on attack detection and isolation

Based on the system model construction in Section 3.3.1, one has learned that the sensor measurements are mixed with some data compromised by attacks. Therefore, the goal of this section is to securely recover the actual operational state of the system (3.25a) from these mixed measurements under the distributed framework by utilizing an attack detection and isolation mechanism. The distributed security monitoring framework is shown in Fig. 3.8, where the solid line indicates the benign sensor channel and the dashed line indicates the wireless channel that may be manipulated by the attacker.

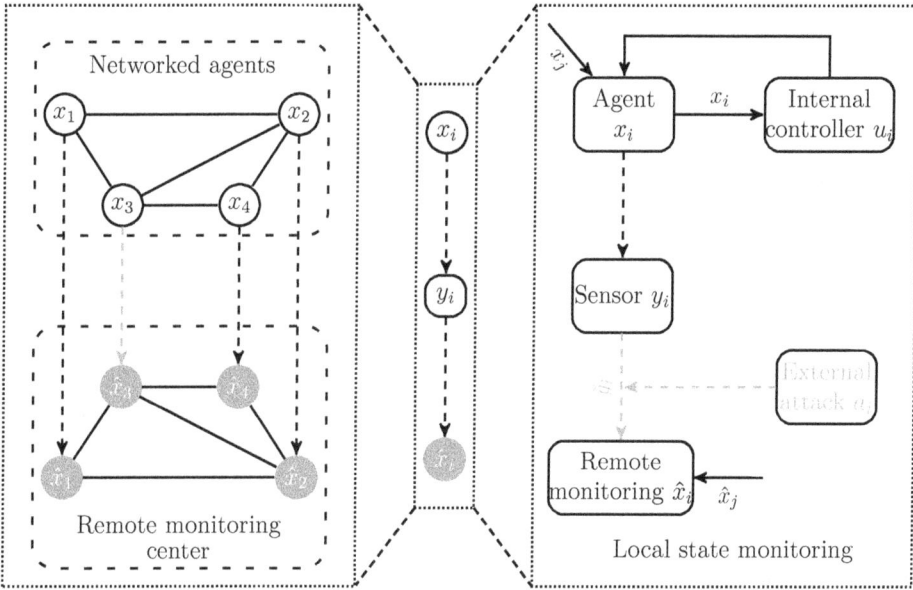

Figure 3.8 The distributed SSM framework against location-varying sparse sensor attacks.

A distributed Luenberger-like observer based on attack detection and isolation mechanism is proposed by borrowing and modifying the SIU algorithm in [37]. The specific distributed observer dynamics are described as follows:

$$\hat{x}_i(k+1) = A\hat{x}_i(k) + Bu(k) + BK\sum_{j=1}^{N} a_{ij}(\hat{x}_j(k) - \hat{x}_i(k)) + \delta_i(k)Lz_i(k), \quad (3.26)$$

where $\hat{x}_i(k)$ is the state monitoring value of $x_i(k)$ generated by the i-th observer at moment k and $z_i(k) = \tilde{y}_i(k) - C\hat{x}_i(k)$ denotes the innovation sequence. The matrix L is the observer gain matrix that needs to be designed, and the term $\delta_i(k)$ is time-varying logical decision gains employed to isolate the localized attack $a_i(k)$, which is designed as

$$\delta_i(k) = \begin{cases} 1, & \|QLz_i(k)\| \leq \tau(k), \\ 0, & \text{otherwise,} \end{cases} \tag{3.27}$$

where $\tau(k)$ is the detection threshold function and the matrix $Q^T Q > 0$ is positive definite that will be designed subsequently. With this attack detection and isolation scheme, the set $\mathcal{S}_k = \{i \mid \delta_i(k) = 0\}$ can be regarded as the sets of malicious attack indicators detected at moment k.

The following assumption is established to refine the driving force of the study.

Assumption 3.5 *The norm of the initial state monitoring error for each observer is bounded by the known bound $\eta > 0$, i.e., $\max_{i \in \mathcal{V}} \|e_i(0)\| \leq \eta$, where $e_i(k) = \hat{x}_i(k) - x_i(k)$ denotes the state monitoring error of the i-th observer.*

Remark 3.3 *In fact, due to the physical modeling limitations of NASs in reality, such as the operating range and the change rate, cannot be arbitrary. Therefore, the initial states of a system usually lie in a bounded known compact set and are publicly known. Hence, Assumption 3.5 is reasonable in practice.*

The role of the gain $\delta_i(k)$ is to ensure that the size of the innovation $z_i(k)$ is limited by the threshold $\tau(k)$, which in turn is identified as an anomaly attack if $\|QLz_i(k)\| \leq \tau(k)$ is violated. This ensures that malicious attacks can only cause limited and repairable damage. As a result, the main challenge is how to design the appropriate thresholds $\tau(k)$ to make sure that all the monitoring values $\hat{x}_i(k)$ converge to the actual system state $x_i(k)$. The key is to make a trade-off between the following two aspects:

i) If $\tau(k)$ is too small, then the gain $\delta_i(k)$ will limit the contribution of the normal and un-compromised measurements so that the RMC may unable to estimate the true state of the agent;

ii) On the contrary, if $\tau(k)$ is too large, then the RMC may fail to detect the malicious attacks in time, which may amplify the performance of the malicious attacks.

Based on the above considerations, the detection threshold function $\tau(k)$ can be designed as follows:

$$\tau(k) = \iota(d+g)\|Q\|(\tau_1(k) + \tau_2(k)), \tag{3.28a}$$

$$\begin{aligned} \tau_1(k+1) = {} & \left(d + \iota(g-d) + \lambda_2(\mathcal{L})\kappa(f-d) + 2\sqrt{s}\iota(d+g)\right)\tau_1(k) \\ & + \sqrt{s}\iota(d+g)\tau_2(k), \end{aligned} \tag{3.28b}$$

$$\tau_2(k+1) = \left(d - \left(1 - \frac{2s}{N}\right)\iota d + \left(1 + \frac{2s}{N}\right)\iota g\right)\tau_2(k) + \frac{2s}{N}\iota(d+g)\tau_1(k), \tag{3.28c}$$

where the initial values are selected as $\tau_1(0) = \sqrt{N}\eta$, $\tau_2(0) = \eta$, and observer parameters d, f, g, κ, ι will be designed and given in the subsequent analysis.

Remark 3.4 *In particular, for the case that the bound η of $\max_{i \in \mathcal{V}} \|e_i(0)\|$ cannot be obtained directly, one can select the maximum value of $\|\hat{x}_i(0) - C^\dagger \tilde{y}_i(0)\|$ to conservatively approximate the value η. However, this approximation method may result in relatively poorer observer monitoring accuracy to some extent in the face of initial attacks with large magnitude.*

3.3.3 Performance of distributed security state monitoring

To simplify the analysis, the following two auxiliary variables are introduced:

$$
\begin{aligned}
\tilde{e}(k) &= e(k) - 1_N \otimes \bar{e}(k) = (\tilde{I}_N \otimes I_n)e(k), \\
\bar{e}(k) &= \frac{1}{N} \sum_{i=1}^{N} e_i(k) = \frac{1}{N}(1_N^T \otimes I_n)e(k),
\end{aligned}
\tag{3.29}
$$

where $e(k) = \operatorname{col}\{e_i(k)\}$, $i \in \mathcal{V}$, $\tilde{I}_N = I_N - O_N$ with $O_N = \frac{1}{N}1_N 1_N^T$.
By the properties of the matrices $\mathcal{L}$ and $\tilde{I}_N$, one has:

$$
\mathcal{L}^T 1_N = \mathcal{L} 1_N = 0_N, \quad \tilde{I}_N \mathcal{L} = \mathcal{L}\tilde{I}_N = \mathcal{L}.
\tag{3.30}
$$

Furthermore, for any $i \in \mathcal{V}$, $\tilde{\delta}_i(k)$ is defined as:

$$
\tilde{\delta}_i(k) = \begin{cases} \delta_i(k), & \text{if } i \in \mathcal{V}_{A,k}, \\ 0, & \text{otherwise.} \end{cases}
$$

Then, denote the matrix $\Delta_{\mathcal{V}_{A,k}} = \operatorname{diag}\{\tilde{\delta}_i(k)\}_{i \in \mathcal{V}}$ and $\Delta_{\mathcal{V}_{N,k}} = \Delta_k - \Delta_{\mathcal{V}_{A,k}}$ with $\Delta_k = \operatorname{diag}\{\delta_1(k), \ldots, \delta_N(k)\}$.
From (3.25), (3.26), (3.29), and (3.30), it follows that:

$$
\begin{aligned}
e(k+1) &= (I_N \otimes A - \mathcal{L} \otimes BK)e(k) + (\Delta_k \otimes L)z(k), \\
&= (I_N \otimes \Phi - \mathcal{L} \otimes BK)e(k) + ((I_N - \Delta_{\mathcal{V}_{N,k}}) \otimes LC)e(k) \\
&\quad + (\Delta_{\mathcal{V}_{A,k}} \otimes L)z(k),
\end{aligned}
\tag{3.31}
$$

$$
\bar{e}(k+1) = \Phi\bar{e}(k) + \frac{1}{N}\bar{z}(k),
\tag{3.32}
$$

where $\Phi = A - LC$, $z(k) = \operatorname{col}\{z_i(k)\}$, $i \in \mathcal{V}$, and

$$
\bar{z}(k) = \left(1_N^T \Delta_{\mathcal{V}_{A,k}}(k) \otimes L\right)z(k) + \left[1_N^T(I_N - \Delta_{\mathcal{V}_{N,k}}) \otimes LC\right]e(k).
$$

Besides, according to Lemma 2.11, there exists a orthogonal matrix U such that $U^T \mathcal{L} U = J$ and

$$
\begin{aligned}
(U^T \otimes I_n)\tilde{e}(k) &= (U^T \tilde{I}_N U U^T \otimes I_n)e(k) \\
&= \begin{bmatrix} I_{(N-1)n} & 0 \\ 0 & 0 \end{bmatrix} \left(\begin{bmatrix} U_1^T \\ U_2^T \end{bmatrix} \otimes I_n \right) e(k) \\
&= \begin{bmatrix} I_{(N-1)n} & 0 \\ 0 & 0 \end{bmatrix} \begin{bmatrix} \check{e}_1(k) \\ \check{e}_2(k) \end{bmatrix}
\end{aligned}
$$

$$= \begin{bmatrix} \breve{e}_1(k) \\ \mathbf{0}_n \end{bmatrix}, \tag{3.33}$$

where U_1 and U_2 denote sub-matrices defined in Lemma 2.11, $\breve{e}_1(k) = (U_1^T \otimes I_n)e(k) \in \mathbb{R}^{(N-1)n}$ and $\breve{e}_2(k) = (U_2 \otimes I_n)e(k) \in \mathbb{R}^n$ denote intermediate auxiliary variables.

In particular, it follows that

$$\begin{aligned} e(k) &= \tilde{e}(k) + 1_N \otimes \bar{e}(k) \\ &= (U \otimes I_n)\breve{e}(k) + 1_N \otimes \bar{e}(k) \\ &= (U_1 \otimes I_n)\breve{e}_1(k) + 1_N \otimes \bar{e}(k), \end{aligned}$$

where $\breve{e}(k) = [\breve{e}_1^T(k) \; \breve{e}_2^T(k)]^T$.

Then, the dynamics of $\breve{e}_1(k)$ can be computed as follows:

$$\begin{aligned} \breve{e}_1(k+1) &= \Lambda \breve{e}_1(k) + \left(U_1^T(I_N - \Delta_{\nu_{N,k}}) \otimes LC \right) e(k) + (U_1^T \Delta_{\nu_{A,k}} \otimes L)z(k) \\ &= \left(\Lambda + (I_N - U_1^T \Delta_{\nu_{N,k}} U_1) \otimes LC \right) \breve{e}_1(k) + (U_1^T \Delta_{\nu_{A,k}} \otimes L)z(k), \end{aligned} \tag{3.34}$$

where $\Lambda = I_{N-1} \otimes \Phi - J_1 \otimes BK$.

Lemma 3.3 For the detection threshold function $\tau(k)$ given in (3.28), if there exist positive parameters $0 < r_1 < 1 \le d$, $\kappa_d > 0$, $\kappa_f > 0$, $\iota_d > 0$ and $\iota_g > 0$ such that the following LMI holds:

$$\begin{bmatrix} -rI_2 & \Psi^T \\ * & -I_2 \end{bmatrix} < 0, \tag{3.35}$$

where

$$\Psi = \begin{bmatrix} \tilde{\varpi}_1 & \sqrt{s}(\iota_d + \iota_g) \\ 2s/N(\iota_d + \iota_g) & \tilde{\varpi}_2 \end{bmatrix},$$

$$\tilde{\varpi}_1 = d + \lambda_2(\mathcal{L})(\kappa_f - \kappa_d) + (1 + 2\sqrt{s})\iota_g + (2\sqrt{s} - 1)\iota_d,$$

$$\tilde{\varpi}_2 = d - (1 - \frac{2s}{N})\iota_d + (1 + \frac{2s}{N})\iota_g.$$

Then, the equilibrium point 0 of systems (3.28) is asymptotically stable by selecting

$$\kappa = \frac{\kappa_d}{d}, \quad \iota = \frac{\iota_d}{d}, \quad f = \frac{\kappa_f}{\kappa}, \quad g = \frac{\iota_g}{\iota}.$$

Proof 3.4 Define the auxiliary variable as $\tilde{\tau}(k) \triangleq [\tau_1^T(k), \; \tau_2^T(k)]^T$, one has $\tilde{\tau}(k+1) = \Psi \tilde{\tau}(k)$.

By the Schur complement in Lemma 2.7, the LMI (3.35) is equivalent to $\Psi^T \Psi < rI_2$. Hence, it follows from $0 < r < 1$ that $\tilde{\tau}(k)$ is asymptotically stable. Furthermore, the equilibrium point 0 of systems $\tau_1(k)$ and $\tau_2(k)$ is concluded to be asymptotically stable, thereby the proof is complete.

Theorem 3.3 For the NAS (3.25) under location-varying sparse attacks, suppose that Assumptions 3.2–3.5 hold. If the LMI (3.35) and the following inequalities holds simultaneously:

$$\lambda_2(\mathcal{L})\kappa_d < d, \quad \iota_d < d,$$

$$(\lambda_2(\mathcal{L}) + \lambda_N(\mathcal{L}))d + (\lambda_N(\mathcal{L}) - \lambda_2(\mathcal{L}))\kappa_f < 2d - 2\iota_d, \tag{3.36}$$

and there exist matrices $P > 0$, $\check{L}$, and $\check{K}$ which satisfy the following LMIs:

$$\begin{bmatrix} -dP & A^T P \\ * & -P \end{bmatrix} < 0,$$

$$\begin{bmatrix} -fP & A^T P - \check{K}^T \\ * & -P \end{bmatrix} < 0, \tag{3.37}$$

$$\begin{bmatrix} -gP & A^T P - C^T \check{L}^T \\ * & -P \end{bmatrix} < 0,$$

then the asymptotically SSM of the system (3.25) can be realized with the distributed observer (3.26) and gain parameters are selected as follows:

$$K = \kappa B^\dagger P^{-1}\check{K}, \quad L = \iota P^{-1}\check{L}, \quad Q = P^{\frac{1}{2}}.$$

Proof 3.5 *Firstly, one can rewrite the matrices Λ and Φ in the following form:*

$$\Lambda = ((1 - \iota)I_{N-1} - \kappa J_1) \otimes A + \iota(A - \tilde{L}C) + \kappa J_1 \otimes (A - B\tilde{K}),$$

$$\Phi = (1 - \iota)A + \iota(A - \tilde{L}C),$$

where $\tilde{L} = \frac{1}{\iota}L$ and $\tilde{K} = \frac{1}{\kappa}K$.

Then, Lyapunov candidate functions for error systems (3.32) and (3.34) are constructed as follows:

$$V_1(k) = \left(\tilde{e}^T(k)P_1\tilde{e}(k)\right)^{\frac{1}{2}} = \left(\check{e}_1^T(k)P_2\check{e}_1(k)\right)^{\frac{1}{2}} = \|(\tilde{I} \otimes Q)e(k)\|,$$

$$V_2(k) = \left(\bar{e}^T(k)P\bar{e}(k)\right)^{\frac{1}{2}} = \frac{1}{N}\|Q(1_N^T \otimes I_n)e(k)\|,$$

where $P_1 = I_N \otimes P$ and $P_2 = I_{N-1} \otimes P$. It follows that

$$V_1(0) = \|(\tilde{I} \otimes Q)e(0)\| \leq \|Q\|\left(\sum_{i=1}^N \|e_i(0)\|^2\right)^{\frac{1}{2}} \leq \sqrt{N}\|Q\|\eta,$$

$$V_2(0) = \frac{1}{N}\|Q(1_N^T \otimes I_n)e(0)\| \leq \frac{1}{N}\|Q\|\sum_{i=1}^N \|e_i(0)\| \leq \|Q\|\eta$$

with $\|\tilde{I}\| = 0$ and $Q^T Q = P$.

From the Schur complement in Lemma 2.7 and LMIs (3.37), one can obtain

$$A^T PA < dP,$$

$$(A - B\tilde{K})^T P (A - B\tilde{K}) < fP,$$
$$(A - \tilde{L}C)^T P (A - \tilde{L}C) < gP.$$

From the inequality (3.36), one has

$$|1 - \iota - \kappa \lambda_N(\mathcal{L})| d + \iota g + \lambda_N(\mathcal{L}) \kappa f < d + \iota(g - d) + \lambda_2(\mathcal{L}) \kappa (f - g).$$

Hence, for any $x_a \in \mathbb{R}^{(N-1)n}$ and $x_b \in \mathbb{R}^n$, one has

$$\|(I_{N-1} \otimes Q) \Lambda x_a\| \leq \|(((1 - \iota) I_{N-1} - \kappa J_1) \otimes QA) x_a\| + \iota \|I_{N-1} \otimes Q(A - \tilde{L}C) x_a\|$$
$$+ \kappa \left\| \left(J_1 \otimes Q(A - B\tilde{K}) \right) x_a \right\|$$
$$< (d + \iota(g - d) + \lambda_2(\mathcal{L}) \kappa(f - d)) \|Q x_a\|,$$
$$\|Q \Phi x_b\| \leq \|(1 - \iota) QA x_b\| + \iota \|Q(A - \tilde{L}C) x_b\|$$
$$\leq ((1 - \iota) d + \iota g) \|Q x_b\|,$$
$$\|Q \tilde{L} C x_b\| < \|QA_0 x_b\| + g \|Q x_b\| < (d + g) \|Q x_b\|,$$
$$\|QLC x_b\| < \iota(d + g) \|Q x_b\|.$$

In the following, the mathematical induction will be used to prove the theorem.
Step 1: Initially, from $\tau_1(0) = \sqrt{N}\eta$, $\tau_2(0) = \eta$, one has

$$V_1(0) \leq \|Q\| \tau_1(0), \quad V_2(0) \leq \|Q\| \tau_2(0).$$

Step 2: At this point, if the following conditions hold:

$$V_1(k) \leq \|Q\| \tau_1(k), \quad V_2(k) \leq \|Q\| \tau_2(k).$$

Then, there holds that $\forall i \in \mathcal{V}$,

$$\|QLC e_i(k)\| = \|QLC(\tilde{e}_i(k) + \bar{e}(k))\|$$
$$\leq \|(I_N \otimes QLC) \tilde{e}(k)\| + \|QLC \bar{e}(k)\|$$
$$\leq \iota(d + g) \|Q\| (\tau_1(k) + \tau_2(k))$$
$$= \tau(k).$$

Hence, from the parameter $\delta_i(k)$ in (3.27), it follows that for all $i \in \mathcal{V}$,

$$\delta_i(k) \|QL z_i(k)\| \leq \tau(k), \quad \Delta_{\mathcal{V}_{N,k}} = I_{\mathcal{V}_{N,k}}(k).$$

Step 3: Furthermore, for any $i \in \mathcal{V}$, one has

$$\|Q \bar{z}(k)\| \leq \left\| \left(1_N^T \Delta_{\mathbf{A}_k^y}(k) \otimes QL \right) z(k) \right\| + \left\| \left(1_N^T I_{\mathbf{A}_k} \otimes QLC \right) e(k) \right\|$$
$$< 2|\mathcal{V}_{A,k}| \tau(k),$$

Based on the orthogonality of the matrices U and $\|U_1\| = 1$, one has

$$V_1(k + 1) \leq \|(I_{N-1} \otimes Q) \Lambda \check{e}_1(k)\| + \|(U_1^T I_{\mathcal{V}_{A,k}} U_1 \otimes QLC) \check{e}_1^x(k)\|$$

$$+ \|(U_1^T \Delta_{\mathcal{V}_{A,k}} \otimes QL)z(k)\|$$
$$< (d + \iota(g - d) + \lambda_2(\mathcal{L})\kappa(f - d - \iota(g - d))) \|Q\|\tau_1(k)$$
$$+ \sqrt{s}(\iota(d + g)\|Q\|\tau_1(k) + \tau(k))$$
$$= (d + \iota(g - d) + \lambda_2(\mathcal{L})\kappa(f - d) + 2\sqrt{s}\iota(d + g)) \|Q\|\tau_1(k)$$
$$+ \sqrt{s}\iota(d + g)\|Q\|\tau_2(k)$$
$$= \|Q\|\tau_1(k + 1),$$
$$V_2(k + 1) < \|Q\Phi\bar{e}^x(k)\| + 2s_k\tau(k)$$
$$\leq (d - (1 - 2s_k)\iota d + (1 + 2s_k)\iota g) \|Q\|\tau_2(k) + 2s_k\iota(d + g)\|Q\|\tau_1(k)$$
$$\leq (d - (1 - 2s_k)\iota d + (1 + 2s_k)\iota g) \|Q\|\tau_2(k) + \frac{2s}{N}\iota(d + g)\|Q\|\tau_1(k)$$
$$= \|Q\|\tau_2(k + 1),$$

where $s_k = |\mathcal{V}_{A,k}|/N$.

Consequently, according to the mathematical induction, one can easily obtain that for all $k \in \mathbb{N}$ with

$$V_1(k) < \|Q\|\tau_1(k), \quad V_2(k) < \|Q\|\tau_2(k).$$

Hence, due to

$$0 \leq \lim_{k \to \infty} \|Qe_i(k)\| \leq \lim_{k \to \infty} \|Q\|(\tau_1(k) + \tau_2(k)) \leq \sqrt{2}\|Q\| \lim_{k \to \infty} \|\tilde{\tau}(k)\| = 0,$$

one can get $\lim_{k \to \infty} \|e_i(k)\| = 0$, $i \in \mathcal{V}$. The proof is complete.

From the LMIs (3.35)–(3.36) and the convergence requirement, both $\tilde{\varpi}_1$ and $\tilde{\varpi}_2$ in Lemma 3.3 should satisfy $|\tilde{\varpi}_1| < 1$ and $|\tilde{\varpi}_2| < 1$. Therefore, it can be concluded that $(1 - 2s/N) > 0$ is necessary, i.e., $s_k < 1/2$ and $|\mathcal{V}_{A,k}| < N/2$. Based on the above results, the secure state monitoring procedure for NASs is summarized as Algorithm 3.2.

It is worth noting that in the above results, the control matrix K of the NAS (3.25) is designed centrally without considering the goal of consensus control. In general, however, the consensus control matrix K can be chosen by using the method such as in Lemma 3.2. In this situation, one has the following result.

Corollary 3.1 For the NAS (3.25) under location-varying sparse attacks, if a consensus control matrix K has been previously designed according to Lemma 3.2 and the corresponding distributed state observer (3.26) is changed into the following form:

$$\hat{x}_i(k + 1) = A_0\hat{x}_i(k) + Bu(k) + \check{K}\sum_{j=1}^{N} a_{ij}(\hat{x}_j(k) - \hat{x}_i(k)) + \delta_i(k)Lz_i(k),$$

where $\check{K} = \kappa P^{-1}\check{K}$ and other parameters are the same as in Theorem 3.3, then the secure state monitoring of the NAS (3.25) can still be realized.

Algorithm 3.2 Distributed security state monitoring algorithm

---

Input: A, B, C, $\tilde{y}_i(k)$, $u(k)$, $\hat{x}_i(0)$, N, s, η;

1: Solving LMIs (3.35)-(3.37) to obtain the gains ι, κ, d, g, f and matrices $\check{K}, \check{L}, P$;

2: $K = \kappa B^\dagger P^{-1}\check{K}$, $L = \iota P^{-1}\check{L}$, $Q = P^{\frac{1}{2}}$;

3: $\tau_1(0) = \sqrt{N}\eta$, $\tau_2(0) = \eta$;

4: **while** $k > 0$ **do**

5:     Communicate with neighbors to get $\hat{x}_j(k)$;

6:     Perform the following auxiliary parameter updates:

$$\tau_1(k+1) = \left(d + \iota(g-d) + \lambda_2(\mathcal{L})\kappa(f-d) + 2\sqrt{s}\iota(d+g)\right)\tau_1(k)$$
$$+ \sqrt{s}\iota(d+g)\tau_2(k),$$

$$\tau_2(k+1) = \left(d - (1 - \frac{2s}{N})\iota d + (1 + \frac{2s}{N})\iota g\right)\tau_2(k) + \frac{2s}{N}\iota(d+g)\tau_1(k).$$

7:     Update the attack detection threshold $\tau(k) = \iota(d+g)\|Q\|(\tau_1(k) + \tau_2(k))$;

8:     $z_i(k) = \tilde{y}_i(k) - C\hat{x}_i(k)$;

9:     **if** $\|QLz_i(k)\| \leq \tau(k)$ **then**

10:         $\delta_i(k) = 1$;

11:     **else**

12:         $\delta_i(k) = 0$;

13:     **end if**

14:     Update the state monitoring as follows:

$$\hat{x}_i(k+1) = A\hat{x}_i(k) + Bu(k) + BK\sum_{j=1}^{N} a_{ij}(\hat{x}_j(k) - \hat{x}_i(k)) + \delta_i(k)Lz_i(k);$$

15: **end while**

Output: The state monitoring value $\hat{x}_i(k)$.

Proof 3.6 *Define the matrix* $\acute{K} = \check{K} - BK$. *In view of the design of the consensus control matrix* K, *one has* $\lim_{k \to \infty} \|x_i(k) - x_j(k)\| = 0$. *Hence, a proof similar to that of Theorem 3.3 can smoothly establish the conclusion assisted by the* $\mathcal{H}_\infty$ *method and the following facts:*

$$\lim_{k \to \infty} \acute{K}\sum_{j=1}^{N} a_{ij}(x_j(k) - x_i(k)) = \mathbf{0}_n.$$

The corollary is proved complete.

3.3.4 Numerical simulations

In order to illustrate the effectiveness of the proposed Algorithm 3.2, similarly carry out simulation validation of the SSM on the IEEE 6-bus power system with the communication topology of Fig. 3.2. With the sampling step taken to be $\Delta t = 0.01$, system matrices of the power network are described in (3.25) with the following parameters:

$$A = I_6 + \Delta t \begin{bmatrix} 0 & I \\ M_g^{-1}\left(L_{Ig}^T L_{II}^{-1} L_{Ig} - L_{gg}\right) & -M_g^{-1}D_g \end{bmatrix},$$

$$B = \Delta t \begin{bmatrix} 0 & 0 \\ M_g^{-1} & -M_g^{-1}L_{Ig}^T L_{II}^{-1} \end{bmatrix}, \quad C = I_6,$$

where matrix parameters and initial states are the same as in Section 3.2.4. Similarly, assume that all remote transmission channels are open to attack, but only up to two channels are compromised at each time instant. It can be seen that Assumption 3.4 is satisfied.

Then, by solving the LMIs (3.35)–(3.37) with the help of the MATLAB Robust Control Toolbox, the following matrices and parameters can be obtained:

$$\kappa = 0.2427, \quad \iota = 0.0433, \quad d = 1.0002, \quad f = 0.0298, \quad g = 0.0026,$$

$$K = \begin{bmatrix} 0.6677 & 0.1738 & 0.2168 & 2.2815 & -0.1543 & -0.0499 \\ -0.2186 & 0.0602 & -0.0626 & -0.5815 & 0.5585 & -0.0520 \\ -0.1567 & -0.0871 & -0.0257 & -0.5766 & -0.1538 & 0.1908 \\ 0.2353 & 0.07964 & 0.0848 & 0.8360 & 0.0281 & 0.0096 \\ 0.0038 & 0.0543 & 0.0114 & 0.0937 & 0.2326 & 0.0023 \\ 0.0309 & 0.0118 & 0.0250 & 0.1234 & 0.0148 & 0.0742 \\ 0.0838 & 0.0630 & 0.0367 & 0.3502 & 0.1620 & 0.0048 \\ 0.1026 & 0.0356 & 0.0460 & 0.3736 & 0.0195 & 0.0515 \\ 0.0196 & 0.0295 & 0.0193 & 0.1110 & 0.1055 & 0.0443 \end{bmatrix},$$

$$L = \begin{bmatrix} 0.0433 & 0 & 0 & 0.0004 & 0 & 0 \\ 0 & 0.0433 & 0 & 0 & 0.0004 & 0 \\ 0 & 0 & 0.0433 & 0 & 0 & 0.0004 \\ -1.02*10^{-4} & 5.84*10^{-5} & 5.02*10^{-5} & 0.0429 & 0 & 0 \\ 1.89*10^{-4} & -3.67*10^{-4} & 1.78*10^{-4} & 0 & 0.0425 & 0 \\ 3.92*10^{-4} & 3.78*10{-4} & -7.71*10^{-4} & 0 & 0 & 0.0303 \end{bmatrix},$$

$$Q = \begin{bmatrix} 0.7387 & -0.1136 & -0.0509 & 0.1014 & -0.0185 & -0.0105 \\ -0.1136 & 0.8087 & -0.0528 & 0.0459 & 0.0512 & -0.0064 \\ -0.0509 & -0.0528 & 0.9176 & 0.0332 & 0.0040 & 0.0303 \\ 0.1014 & 0.0459 & 0.0332 & 0.9127 & 0.0071 & -0.0020 \\ -0.0185 & 0.0512 & 0.0040 & 0.0071 & 0.9065 & 0.0048 \\ -0.0105 & -0.0064 & 0.0303 & -0.0020 & 0.0048 & 0.9175 \end{bmatrix}.$$

Set the attack vector $a_i(k) = -e^{-\frac{5}{k}}y_i(k) + 0.3\epsilon_i(k)$ with $\epsilon_i(k)$ obeying a uniform distribution on the interval $[0, 1]$, and suppose that the attacker randomly switches the attack target every 50 steps. Then, by applying the proposed Algorithm 3.2, actual attack locations and the detected attack locations are displayed in Fig. 3.9, where the circle represents the locations picked by the attacker at each moment, the crosses indicate attack channels detected and isolated by local observers at the corresponding time instant. For viewing purposes, the results are visualized only every $0.8s$.

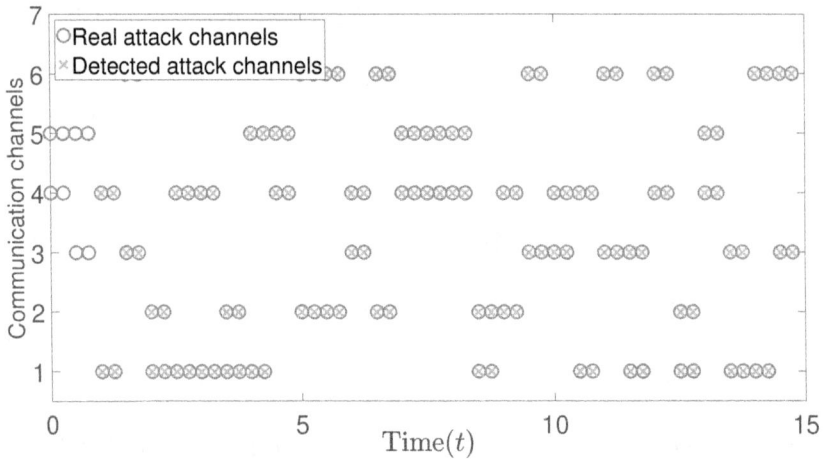

Figure 3.9 Compromised transmission channels and malicious channels identified by observers.

The SSM result is shown in Fig. 3.10, where solid lines represent the actual state trajectories of the power system and dashed lines indicate state monitoring trajectories produced by the distributed observer (3.26).

In addition, the performance of secure asymptotic state monitoring under location-varying sparse attacks is verified by plotting the local monitoring error norm curve of each observer in Fig. 3.11. Fig. 3.12 reveals that the residual norm $\tilde{V}_i(k) = \|QLCe_i(k)\|$ of a benign measurement remains within the threshold function $\tau(k)$, as shown in (3.28). It can be seen that the benign transmission channel does not trigger the attack detection mechanism, and all data that violate the detection rules are subject to isolation.

Next, consider the case where the NAS has designed the consensus control gain matrix K in advance to show the applicability and validity of Corollary 3.1. Assume that the synchronization control matrix K_0 has been designed as in (3.24), and the matrix $\check{K}$ is given as:

$$\check{K} = \begin{bmatrix} 0.2579 & 0 & 0 & 0.0026 & 0 & 0 \\ 0 & 0.2579 & 0 & 0 & 0.0026 & 0 \\ 0 & 0 & 0.2579 & 0 & 0 & 0.0026 \\ 0.0006 & 0.0003 & 0.0003 & 0.2553 & 0 & 0 \\ 0.0011 & -0.0022 & 0.0011 & 0 & 0.2527 & 0 \\ 0.0023 & 0.0023 & -0.0046 & 0 & 0 & 0.1805 \end{bmatrix}.$$

Then, the corresponding system states trajectories and local state monitoring are shown in Fig. 3.13, and the monitoring error norm of each observers are exhibited in Fig. 3.14, which show that effective and secure state monitoring can still be achieved with the matrix K_0 given in advance.

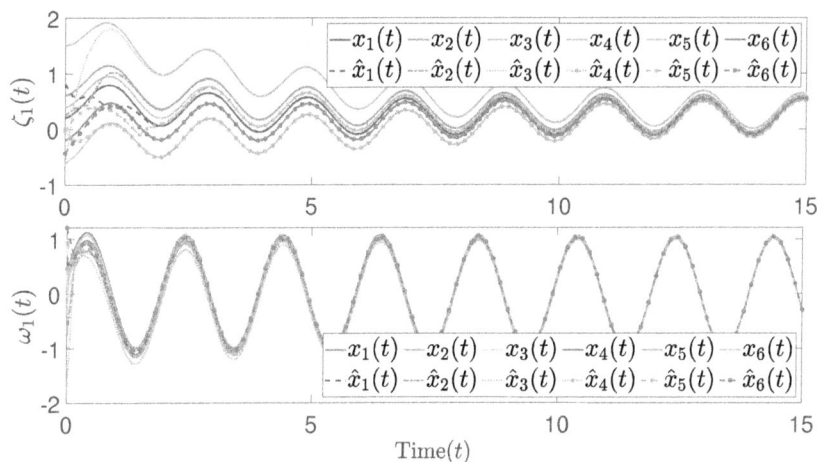

(a) $j = 1$

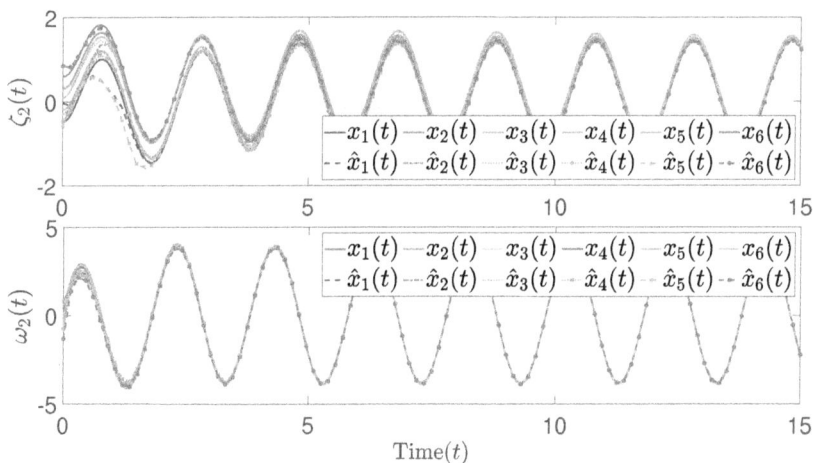

(b) $j = 2$

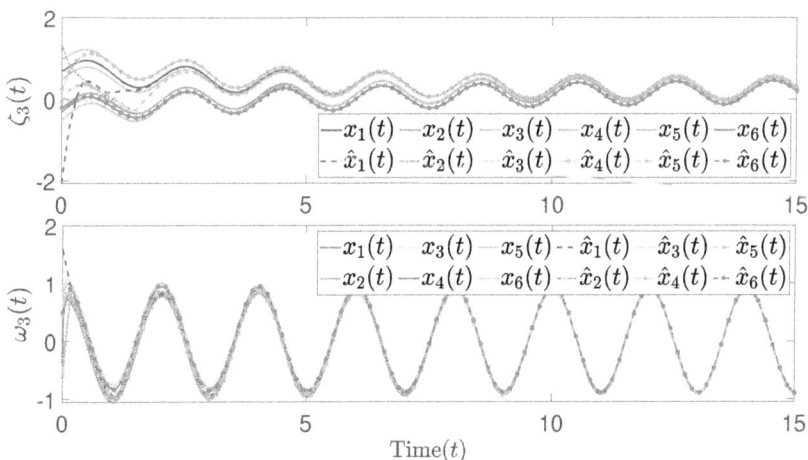

(c) $j = 3$

Figure 3.10 System states trajectories of $\zeta_j(t)$, $\omega_j(t)$, $j \in [3]$ and their monitoring values $\hat{x}_i(t)$ with the control gain matrix K.

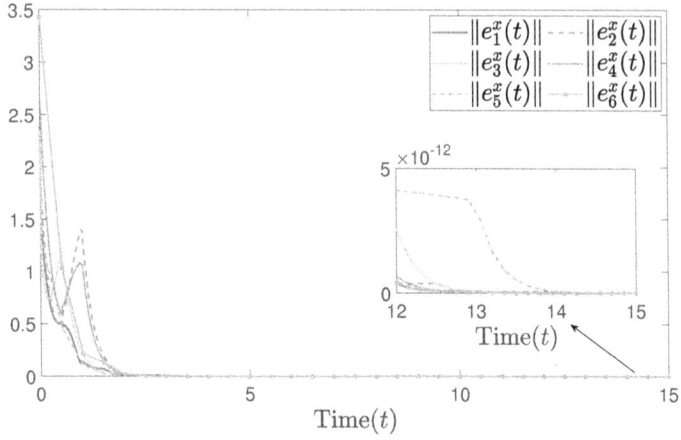

Figure 3.11   The curve of local state monitoring error norm.

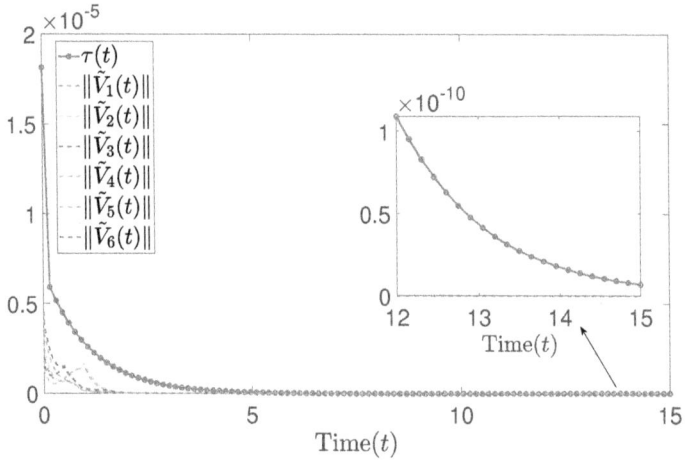

Figure 3.12   The real residual norm $\tilde{V}_i(t)$ and the threshold function $\tau(t)$.

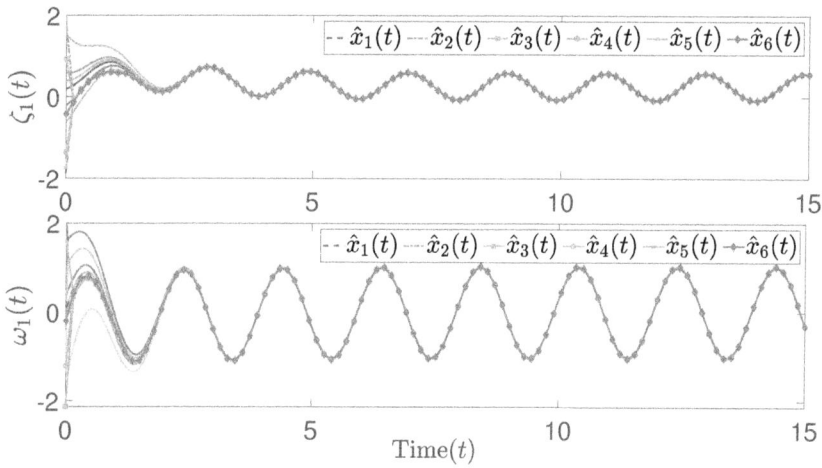

(a) $j = 1$

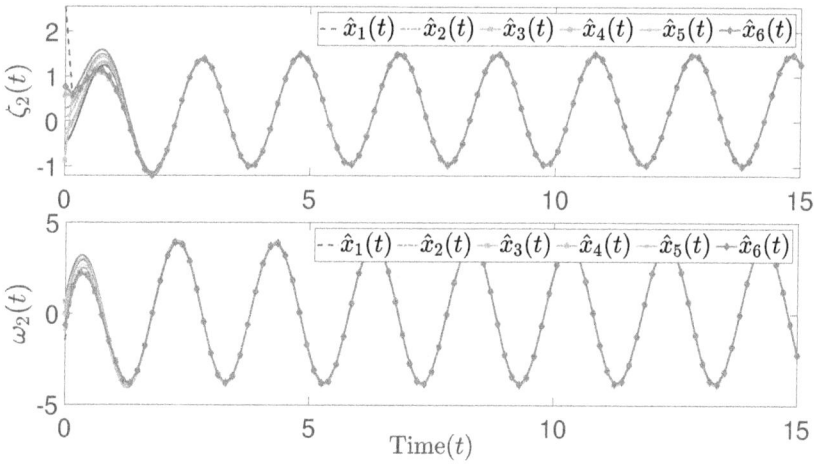

(b) $j = 2$

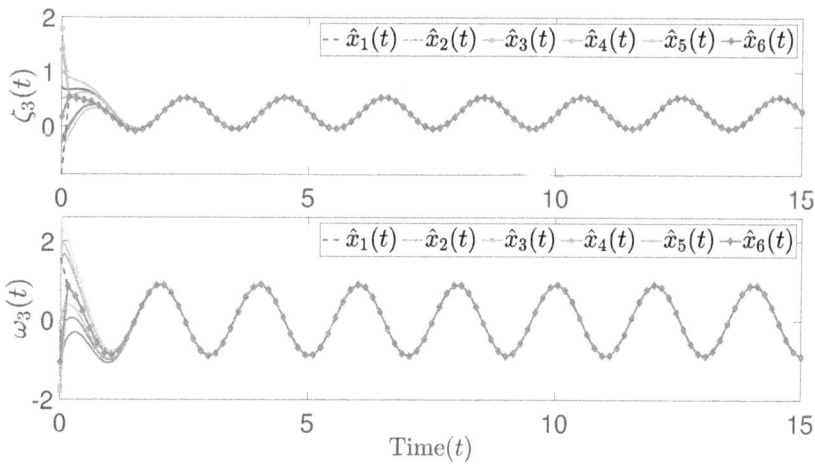

(c) $j = 3$

Figure 3.13   System states trajectories of $\zeta_j(t)$, $\omega_j(t)$, $j \in [3]$ and their monitoring values $\hat{x}_i(t)$ with the control gain matrix K_0.

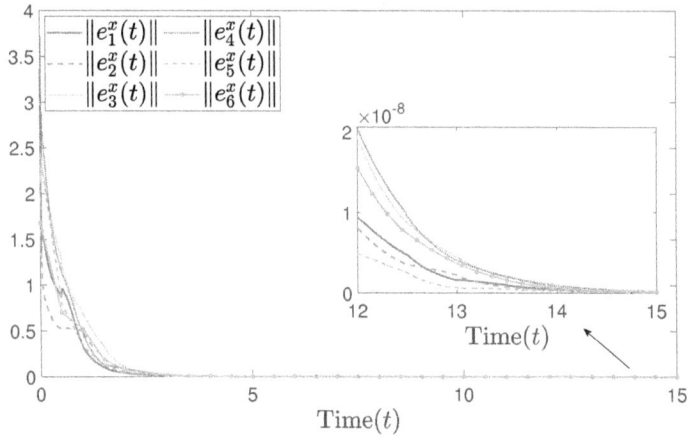

Figure 3.14 The curve of local state monitoring error norm with the matrix K_0 given in (3.24).

3.4 CONCLUSIONS

This chapter has studied the distributed SSM problem for NASs in the presence of location-fixed or location-varying sparse sensor attacks, respectively. For continuous-time NASs under location-fixed sparse attacks, a dual-layer dynamic observation framework has been proposed for distributed SSM of NASs. The decentralized and distributed observers have been utilized to decouple the measurement residual feedback estimation from the collaborative attack suppression effort, and a dynamic gain adjustment strategy has been proposed to assist in achieving attack suppression and isolation. Based on the proposed algorithm, the distributed design of the dynamic threshold function has been improved without losing the security performance of state monitoring, which broadens the application of the SSM algorithm.

For discrete-time NAS under location-varying sparse sensor attacks, a logic-verified attack detection mechanism has been utilized by the designed observers to block attack channels so that the remote monitoring center can effectively identify and precisely isolate the measurement data under malicious attacks. Compared to previous works in [37, 40, 42], where the threshold detection function is designed based on ambiguous parameters, an explicit detection threshold function design has been given in this chapter based on the analysis of the monitoring system dynamics, which can further regulate the convergence rate of state monitoring without losing the security performance.

CHAPTER 4

Distributed security monitoring of heterogeneous sensor networks against sparse sensor attacks

This chapter investigates distributed SSM for heterogeneous sensor networks under adversarial attack scenarios. Following a critical review of existing approaches and motivation analysis, two principal contributions are presented. Section 4.2 addresses sparsity-varying sensor attacks that relax conventional sparsity constraints, posing heightened security challenges. A novel resilient distributed observer is developed using sliding window estimation techniques, establishing sufficient defense conditions while preserving original measurement redundancy requirements. Notably, the proposed solution maintains system observability without augmenting sensor configurations. Building upon the attack isolation framework from preceding chapters, Section 4.3 proposes a hybrid SSM architecture countering vulnerability-based sparse attacks. The analysis systematically differentiates defense mechanisms for heterogeneous trusted sensors, incorporating adaptive detection thresholds that balance false alarm suppression and attack sensitivity. Through rigorous Lyapunov-based stability proofs and comparative simulations, the chapter demonstrates quantifiable performance improvements in attack resilience while maintaining computational efficiency.

4.1 INTRODUCTION

In the homogeneous sensor network discussed in the previous chapter, each sensor has an equal probability of being affected by an attack [29,37,40,42,111,112]. However, in large-scale industrial applications, the types of sensor and network device nodes are diverse, and the attack intentions at different moments can vary, resulting in different sensors having varying levels of resistance to malicious attacks at different times [113, 114]. Moreover, as attackers become more sophisticated, they tend to target sensors with evident vulnerabilities first. In light of this, Faiq et al. [115] proposed enhancing

DOI: 10.1201/9781003669913-4

the robustness of the underlying network by leveraging heterogeneous nodes, which utilizes the disjointness vulnerability characteristics of heterogeneous nodes against the attack so that the sparse network can also have sufficient robustness. Furthermore, Mitra et al. [85] augmented the network structure with trusted nodes, which, along with heterogeneous nodes, enhanced the robust performance of the network.

The above ideas and some of the research works have broadened our horizon on the resistance of sensor networks against sparse attacks. Therefore, in this chapter, inspired by the above works and the diversity and trust of sensors, the efficacy of this characteristic in enhancing the resilience of sensor networks against sparse attacks is investigated. In other words, the susceptibility of different sensors against attacks is heterogeneous, which makes it possible to label sensors and improve algorithms targetedly. Firstly, with the idea of the sliding window, an improved resilient distributed state observer is constructed to effectively address the performance degradation problem of observer-based algorithms developed in [37, 40, 42, 43] against sparse-varying attacks. In addition, the potential performance of distributed security state monitoring algorithms against sparse attacks is motivated by the accessibility of global historical attack detection information. Secondly, leveraging the vulnerabilities of the heterogeneous sensor network, different weights are constructed for each distributed observer, effectively resisting the sparse attacks related to the vulnerabilities in this heterogeneous network. Finally, with the assistance of trusted sensor nodes, a simplified design of the distributed security monitoring algorithm is further explored, enhancing the resistance of the sensor network against sparse attacks. The results indicate that by considering the heterogeneity of the sensor nodes, the network's tolerance to sparse attacks can be increased, further enhancing the security performance of the networked agent systems.

4.2 DISTRIBUTED SECURITY MONITORING AGAINST SPARSE-VARYING SENSOR ATTACKS

4.2.1 Model formulation

Consider an unknown static signal $\theta^* \in \mathbb{R}^n$ monitored by N wireless sensors operating in a vulnerable cyber environment:

$$y_i(k) = C_i \theta^* + d_i(k) + a_i(k), \quad i \in \mathcal{V} = [N], \tag{4.1}$$

where $y_i(k)$, $d_i(k) \in \mathbb{R}$ are the measurement output and the unknown measurement disturbance of the i-th sensor at time k, respectively, and $a_i(k)$ denotes the malicious anomalous data on sensor i injected by the attacker at time k. Assume that all the observation matrix $C_i \in \mathbb{R}^{1 \times n}$ are normalized [40], i.e., $\|C_i\| = 1$, $\forall i \in \mathcal{V}$. Denote the measurement matrix as $C = [C_1^T, \ldots, C_N^T]^T$.

Consider a cyber observer deployed on each physical sensor, which altogether forms a NAS, as shown in Fig. 4.1. The dotted line represent the measurement channel that have been compromised by attacks, while the rest refer to benign data or channels.

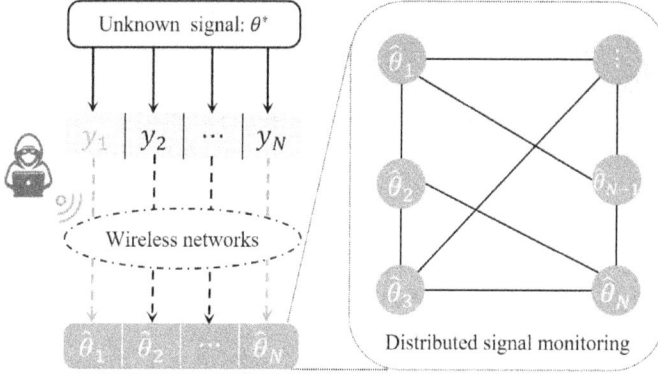

Figure 4.1   Distributed security signal monitoring framework.

Let $\hat{\theta}_i(k)$ be the estimation of the signal θ^* generated by the observer i at time k. To advance the research process, it is natural to impose the following boundedness assumption given the constraints of practical physical devices.

Assumption 4.1 *The following conditions hold:*

$$\max_{i \in \mathcal{V}} \sup_k \|d_i(k)\| \leq \tau, \quad \|\hat{\theta}_i(0) - \theta^*\| \leq \eta_0,$$

where τ and η_0 are known positive constants.

It is worth noting that the attack $a_i(k)$ and the disturbance $d_i(k)$ both affect the same measurement channel. Consequently, when both are present, the attack $a_i(k)$ can be identified as malicious if and only if $\|d_i(k) + a_i(k)\| > \tau$; otherwise, it is deemed part of the disturbance [28]. In subsequent discussions, for simplicity, it is assumed that all sparse attacks are malicious.

With the above setting, the task of each observer is to ensure that

$$\lim_{k \to \infty} \|e_i(k)\| \leq g(\tau), \quad \forall i \in \mathcal{V}, \tag{4.2}$$

where $e_i(k) = \hat{\theta}_i(k) - \theta^*$ denotes the signal estimation error of i-th observer and $g(\tau)$ is a bounded function related only to τ with $g(0) = 0$.

In this section, the measurement from each sensor can be unobservable, which implies that local observability is not required. Furthermore, in the absence of disturbance $d_i(k)$ and attacks $a_i(k)$, if the global matrix C is of full column-rank, each observer can asymptotically generate consensus signal estimation based on its local measurement flow and distributed cooperative control, i.e., $\lim_{k \to \infty} \|\hat{\theta}_i(k) - \theta^*\| = 0$ and $\lim_{k \to \infty} \|\hat{\theta}_i(k) - \hat{\theta}_j(k)\| = 0$, $\forall i, j \in \mathcal{V}$.

A malicious attacker, by disrupting the limited measurement outputs, aims to invalidate the performance of signal monitoring (4.2). Thus, if the i-th sensor is targeted by an adversary at time k, then $a_i(k)$ can take any non-zero value satisfying $\|d_i(k) + a_i(k)\| > \tau$; otherwise, for simplicity, it is treated as a disturbance and integrated into $d_i(k)$, with $a_i(k) = 0$ being recorded. Further, divide the sensor set into two subsets based on whether they are compromised: one is the compromised subset $\mathbf{A}_k = \{i \mid \|d_i(k) + a_i(k)\| > \tau\}$, and the other is the compromised-free subset $\mathbf{N}_k = \mathcal{V} \setminus \mathbf{A}_k$. Note that the information of the compromised subset $\mathbf{A}_k$ is unavailable to any node in the network.

Constrained by limited attack energy of the adversary, s-sparse attacks (or location-varying ones) have been explored as early as in the previous section and the works [17–21, 27, 31–33]. However, the attack model considered above assumes that attacks at each moment are s-sparse, and any attack that violates this property may lead to the failure of existing security state monitoring algorithms. This subsection addresses the risk associated with an attack model that deviates from s-sparse attacks, especially in scenarios involving sophisticated adversaries capable of concentrating their attack energy in a few moments. Such deviations can render existing security state monitoring algorithms ineffective. In the following, the sparsity requirement will be weakened for sensor attacks to time-dependent sparsity.

The work of [17, 33] shows that when measurements fall under s-sparse attacks, the SSM of the signal θ^* can be carried out within a centralized framework and under the assumption of fixed attack locations if and only if the global matrix C is conformed to be $2s$-sparse full column-rank, which is defined in Remark 2.2. This means that the measurement matrix retains the full column-rank after removing any s local measurements. However, from a distributed perspective, in the absence of global measurement information, the s-sparse full column-rank has been extended in [37,40,42,93] to being s-sparse redundant defined in Definition 2.10. This extension removes the location-fixed constraint of sparse attacks, thereby ensuring the security and feasibility of the distributed SSM against location-varying s-sparse sensor attacks.

Assumption 4.2 *The sensor network* (4.1) *is assumed to be $2s$-sparse redundant.*

Note that (location-fixed/varying) s-sparse attacks have been well addressed in [17, 20, 21, 31–33, 37, 40–43]. However, the attack models considered above all assume that the attack at each moment is s-sparse. Therefore, under Assumption 4.2, any violation of this assumption could lead to the failure of existing SSM algorithms. This chapter addresses the risk associated with an attack model that deviates from s-sparse attacks, especially in scenarios involving sophisticated adversaries capable of concentrating their attack energy in a few moments. Such deviations can render the existing SSM algorithms ineffective. In the following, the sparsity requirement will be weakened for sensor attacks to a time-dependent sparsity.

Assumption 4.3 *The sensor attack $a(k)$ is assumed to be (f, s)-sparse-varying.*

Compared to location-varying s-sparse sensor attacks, (f, s)-sparse-varying sensor attacks are more flexible but challenging to resist. This is due to its ability to disrupt the sparsity of attacks at a certain moment, which, in extreme cases, may destroy the measurement outputs of all sensors instantaneously.

4.2.2 Improved attack detection mechanism for sparse-varying sensor attacks

To resist the sparse-varying attacks mentioned above, the idea of a sliding window observer is introduced, based on which distributed security signal observers are constructed over the communication network $\mathcal{G}$, as follows:

$$\hat{\theta}_i(k+1) = \hat{\theta}_i(k) + \alpha \sum_{j=1}^{N} a_{ij}(\hat{\theta}_j(k) - \hat{\theta}_i(k)) + \frac{\beta}{f}C_i^T \sum_{l=0}^{f-1} \delta_i(k,l)z_i(k,l), \qquad (4.3)$$

where the parameters α and β are the positive gains to be designed, $z_i(k,l) = y_i(k - l) - C_i\hat{\theta}_i(k)$ denotes the innovation item with $y_i(k - l) = y_i(0)$ if $l \geq k$, and

$$\delta_i(k,l) = \begin{cases} 1, & \text{if } \|z_i(k,l)\| \leq \gamma(k) + \tau, \\ 0, & \text{otherwise,} \end{cases}$$

where $\gamma(k)$ is a dynamic detection threshold function that needs to be designed. The significance of designing $\gamma(k)$ lies in ensuring the effective utilization of all innovation $z_i(k,l)$ based on benign measurement data while also maximizing the detection and isolation of malicious attacks to the greatest possible extent. Therefore, the selection of $\gamma(k)$ becomes a crucial factor in balancing the feedback performance of the innovation term and the impact of attacks. Following the design ideas in Chapter 3, the threshold $\gamma(k)$ needs to be designed according to the following regulations:

$$\begin{cases} \|e_i(k)\| \leq \gamma(k), & \forall i \in \mathcal{V}, \\ \lim_{k \to \infty} \gamma(k) \leq g(\tau). \end{cases} \qquad (4.4)$$

The regulations (4.4) ensure that, for any benign sensor $i \in \mathbf{N}_{k-l}$, one has $\|z_i(k,l)\| = \|C_ie_i(k) + d_i(k - l)\| \leq \gamma(k) + \tau$, where $\|C_i\| = 1$. Additionally, the second condition in (4.4) guarantees that the estimation error is ultimately uniformly bounded. Hence, if any sensor i violates $\|z_i(k,l)\| \leq \gamma(k) + \tau$, it will be inferred that this sensor has been subjected to malicious attacks at time $k - l$.

Remark 4.1 *Under the setting of sparse-varying attacks, the feedback method relying on a single innovation item $z_i(k,0)$, as employed in [37, 40, 42, 43], is no longer applicable due to the possibility of a momentary violation of the attack sparsity. To overcome this limitation, the observer (4.3) draws inspiration from the concept of a sliding window. Then, the innovation terms $z_i(k,l)$ from multiple time instants are used to enhance the flexibility of the distributed SSM algorithm.*

In this section, the following problem is addressed.

Problem 1: For the given sensor networks (4.1) and distributed observers (4.3), under Assumptions 4.1–4.2, how can one devise the observation gains α, β, and threshold function $\gamma(k)$ to resist (f, s)-sparse-varying sensor attacks while ensuring that the output of the observer (4.3) securely achieves its signal monitoring objective (4.2).

4.2.3 Performance of distributed security signal monitoring

In this subsection, a candidate design solution for the threshold function $\gamma(k)$ is presented. Subsequently, based on theoretical analyses, the utility and performance of the proposed observer (4.3) against (f, s)-sparse-varying sensor attacks are analyzed. Furthermore, analysis is conducted to explore the potential performance enhancement of the security monitoring algorithm through the utilization of historical attack detection information.

● *Threshold function construction*

The design of the threshold function $\gamma(k)$ primarily aims to mitigate the negative effects of attacks while facilitating the state feedback of normal measurements. To strike a balance between the two, the following design process is carried out.

Under the Assumption 4.1, an iteration rule for the threshold function $\gamma(k) = \gamma_1(k) + \gamma_2(k)$ is proposed as follows:

$$\gamma_1(k+1) = \varrho_s \gamma_1(k) + \sqrt{N}\beta\gamma_2(k) + \sqrt{N}\beta\tau, \tag{4.5a}$$

$$\gamma_2(k+1) = \beta\gamma_1(k) + \rho_s\gamma_2(k) + \beta\tau, \tag{4.5b}$$

where $\varrho_s = 1 - \alpha\lambda_2(\mathcal{L}) + \beta\sqrt{s}$, $\rho_s = 1 - \frac{\beta}{N}(\lambda_{min}(G) - 2s)$, and the initial values are chosen as $\gamma_1(0) = \sqrt{N}\eta_0$ and $\gamma_2(0) = \eta_0$. The parameters α, β are taken to satisfy the regulations (4.4), for which an alternative design scheme is given below.

The positive parameters α, β are designed as $\alpha = \tilde{\alpha}/\tilde{\beta}$ and $\beta = 1/\tilde{\beta}$, where the scalars $\tilde{\alpha}$, $\tilde{\beta}$ are chosen as a solution to the following LMIs:

$$0 < \tilde{\alpha} < \frac{2\tilde{\beta}}{\lambda_2(\mathcal{L}) + \lambda_N(\mathcal{L})}, \tag{4.6}$$

$$\frac{\lambda_{max}(G) + \lambda_{min}(G)}{2N} < \tilde{\beta}, \tag{4.7}$$

$$\begin{bmatrix} -hI_2 & \Psi^T \\ * & -\tilde{\beta}I_2 \end{bmatrix} < 0, \tag{4.8}$$

where $0 < h < \tilde{\beta}$ and

$$\Psi = \begin{bmatrix} \tilde{\beta} - \tilde{\alpha}\lambda_2(\mathcal{L}) + \sqrt{s} & \sqrt{N} \\ 1 & \tilde{\beta} - \frac{1}{N}(\lambda_{min}(G) - 2s) \end{bmatrix}.$$

Compared to the threshold design in the saturated adaptive gain estimator [42], a more intuitive and explicit iterative form (4.5) is presented here. In [42], an appropriate parameter selection may be crucial, whereas the parameter design relying on LMIs (4.6)–(4.8) reduces much of the tuning effort. Furthermore, the design concept of dynamic thresholds from [43] is extended by selecting different coefficients ϱ_s and ρ_s to accommodate the distributed SSM under locally unobservable measurements.

Lemma 4.1 For the threshold function $\gamma(k)$ designed in (4.5), if the parameters α and β are picked to conform to the solution of the LMI (4.8), then it can be ensured that $\gamma(k)$ is uniformly eventually bounded, i.e.,

$$\lim_{k \to \infty} \gamma(k) \leq g(\tau) = \frac{\sqrt{2(N+1)}}{1 - \sqrt{h\beta}} \beta \tau. \tag{4.9}$$

Proof 4.1 *Denote* $\tilde{\gamma}(k) = [\gamma_1(k), \gamma_2(k)]^T$, *then (4.5) can be rewritten in the following compact form:*

$$\tilde{\gamma}(k+1) = \beta \Psi \tilde{\gamma}(k) + \begin{bmatrix} \sqrt{N}\beta \\ \beta \end{bmatrix} \tau.$$

It follows from LMI (4.8) and Schur complement Lemma that $\Psi^T \Psi < \tilde{\beta} h I_2$. *Thus, for the Lyapunov function* $W(k) = \|\tilde{\gamma}(k)\|$, *one has*

$$\begin{aligned} W(k+1) &= \|\tilde{\gamma}(k+1)\| \\ &\leq \|\beta \Psi \tilde{\gamma}(k)\| + \sqrt{N+1}\beta \tau \\ &< \sqrt{h\tilde{\beta}}\beta W(k) + \sqrt{N+1}\beta \tau \\ &= \sqrt{h\beta}W(k) + \sqrt{N+1}\beta \tau. \end{aligned}$$

From the fact that $h\beta < 1$, *it can be verified that* $W(k+1)$ *is uniformly ultimately bounded and*

$$\lim_{k \to \infty} \gamma(k) \leq \lim_{k \to \infty} \sqrt{2}W(k) \leq \frac{\sqrt{2(N+1)}}{1 - \sqrt{h\beta}} \beta \tau = g(\tau).$$

This completes the proof.

- *Security signal monitoring performance analysis*

In this subsection, a performance analysis is provided of solving the distributed SSM problem under the threshold function $\gamma(k)$ designed in (4.5).

It follows from (4.1) and (4.3) that

$$e_i(k+1) = e_i(k) + \alpha \sum_{j=1}^{N} a_{ij}(e_j(k) - e_i(k)) + \frac{\beta}{f} C_i^T \sum_{l=0}^{f-1} \delta_i(k,l) z_i(k,l).$$

Let $e(k) = \text{col}\{e_i(k)\}$, $i \in \mathcal{V}$ be the global form of the estimation error $e_i(k)$, $d(k-l) = \text{col}\{d_i(k-l)\}$, and $z(k,l) = \text{col}\{z_i(k,l)\}$, then one has

$$e(k+1) = (I_{Nn} - \alpha \mathcal{L} \otimes I_n)e(k) + \frac{\beta}{f} \tilde{C}^T \sum_{l=0}^{f-1} \sigma(k,l), \tag{4.10}$$

where $\sigma(k,l) = \tilde{\Delta}_{\mathbf{A}_{k-l}} z(k,l) - \tilde{\Delta}_{\mathbf{N}_{k-l}}(\tilde{C}e(k) - d(k-l))$ with the matrix $\tilde{C} = \text{diag}\{C_i\}$, $i \in \mathcal{V}$, and

$$\tilde{\Delta}_{\mathbf{A}_{k-l}} = \text{diag}\{\tilde{\delta}_i(k,l)\} \otimes I_n,$$

$$\tilde{\Delta}_{\mathbf{N}_{k-l}} = \operatorname{diag}\{\delta_i(k,l)\} \otimes I_n - \tilde{\Delta}_{\mathbf{A}_{k-l}},$$

$$\tilde{\delta}_i(k,l) = \begin{cases} \delta_i(k,l), & \text{if } i \in \mathbf{A}_{k-l}, \\ 0, & \text{otherwise}. \end{cases}$$

Introduce the following two variables:

$$\bar{e}(k) = \frac{1}{N}\sum_{i=1}^{N} e_i(k) = \frac{1}{N}(\mathbf{1}_N^T \otimes I_n)e(k),$$

$$\tilde{e}(k) = e(k) - \mathbf{1}_N \otimes \bar{e}(k) = \tilde{\mathcal{I}}e(k),$$

where $\tilde{\mathcal{I}} = (I_N - \frac{1}{N}\mathbf{1}_{N\times N}) \otimes I_n$.

It follows from (4.10), $\tilde{\mathcal{I}}(\mathcal{L}\otimes I_n) = (\mathcal{L}\otimes I_n)\tilde{\mathcal{I}} = \mathcal{L}\otimes I_n$, and $e(k) = \tilde{e}(k)+\mathbf{1}_N\otimes\bar{e}(k)$ that

$$\tilde{e}(k+1) = [I_{Nn} - \alpha(\mathcal{L}\otimes I_n)]\,\tilde{e}(k) + \frac{\beta}{f}\tilde{\mathcal{I}}\tilde{C}^T \sum_{l=0}^{f-1} \sigma(k,l)$$

$$= \left[I_{Nn} - \alpha(\mathcal{L}\otimes I_n) - \frac{\beta}{f}\sum_{l=0}^{f-1}\tilde{\mathcal{I}}\tilde{\Delta}_{\mathbf{N}_{k-l}}\tilde{G}\right]\tilde{e}(k) + \frac{\beta}{f}\sum_{l=0}^{f-1}\tilde{\sigma}(k,l), \qquad (4.11)$$

$$\bar{e}(k+1) = \bar{e}(k) + \frac{\beta}{Nf}\sum_{l=0}^{f-1}\left(\sum_{i\in\mathbf{A}_{k-l}}\delta_i(k,l)C_i^T z_i(k,l) - \sum_{i\in\mathbf{N}_{k-l}}C_i^T C_i e_i(k)\right)$$

$$= \left(I - \frac{\beta}{N}G\right)\bar{e}(k) + \frac{\beta}{Nf}\sum_{l=0}^{f-1}\bar{\sigma}(k,l), \qquad (4.12)$$

where $\tilde{G} = \operatorname{blkdiag}\{C_i^T C_i\}$ and

$$\tilde{\sigma}(k,l) = \tilde{\mathcal{I}}\tilde{\Delta}_{\mathbf{A}_{k-l}}\tilde{C}^T z(k,l) - \tilde{\mathcal{I}}\tilde{\Delta}_{\mathbf{N}_{k-l}}[\tilde{G}(\mathbf{1}_N \otimes \bar{e}(k)) - \tilde{C}^T d(k-l)],$$

$$\bar{\sigma}(k,l) = \sum_{i\in\mathbf{A}_{k-l}}(C_i^T C_i\bar{e}(k) + \delta_i(k,l)C_i^T z_i(k,l))$$

$$- \sum_{i\in\mathbf{N}_{k-l}}(C_i^T C_i\tilde{e}_i(k) + C_i^T d_i(k-l)).$$

Theorem 4.1 Under Assumptions 4.1–4.2, consider the distributed observer (4.3) with threshold function designed as (4.5). If the attack signal $a(k)$ is (f,s)-sparse-varying, then the signal monitoring objective (4.2) can be achieved with $g(\tau)$ given in (4.9).

Proof 4.2 *Given the following Lyapunov function:*

$$V_1(k) = \|\tilde{e}(k)\|, \quad V_2(k) = \|\bar{e}(k)\|.$$

Then, a mathematical induction method is applied to prove that $V_1(k) \le \gamma_1(k)$ and $V_2(k) \le \gamma_2(k)$, $\forall k \ge 0$.

At the initial step, one has that $V_1(0) = \|\tilde{e}(0)\| \leq \sqrt{N}\eta_0 = \gamma_1(0)$ and $V_2(0) = \|\bar{e}(0)\| \leq \eta_0 = \gamma_2(0)$.

Next, assume that $V_1(k) \leq \gamma_1(k)$ and $V_2(k) \leq \gamma_2(k)$ hold at time k. From LMI (4.6) and $\alpha = \tilde{\alpha}/\tilde{\beta}$, one has

$$|1 - \alpha\lambda_N(\mathcal{L})| < 1 - \alpha\lambda_2(\mathcal{L}).$$

Then, it follows from Lemma 2.11 and $(\mathbf{1}_N^T \otimes I_n)\tilde{e}(k) = \mathbf{0}_n$ that

$$\|(I_{Nn} - \alpha(\mathcal{L} \otimes I_n))\tilde{e}(k)\| \leq (1 - \alpha\lambda_2(\mathcal{L})) \|\tilde{e}(k)\|.$$

In addition, based on the semi-positive definiteness of matrices $\tilde{I} \geq 0$, $\tilde{\Delta}_{\mathbf{N}_{k-l}} \geq 0$, and $\tilde{G} \geq 0$, it can be concluded that $\tilde{I}\tilde{\Delta}_{\mathbf{N}_{k-l}}\tilde{G} \geq 0$.

Hence, according to (4.11), one has

$$V_1(k+1) \leq \left\| (I_{Nn} - \alpha(\mathcal{L} \otimes I_n)) - \frac{\beta}{f} \sum_{l=0}^{f-1} \tilde{I}\tilde{\Delta}_{\mathbf{N}_{k-l}}\tilde{G})\tilde{e}(k) \right\| + \frac{\beta}{f} \sum_{l=0}^{f-1} \|\tilde{\sigma}(k,l)\|$$

$$\leq (1 - \alpha\lambda_2(\mathcal{L})) V_1(k) + \frac{\beta}{f} \sum_{l=0}^{f-1} \|\tilde{\sigma}(k,l)\|.$$

Since $\delta_i(k,l)\|C_i^T z_i(k,l)\| \leq \gamma(k)+\tau$, $\forall i \in \mathcal{V}$, $\|\tilde{I}\| = 1$ and $\max_{i \in \mathcal{V}} \sup_{k \geq 0} \|d_i(k)\| \leq \tau$, one gets

$$\begin{aligned}
\|\tilde{\sigma}(k,l)\| &\leq \|\tilde{\Delta}_{\mathbf{A}_{k-l}}\tilde{C}^T z(k-l) - \tilde{\Delta}_{\mathbf{N}_{k-l}}\tilde{G}(\mathbf{1}_N \otimes \bar{e}(k)) \\
&\quad + \tilde{\Delta}_{\mathbf{N}_{k-l}}\tilde{C}^T d(k-l)\| \\
&\leq \|\mathbf{1}_{\mathbf{A}_{k-l}}\gamma(k) + \mathbf{1}_{\mathbf{N}_{k-l}}\gamma_2(k) + \mathbf{1}_N\tau\| \\
&= \|\mathbf{1}_{\mathbf{A}_{k-l}}\gamma_1(k) + \mathbf{1}_N\gamma_2(k) + \mathbf{1}_N\tau\| \\
&= \sqrt{|\mathbf{A}_{k-l}|}\gamma_1(k) + \sqrt{N}(\gamma_2(k) + \tau),
\end{aligned} \tag{4.13}$$

where the i-th element of the vector $\mathbf{1}_{\mathbf{A}_{k-l}}$ is 1 if $i \in \mathbf{A}_{k-l}$, otherwise it is 0, and $\mathbf{1}_{\mathbf{N}_{k-l}} = \mathbf{1}_N - \mathbf{1}_{\mathbf{A}_{k-l}}$.

Since for $\forall a_i \geq 0$, $\sum_{i=1}^{f} \sqrt{a_i} \leq (f \sum_{i=1}^{f} a_i)^{\frac{1}{2}}$ holds, one can infer that

$$\sum_{l=0}^{f-1} \sqrt{|\mathbf{A}_{k-l}|} \leq \left(f \sum_{l=0}^{f-1} |\mathbf{A}_{k-l}| \right)^{\frac{1}{2}} \leq f\sqrt{s}.$$

Hence,

$$\begin{aligned}
V_1(k+1) &\leq (1 - \alpha\lambda_2(\mathcal{L}))\gamma_1(k) + \sqrt{N}\beta\gamma_2(k) \\
&\quad + \frac{\beta}{f} \sum_{l=0}^{f-1} \sqrt{|\mathbf{A}_{k-l}|}\gamma_1(k) + \sqrt{N}\beta\tau \\
&\leq \varrho_s\gamma_1(k) + \sqrt{N}\beta\gamma_2(k) + \sqrt{N}\beta\tau \\
&= \gamma_1(k+1).
\end{aligned} \tag{4.14}$$

In addition, based on the fact that $\beta = 1/\tilde{\beta}$ and LMI (4.7), one has $\|I - \frac{\beta}{N}G\| \leq 1 - \beta/N\lambda_{min}(G)$. Hence, it follows from (4.12) that

$$V_2(k+1) \leq \left(1 - \frac{\beta}{N}\lambda_{min}(G)\right) V_2(k) + \frac{\beta}{Nf} \sum_{l=0}^{f-1} \|\bar{\sigma}(k,l)\|.$$

It then follows from (4.4) and (4.12) that

$$\begin{aligned}
\|\bar{\sigma}(k,l)\| &\leq \sum_{i \in \mathbf{A}_{k-l}} \|C_i^T C_i \bar{e}(k) + \delta_i(k,l) C_i^T z_i(k,l)\| \\
&\quad + \sum_{i \in \mathbf{N}_{k-l}} \left(\|C_i^T C_i \tilde{e}_i(k)\| + \|C_i^T d_i(k-l)\|\right) \\
&\leq (V_2(k) + \gamma(k) + \tau)|\mathbf{A}_{k-l}| + (\gamma_1(k) + \tau)|\mathbf{N}_{k-l}| \\
&\leq N\gamma_1(k) + 2|\mathbf{A}_{k-l}|\gamma_2(k) + N\tau.
\end{aligned} \tag{4.15}$$

Hence,

$$V_2(k+1) \leq \rho_s \gamma_2(k) + \beta\gamma_1(k) + \beta\tau = \gamma_2(k+1). \tag{4.16}$$

Combining (4.14), (4.16) and $e_i(k) = \tilde{e}_i(k) + \bar{e}(k)$, one has

$$\begin{aligned}
\|e_i(k)\| &\leq \|\tilde{e}(k)\| + \|\bar{e}(k)\| \\
&\leq \gamma_1(k) + \gamma_2(k) \\
&= \gamma(k), \ \forall i \in \mathcal{V}.
\end{aligned} \tag{4.17}$$

Finally, it follows from Lemma 4.1 that

$$\lim_{k\to\infty} \|e_i(k)\| \leq \lim_{k\to\infty} \gamma(k) \leq g(\tau), \ \forall i \in \mathcal{V}.$$

This completes the proof.

Remark 4.2 *If all observers have their initial estimations $\hat{\theta}_i(0)$ chosen to be equal, then one has $V_1(0) = 0$. Hence, in this case, one can set the initial value of $\gamma_1(k)$ to be $\gamma_1(0) = 0$.*

Remark 4.3 *Theorem 4.1 demonstrates that an effective distributed SSM algorithm can be developed for sparse-varying attacks without necessitating any additional redundancy conditions on the sensor network beyond Assumption 4.2. The key to the matter is to tackle the sparsity of attacks by establishing sliding-window observers at particular time scales and to integrate an attack detection and isolation mechanism to maintain robust estimation performance.*

● *Distributed SSM with attack detection information*

This subsection explores a more comprehensive analysis of the potential performance enhancements against sparse-varying attacks that can be achieved by leveraging historical attack detection information.

For clarity, let $\hat{\mathbf{A}}_{k-l} \triangleq \{i \in \mathcal{V} \mid \delta_i(k,l) = 0, \ 0 \leq l \leq k\}$ denote the set of historical attack detection results. By following regulations (4.4), it follows that $\delta_i(k,l) = 1, \forall i \in \mathbf{N}_{k-l}$, thus one can conclude that $\hat{\mathbf{A}}_{k-l} \subseteq \mathbf{A}_{k-l}$ and $\hat{\mathbf{A}}_{k-l} \cap \mathbf{N}_{k-l} = \emptyset$.

Note that the defender has access to the global historical information on the number of attacks that have been detected and isolated, i.e., $|\hat{\mathbf{A}}_{k-l}|$. Hence, based on the accessibility of $|\hat{\mathbf{A}}_{k-l}|$, $\tilde{\sigma}(k,l)$ in (4.13) and $\bar{\sigma}(k,l)$ in (4.15) can be reformulated as follows:

$$
\begin{aligned}
\|\tilde{\sigma}(k,l)\| &\leq \|\tilde{\Delta}_{\mathbf{A}_{k-l}} \tilde{C}^T z(k-l) - \tilde{\Delta}_{\mathbf{N}_{k-l}} \tilde{G}(\mathbf{1}_N \otimes \bar{e}(k)) + \tilde{\Delta}_{\mathbf{N}_{k-l}} \tilde{C}^T d(k-l)\| \\
&\leq \|\mathbf{1}_{\bar{\mathbf{A}}_{k-l}} \gamma(k) + \mathbf{1}_{\mathbf{N}_{k-l}} \gamma_2(k) + \mathbf{1}_{\bar{N}_{k-l}} \tau\| \\
&= \|\mathbf{1}_{\bar{\mathbf{A}}_{k-l}} \gamma_1(k) + \mathbf{1}_{\bar{N}_{k-l}} \gamma_2(k) + \mathbf{1}_{\bar{N}_{k-l}} \tau\| \\
&= \sqrt{|\bar{\mathbf{A}}_{k-l}|} \gamma_1(k) + \sqrt{N - |\hat{\mathbf{A}}_{k-l}|}(\gamma_2(k) + \tau), \qquad (4.18)
\end{aligned}
$$

$$
\begin{aligned}
\|\bar{\sigma}(k,l)\| &\leq \sum_{i \in \mathbf{A}_{k-l}} \|C_i^T C_i \bar{e}(k) + \delta_i(k,l) C_i^T z_i(k,l)\| \\
&\quad + \sum_{i \in \mathbf{N}_{k-l}} (\|C_i^T C_i \tilde{e}_i(k)\| + \|C_i^T d_i(k-l)\|) \\
&\leq V_2(k)|\mathbf{A}_{k-l}| + (\gamma(k) + \tau)|\bar{\mathbf{A}}_{k-l}| + (\gamma_1(k) + \tau)|\mathbf{N}_{k-l}| \\
&\leq (N - |\hat{\mathbf{A}}_{k-l}|)(\gamma_1(k) + \tau) + (2|\mathbf{A}_{k-l}| - |\hat{\mathbf{A}}_{k-l}|)\gamma_2(k). \qquad (4.19)
\end{aligned}
$$

where $\bar{\mathbf{A}}_{k-l} = \mathbf{A}_{k-l} \setminus \hat{\mathbf{A}}_{k-l}$, $\bar{N}_{k-l} = \mathcal{V} \setminus \hat{\mathbf{A}}_{k-l}$.

Furthermore, the dynamics of the threshold function $\gamma(k) = \gamma_1(k) + \gamma_2(k)$ designed in (4.5) can be further modified as

$$
\gamma_1(k+1) = \tilde{\varrho}_{s,k} \gamma_1(k) + \sqrt{N - \psi_k} \beta(\gamma_2(k) + \tau), \qquad (4.20a)
$$

$$
\gamma_2(k+1) = \left(1 - \frac{\psi_k}{N}\right) \beta(\gamma_1(k) + \tau) + \tilde{\rho}_{s,k} \gamma_2(k), \qquad (4.20b)
$$

where $\tilde{\varrho}_{s,k} = 1 - \alpha\lambda_2(\mathcal{L}) + \beta\sqrt{s - \psi_k}$, $\tilde{\rho}_{s,k} = 1 - \frac{\beta}{N}(\lambda_{min}(G) - 2s + \psi_k)$ with $\psi_k = \frac{1}{f}\sum_{l=0}^{f-1} |\hat{\mathbf{A}}_{k-l}|$, and the parameters α, β are taken same as (4.5) because $\psi_k \geq 0$.

Theorem 4.2 *Consider the sensor network (4.1) and the distributed observer given by (4.3), operating under (f,s)-sparse-varying sensor attacks. Suppose that Assumptions 4.1–4.2 hold and $|\hat{\mathbf{A}}_{k-l}|$ is accessible. Then, the signal monitoring objective (4.2) can be attained if the threshold function $\gamma(k)$ is designed according to (4.20).*

Proof 4.3 *The proof is similar to the proof of Theorem 4.1, which is briefly outlined below.*

With the same Lyapunov function chosen as in Theorem 4.1, it follows from (4.18) and (4.19) that

$$
\begin{aligned}
V_1(k+1) \leq{}& (1 - \alpha\lambda_2(\mathcal{L}))\gamma_1(k) + \frac{\beta}{f}\sum_{l=0}^{f-1}\Big(\sqrt{|\bar{\mathbf{A}}_{k-l}|}\gamma_1(k) \\
&+ \sqrt{N - |\hat{\mathbf{A}}_{k-l}|}\beta(\gamma_2(k) + \tau)\Big)
\end{aligned}
$$

$$\leq \tilde{\varrho}_{s,k}\gamma_1(k) + \sqrt{N - \psi_k}\beta(\gamma_2(k) + \tau),$$

$$V_2(k+1) \leq \left(1 - \frac{\psi_k}{N}\right)\beta(\gamma_1(k) + \tau) + \tilde{\rho}_{s,k}\gamma_2(k).$$

Then, the same conclusion can be obtained similarly. Hence, details are omitted here.

Remark 4.4 *Building on the previous analysis, it becomes apparent that a critical condition for achieving viable security signal monitoring is to meet the requirement $\frac{1}{f}\sum_{l=0}^{f-1}(2|\mathbf{A}_{k-l}| - |\hat{\mathbf{A}}_{k-l}|) = \frac{2}{f}\sum_{l=0}^{f-1}|\mathbf{A}_{k-l}| - \psi_k \leq 2s$. Consequently, under Assumption 4.2, the maximum number of (f, s'_k)-sparse-varying sensor attacks that the distributed SSM algorithm can withstand is given by $s'_k \leq s + \frac{\psi_k}{2}$. In particular, due to the irreversibility of time, it can be deduced that at time k, the maximum number of attacks that the distributed SSM algorithm can handle is $f(s + \psi_k/2) - \sum_{l=1}^{f-1}|\mathbf{A}_{k-l}|$. Thus, for every two additional attacks detected in the $f - 1$ steps prior to moment k, the system can tolerate one more attack injection at time k.*

Remark 4.5 *Obviously, when ψ_k reaches its maximum value $\frac{1}{f}\sum_{l=0}^{f-1}|\mathbf{A}_{k-l}|$, the system can withstand the maximum number of attacks, satisfying $\frac{1}{f}\sum_{l=0}^{f-1}|\mathbf{A}_{k-l}| \leq 2s$. In other words, at this moment, for any $(f, 2s)$-sparse-varying attack, the distributed SSM algorithm provided in this subsection can still maintain its resilience.*

4.2.4 Numerical simulations

In this section, two numerical simulation examples are presented to verify the theoretical results.

Consider an unknown two-dimensional signal vector $\theta^* = [3, -2]^T$ monitored by a sensor network consisting of 10 agents, where $C_i = [0, \ 1]$ for sensors $\{1, \ldots, 5\}$ and $C_i = [1, \ 0]$ for the other sensors. The static signal θ^* may refer to, for example, a remote command signal, the pixel values of a monitoring image, the infrastructure information of a city, and so on. In addition, all unknown measurement disturbances $d_i(k)$ follow the uniform distribution in $[-0.05, 0.05]$, thus one has $\tau = 0.05$. The communication topology among the sensors is shown in Fig. 4.2.

Assume that each element of $\hat{\theta}_i(0)$, $i \in [N]$ follows the uniform distribution on the interval $[-5, 5]$ with $\eta_0 = 10.5$. The attack injection signals are set to be $a_i(k) = -0.5C_i\theta^* + b\sqrt{k}$ if the i-th sensor is selected by the adversary as the attack target, where the coefficient b is a random variable that also follows the uniform distribution on the interval $[0, 1]$.

Example 4.1: To demonstrate the effectiveness of the algorithm proposed in this section, the signal monitoring performance is evaluated for both the traditional resilient distributed observer and the sliding window-based observer proposed in (4.3) against sparse-varying attack settings.

According to the above sensor setting, $\lambda_{min}(G) = 5$. Then, to verify the effectiveness of the proposed design, assume that the number of attack injections follows the

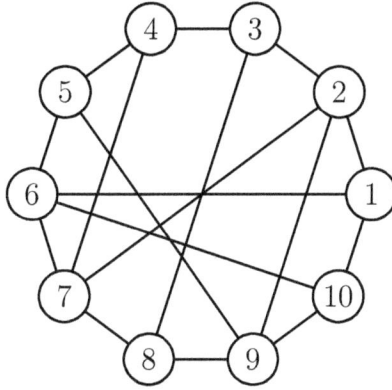

Figure 4.2   Communication topology of 10 sensors.

following recurring rule:

$$|\mathbf{A}_k| = \begin{cases} 0, & \text{if } k = 4r, \\ 3, & \text{if } k = 4r + 1, \\ 4, & \text{if } k = 4r + 2, \\ 1, & \text{if } k = 4r + 3, \end{cases} \quad \forall r \in \mathbb{N}.$$

It can be seen that the attack is $(4, 2)$-sparse-varying, where the attack locations can be selected at any position among the 10 sensors.

By selecting $\alpha = 0.3460$ and $\beta = 0.0024$ based on LMIs (4.6)–(4.8), the sliding window-based observer (4.3) and the following conventional distributed observer (4.21) proposed in [37, 40, 42, 43] are used for distributed SSM simulations, respectively:

$$\hat{\theta}_i(k + 1) = \hat{\theta}_i(k) + \alpha \sum_{j=1}^{N} a_{ij}(\hat{\theta}_j(k) - \hat{\theta}_i(k)) + \beta \delta_i(k, 0)C_i^T z_i(k, 0). \tag{4.21}$$

To reflect the efficacy of regulations (4.4), under the sliding window-based observer (4.3), the comparison results of the functions $\|e_i(k)\|$ with the threshold function $\gamma(k)$ are depicted in Fig. 4.3. For clarity of the attack detection performance, the positions of sparse-varying attacks and the corresponding attack detection results at some moments are presented in Fig. 4.4. Obviously, the constraints outlined in regulations (4.4) are always met, and the accurate detection and isolation of attacks can be achieved as time progresses.

The signal monitoring results with distributed sliding window-based observer (4.3) are shown in Fig. 4.5. It can be seen that the sliding window-based observer (4.3) is capable of achieving distributed SSM under the aforementioned harsh attack setting.

In addition, to evaluate the performance of SSM under different algorithms, the average error performance function is defined as $\eta(k) = \frac{1}{N} \sum_{i=1}^{N} \|e_i(k)\|$. And Fig. 4.6 exhibits the curve of the function $\eta(k)$ for different observers, i.e., (4.3), (4.21), and the upper bound $g(\tau) = 3.2761$. Note that, due to the consideration of the worst-case scenario, the error upper bound $g(\tau)$ provided here is somewhat conservative.

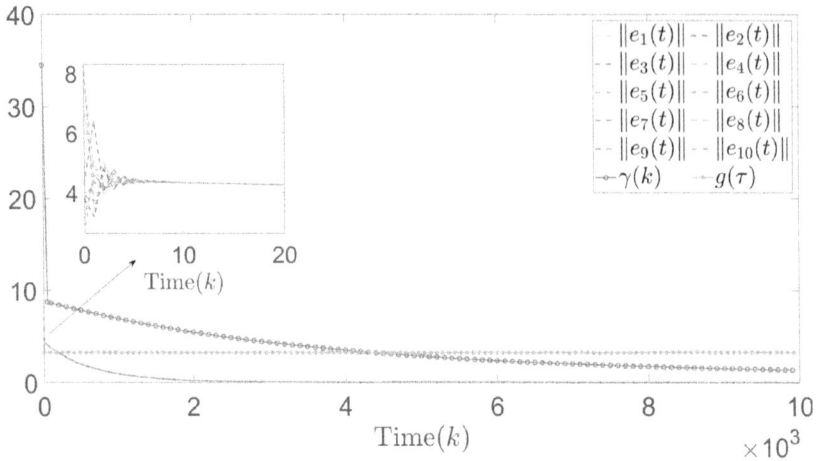

Figure 4.3 The comparison results of $\|e_i(k)\|$ with the threshold function $\gamma(k)$.

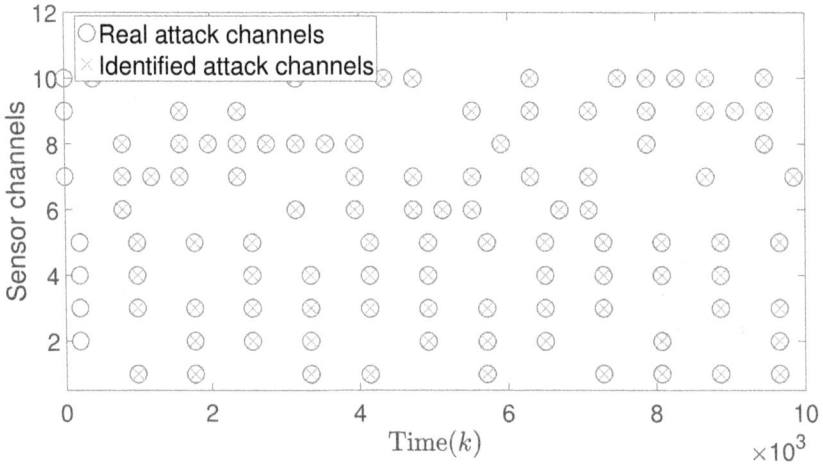

Figure 4.4 The attack locations and the identification results for some moments.

Obviously, the performance of the sliding window-based distributed SSM is superior to those designed in [37, 40, 42, 43].

Example 4.2: This example further demonstrates the potential performance improvement of the proposed distributed SSM algorithm based on conditions where historical attack detection information is available.

For simplicity, the same parameter configuration is taken as in Example 4.1, except that the number of sparse-varying attacks follows the following recursion:

$$|\mathbf{A}'_k| = \begin{cases} \frac{\psi_{k-1}}{2}, & \text{if } k = 4r, \\ 3 + \frac{\psi_{k-1}}{2}, & \text{if } k = 4r + 1, \\ 4 + \frac{\psi_{k-1}}{2}, & \text{if } k = 4r + 2, \\ 1 + \frac{\psi_{k-1}}{2}, & \text{if } k = 4r + 3, \end{cases} \quad \forall r \in \mathbb{N}, \tag{4.22}$$

where $\psi_k = 0$ if $k < 4$.

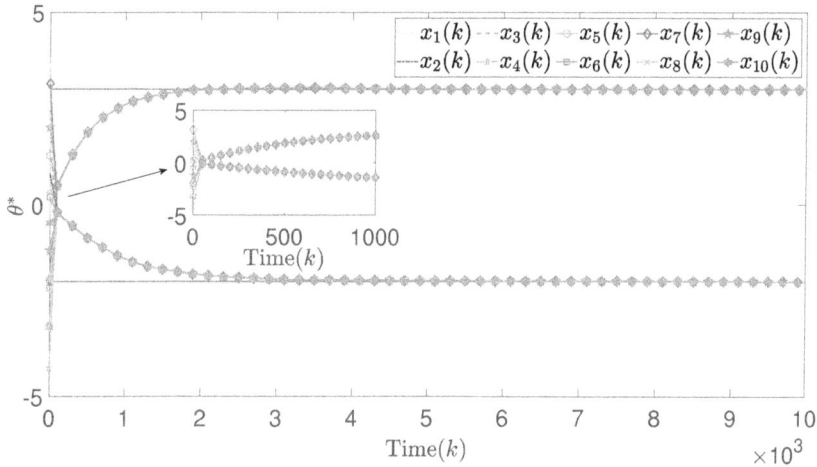

Figure 4.5   The signal monitoring results $\hat{\theta}_i(k)$ generated by observer (4.3).

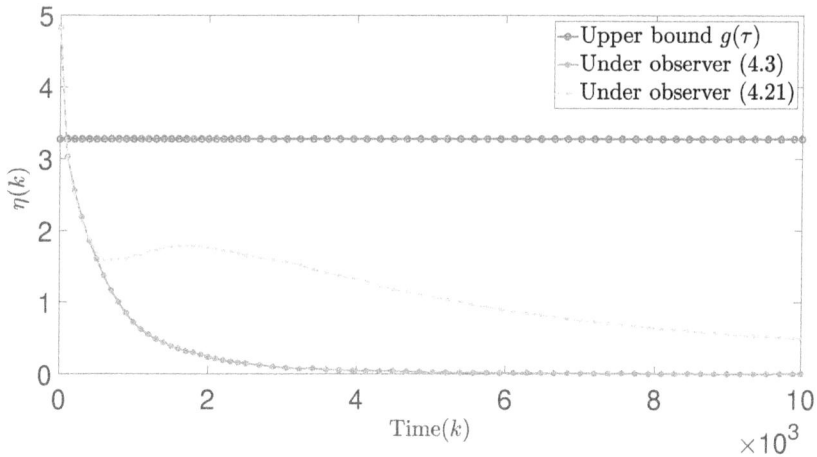

Figure 4.6   The profile of the average error function $\eta(k)$ for different observers and the upper bound $g(\tau)$.

Numerical comparative simulations of parameter estimation based on the distributed observer (4.3) are conducted under the threshold function $\gamma(k)$ designed in (4.5) and (4.20), respectively. Under the distributed SSM algorithm with the historical detection information, the specific attack locations and detection results are depicted in Fig. 4.7. The results indicate that, even with a large number of attacks given in (4.22), the detection threshold generated based on (4.20) still accurately identifies malicious attacks. Fig. 4.8 depicts the dynamics of the variation in the attack sparsity magnitude $s'_k = \frac{1}{f}\sum_{l=0}^{f-1}|\mathbf{A}_{k-l}|$, revealing that the attacks sustained by the system over time are gradually transformed from $(4, 2)$-sparse-varying attacks to $(4, 3.75)$-sparse-varying attacks.

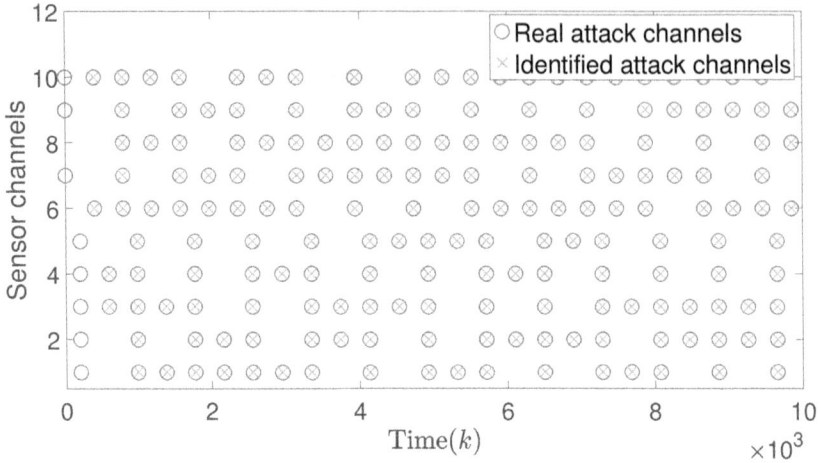

Figure 4.7 The actual attack locations and detection results under the threshold designed in (4.20).

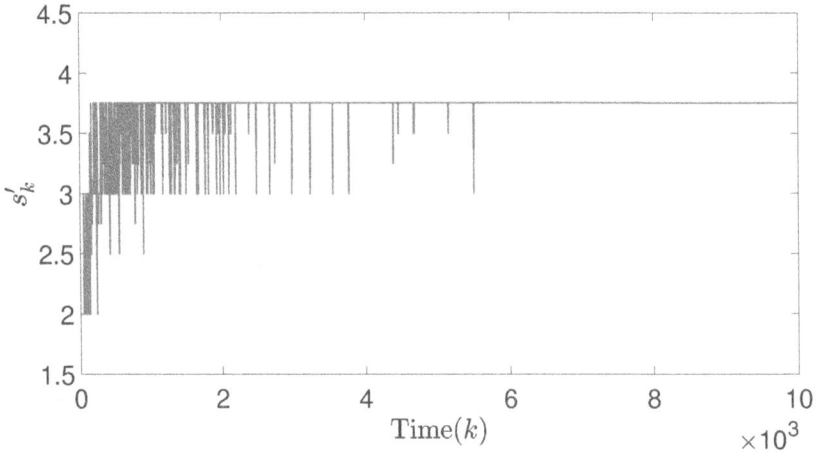

Figure 4.8 The profile of the attack sparsity magnitude s'_k in Example 4.2.

The signal monitoring results obtained for different threshold forms are presented in Fig. 4.9. It can be observed that, under the attack setting in this example, the detection thresholds derived from (4.5) fail to address all attacks, resulting in estimation results that are affected by the attacks and deviate from the actual values, in comparison to those generated based on (4.20). To visually highlight the disparity in the signal monitoring performance under different detection thresholds, Fig. 4.10 shows the curve of the average error function $\eta(k)$. The results indicate that the performance of the security signal monitoring can be significantly improved after utilizing the historical attack detection information.

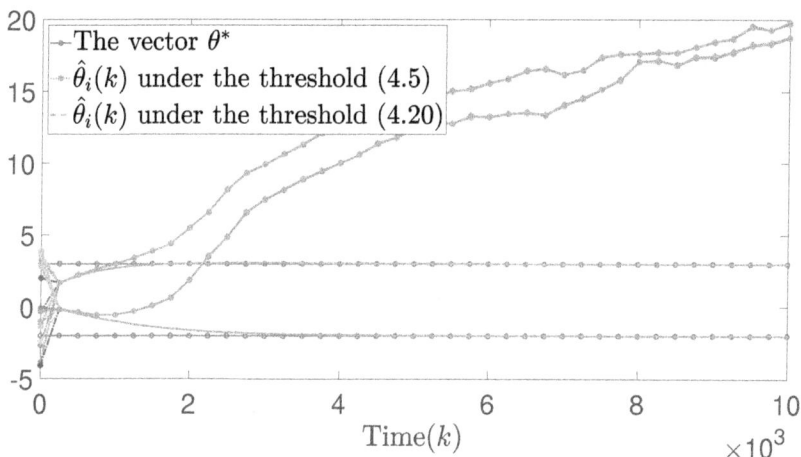

Figure 4.9 The signal monitoring results by observer (4.3) under different detection thresholds.

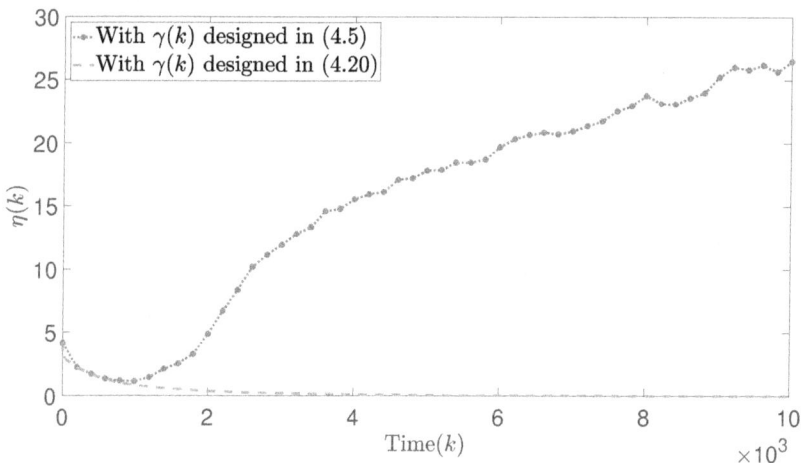

Figure 4.10 The profile of the function $\eta(k)$ under different thresholds.

4.3 DISTRIBUTED SECURITY MONITORING AGAINST VULNERABILITY-RELATED SPARSE SENSOR ATTACKS

4.3.1 Model formulation

In an attack-prone environment, consider a set of heterogeneous wireless sensor networks for monitoring a continuous LTI system, with the system dynamics and the measurement model of the i-th sensor given as follows:

$$
\begin{aligned}
\dot{x}(t) &= Ax(t) + Bu(t), \\
\tilde{y}_i(t) &= C_i x(t) + a_i(t), \quad i \in \mathcal{V} = [N],
\end{aligned}
\tag{4.23}
$$

where $x(t) \in \mathbb{R}^n$, $u(t) \in \mathbb{R}^p$ are the system state and the control input, respectively. $\tilde{y}_i(t) \in \mathbb{R}^{m_i}$ is the measurement output of i-th sensor that might be damaged by the attack $a_i(t)$, where $a_i(t)$ can be arbitrarily sized for malicious attack injection. A, B are system matrices with appropriate dimensions, $C_i \in \mathbb{R}^{m_i \times n}$ is the local measurement matrix of the sensor i, and $C = [C_1^T, \ldots, C_N^T]^T$ denotes the collective observation matrix.

Remark 4.6 *To intuitively reflect the performance of SSM and simplify the analysis, this section temporarily removes the noise or disturbance terms from the model. However, it is noteworthy that by sacrificing some estimation performance to a certain extent, the results of this chapter can also extend to noisy systems. Furthermore, the essence of detecting and isolating attacks lies in the analysis and treatment conducted after decoupling the system from noise. The significance of distinguishing between attacks and noise in practical SSM has been discussed in [24, 28]. The results suggest that under worst-case scenarios, the estimation error is linearly related to the magnitude of the noise, and attacks can be regarded as noise when distinguishing between attack and noise becomes challenging.*

Consider the global attack $a(t) = \text{col}\{a_i(t)\}$, $i \in \mathcal{V}$ is location-varying s-sparse, i.e., $|\mathcal{V}_{A,t}| \leq s$, $\forall t \geq 0$ with $\mathcal{V}_{A,t} = \{i \mid a_i(t) \neq \mathbf{0}_{m_i}\}$. Recalling previous centralized and distributed SSM frameworks to target such s-sparse attacks, some definitions such as s-sparse observable in Definition 2.9 or s-sparse full column-rank are usually necessary. From the works of [17, 23, 24, 28, 32, 37, 40–43, 109], the following lemma can be obtained to derive the SSM algorithm.

Lemma 4.2 *For the NAS (4.23) subjected to any location-varying (fixed) s-sparse attacks. If the network with the global measurement matrix C is $2s$-sparse full column-rank, a distributed (or centralized) algorithm exists that can achieve SSM for NAS (4.23) with any system matrix A, i.e.,*

$$\lim_{t \to \infty} \|e_i(t)\| = 0, \quad \forall i \in \mathcal{V},$$

where $e_i(t) = \hat{x}_i(t) - x(t)$ denote the local state monitoring error.

It follows from Lemma 4.2 that the $2s$-sparse full column-rank of measurement matrix C can achieve SSM under arbitrary s-sparse attacks. However, this guarantee may not be established once the number of sensors under attack exceeds the value s. To overcome this limitation, the following reflections are prompted.

Note that the SSM problem for homogeneous sensor networks was addressed in the previous chapter and the works [17, 22–24, 28, 29, 32–42, 109, 111, 116], where the vulnerabilities of each sensor to attacks are treated identically. However, in fact, the vulnerabilities of different sensors tend to be heterogeneous. Take Fig. 4.11 as an example, in vehicle localization, on-board sensors are relatively vulnerable to attacks compared to satellites and ground stations [117]. How to leverage this diverse vulnerability to positively impact the SSM approaches has caught our attention. Therefore, there are two interesting questions to motivate the investigation of this chapter.

i) Whether the upper bound of tolerable attacks can be increased by assigning different weights to sensors based on their vulnerabilities against attacks?

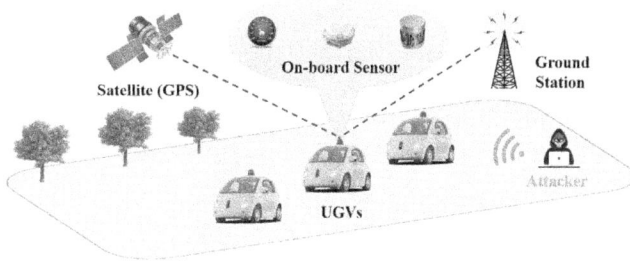

Figure 4.11   Vehicle localization with heterogeneous sensors.

ii) Whether the upper bound can be further increased if the attacker tends to compromise the more vulnerable sensors (those with smaller vulnerability metric)?

• *Heterogeneous sensor network*

Drawing inspiration from the above analysis and the diversity modeling in [85, 115], we consider the heterogeneity in the vulnerability of sensors against specific attacks and assign different weights to each sensor to reflect its relative vulnerability. The vulnerability of each sensor is an intrinsic property that may include attributes such as physical defense capabilities and human-made protective features. In particular, these attributes are known to the defender and have not hindered adversaries' access.

As is commonly recognized, due to cost and energy constraints, inexpensive sensors with weaker protection are often present in more significant numbers. They are more susceptible to being targeted by adversaries in practice. Hence, building on these concepts, the vulnerability metric of each sensor is described as follows.

The vulnerability metric model: As described in Chapter 2, there exists a unique mapping relation can be constructed from the sensor set $\mathcal{V}$ to the weight metric set $\Upsilon = \{\rho_1, \ldots, \rho_N\}$: $\Lambda(i) = \rho_i$ with $\rho_i \in \mathbb{R}^+$. In which, as ρ_i approaches closer to 0, it indicates that the adversary can more easily launch a successful attack on sensor i.

Assumption 4.4 The average vulnerability metric $\bar{\rho} = \frac{1}{N} \sum_{i=1}^{N} \rho_i$ is bounded and known.

Remark 4.7 *The vulnerability metrics of heterogeneous sensors are considered to be assigned offline, and the average vulnerability metrics are counted in a global perspective, therefore, the feasibility of Assumption 4.4 is guaranteed. Furthermore, the boundedness condition ensures the tractability of the SSM algorithm; otherwise, unbounded metrics could lead to biases in trust. This specific scenario with trusted sensors will be discussed in detail in the following subsection.*

The weighted metric ρ_i is determined by the defender based on an assessment of sensor hardware attributes, software protection configurations, and environmental factors. Since the attack information is random and unpredictable, the following vulnerability-related sparse attack model is considered based on the maximum attack energy that can be tolerated by the defender at the beginning of the deployment.

Denote $\rho = \text{col}\{\rho_i\}$, $i \in [N]$ and $G(\rho) = \frac{1}{\bar{\rho}} \sum_{i=1}^{N} \rho_i C_i^\dagger C_i$. Inspired by [42], the following definition is introduced to facilitate observers (4.5) against vulnerability-related sparse attacks.

Definition 4.1 The NAS (4.23) is said to be (ρ, r)-weighted sparse redundant, if the following condition holds

$$\lambda_{min}(G(\rho)) > r, \quad r \in \mathbb{R}^+.$$

The notion of Definition 4.1 is also established globally, with the matrix $G(\rho)$ representing the global category. Again, from the time invariance of C_i and ρ, it can be inferred that the weighted sparsity observable of the NAS (4.23) can be assessed offline.

Sort the elements of vector ρ in a descending order, i.e., $\rho_1^d \geq \rho_2^d \geq \cdots \geq \rho_N^d$. The following lemma builds a bridge between weighted sparse redundant and global sparse full column-rank.

Lemma 4.3 *If the NAS (4.23) is (ρ, r)-weighted sparse redundant with $r = \frac{1}{\bar{\rho}} \sum_{j=1}^{s} \rho_j^d$, it is s-sparse full column-rank for the global matrix C.*

Proof 4.4 *The result is proved by the contrapositive method. Suppose that the global matrix C is not s-sparse full column-rank. Then, there exists a subset $\Gamma_1 \subset [N]$ with $|\Gamma_1| = s$ such that $C_{\bar{\Gamma}_1}$ is column rank-deficient, i.e., $\tilde{G}_1 = \frac{1}{\bar{\rho}} \sum_{i \in \bar{\Gamma}_1} \rho_i C_i^\dagger C_i$ is not invertible. Hence, there exists a nonzero vector $\nu \in \mathbb{R}^n$ such that $\tilde{G}_1 \nu = \mathbf{0}_n$ holds. Besides, one has that*

$$\lambda_{min}(G(\rho)) \leq \mu^T G(\rho)\mu, \quad \mu = \nu/\|\nu\|.$$

Note that $G(\rho) = \frac{1}{\bar{\rho}} \sum_{i \in \Gamma_1} \rho_i C_i^\dagger C_i + \tilde{G}_1$, the following inequality can be derived

$$\mu^T G(\rho)\mu = \frac{1}{\bar{\rho}} \sum_{i \in \Gamma_1} \rho_i \mu^T C_i^\dagger C_i \mu \leq \frac{1}{\bar{\rho}} \sum_{i \in \Gamma_1} \rho_i \leq \frac{1}{\bar{\rho}} \sum_{j=1}^{s} \rho_j^d = r.$$

It suggests that $\lambda_{min}(G(\rho)) \leq r$, which is a contradiction.

Remark 4.8 *It can be directly derived from Lemma 4.3 that the global matrix C is s-sparse full column-rank if the NAS (4.23) is $(\bar{\rho}\mathbf{1}_N, s)$-weighted sparse redundant. This statement is consistent with that in Theorem 2 in [42].*

● *Vulnerability-related sparse sensor attacks*

Given the randomness and unpredictability of cyber-attacks, the following vulnerability-related sparse attack model can be constructed during the initial deployment phase, where the sparsity is optimally selected through statistical analysis of historical data, predictive modeling, or by referencing the maximum tolerable attack intensity within the resilience threshold of the defender. Therefore, a vulnerability-related sparse attack model that the system can withstand is assumed to be as follows.

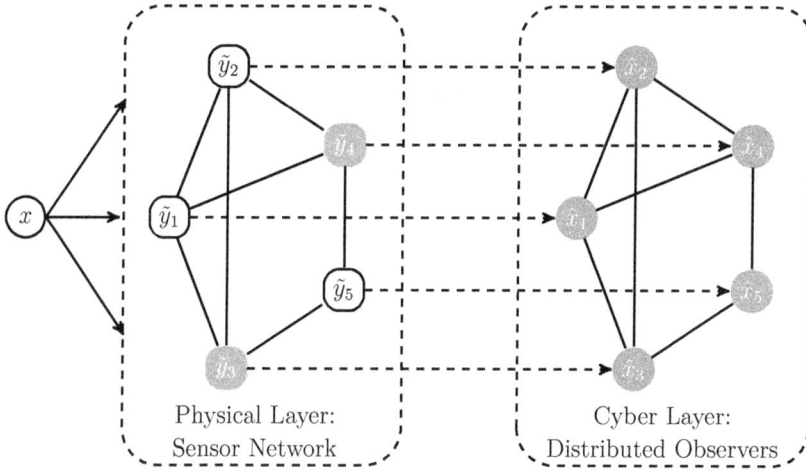

Figure 4.12 An illustration framework for distributed SSM.

Assumption 4.5 *The NAS (4.23) is compromised by vulnerability-related s-sparse attacks, i.e., $\frac{1}{\rho}\sum_{i\in\mathcal{V}_{A,t}}\rho_i \leq s$.*

Remark 4.9 *Note that $s \leq N$ due to $\frac{1}{\rho}\sum_{i\in\mathcal{V}}\rho_i = N$. Specifically, when $\rho_i = \bar{\rho}$, $\forall i \in \mathcal{V}$, a vulnerability-related s-sparse attack is equivalent to an s-sparse attack.*

According to the attack suffered by the i-th sensor, the sensors set can be similarly divided into two subsets, including the attack set $\mathcal{V}_{A,t}$ and the normal set $\mathcal{V}_{N,t} = \{i \mid a_i(t) = \mathbf{0}_{m_i},\ t \geq 0\}$, with $\mathcal{V} = \mathcal{V}_{A,t} \cup \mathcal{V}_{N,t}$.

4.3.2 Distributed SSM against vulnerability-related attacks

The SSM architecture established in [37, 40–43] consists of a set of distributed observers with attack-isolation-based mechanisms. A simple illustration example is shown in Fig. 4.12, where the gray and the white squares in the physical layer denote the compromised sensors and the normal ones, respectively, and the circles in the cyber layer indicate the distributed observers.

Under the established distributed observer architecture, the dynamics of the i-th observer is described as

$$\dot{\hat{x}}_i(t) = A\hat{x}_i(t) + Bu(t) + \theta\mu_i(t) + \beta_i\delta_i(t)C_i^{\dagger}\tilde{z}_i(t), \tag{4.24}$$

where $\hat{x}_i(t)$ is the estimation of $x(t)$ generated by the i-th observer, the parameters θ and β_i are the positive observer gains to be designed. $\mu_i(t)$ and $\tilde{z}_i(t)$ are the consensus control input and the innovation sequence, respectively, which are expressed as

$$\mu_i(t) = \sum_{j=1}^{N} a_{ij}(\hat{x}_j(t) - \hat{x}_i(t)),$$

$$\tilde{z}_i(t) = \tilde{y}_i(t) - C_i\hat{x}_i(t).$$

In contrast to the unified feedback gain, β in [37, 40–43], the different β_i respective to each observer is considered in (4.24) for weighted construction, aiming to improve the security of the SSM strategy to more detrimental attacks.

In addition, $\delta_i(t)$ denotes the isolation factor for the i-th observer, representing the recognition response to the attacks, and its design architecture can be given as follows:

$$\delta_i(t) = \begin{cases} 1, & \|C_i^\dagger \tilde{z}_i(t)\| \leq \gamma(t), \\ 0, & \text{otherwise}, \end{cases} \tag{4.25}$$

where $\gamma(t)$ is the detection threshold function to be designed. The terms $\delta_i(t)$ and $\gamma(t)$ help halt the negative impact of extreme innovation entries (due to attacks) on the estimation. $\delta_i(t) = 1$ means that $\tilde{y}_i(t)$ can be used for estimation, otherwise, $\tilde{y}_i(t)$ should be discarded. The threshold function $\gamma(t)$ relates to the accuracy of attack isolation and further determines the performance of state monitoring. The selection of $\gamma(t)$ is a major challenge in the design of SSM approaches. To this end, the following constraint rule is given.

For the sake of security guarantees of estimation, the following constraints on the threshold function $\gamma(t)$ are noteworthy:

$$\lim_{t \to \infty} \gamma(t) = 0, \quad \gamma(t) \geq \|C_i^\dagger C_i e_i(t)\|, \ \forall i \in \mathcal{V}. \tag{4.26}$$

The goal of this section is to utilize distributed techniques to counter arbitrary vulnerability-related s-sparse sensor attacks and achieve asymptotic SSM, i.e.,

$$\lim_{t \to \infty} \|e_i(t)\| = \lim_{t \to \infty} \|\hat{x}_i(t) - x(t)\| = 0.$$

4.3.3 Performance of distributed security state monitoring

To effectively counter vulnerability-related sparse sensor attacks, the primary task is to achieve accurate detection against malicious attacks. Setting an appropriate detection threshold is crucial, as it not only limits the negative impact of attacks but also directly influences the security of state monitoring. Therefore, in the following discussion, we first design a set of feasible threshold detection functions tailored to the different characteristics of sensor networks and then analyze the performance of distributed SSM by means of the Lyapunov theory.

• Distributed SSM with heterogeneous sensors

This subsection focuses on the design of the threshold function $\gamma(t)$ and the gains θ, β_i in (4.24), in order to enable the observer (4.24) to reach SSM under any vulnerability-related s-sparse attacks.

Assume that the initial value of the estimation error is limited by a constant η, i.e., $\max_{i \in \mathcal{V}} \|e_i(0)\| \leq \eta$, and each observer knows this constant η. Then, the update rule of the threshold function is designed as $\gamma(t) = \gamma_1(t) + \gamma_2(t)$ with

$$\dot{\gamma}_1(t) = (\alpha - \theta \lambda_2(\mathcal{L}) + \bar{\beta}s)\gamma_1(t) + N\bar{\beta}\gamma_2(t), \tag{4.27a}$$

$$\dot{\gamma}_2(t) = \left(\alpha - \rho_s\bar{\beta}\right)\gamma_2(t) + \bar{\beta}\gamma_1(t), \tag{4.27b}$$

where $\alpha \geq \frac{1}{2}\lambda_{max}(A + A^T)$ and $\rho_s = (\lambda_{min}(G(\rho)) - 2s)/N$. The initial values of (4.27a) and (4.27b) satisfy $\gamma_1(0) = \sqrt{N}\eta$ and $\gamma_2(0) = \eta$, the parameters $\theta, \bar{\beta}$ are taken such that constraint (4.26) holds, and the selection criteria are provided in the following.

The parameters θ, $\bar{\beta}$ are designed as $\theta = \tilde{\theta}/\check{\beta}$, and $\bar{\beta} = 1/(N\check{\beta})$, where the variables $\tilde{\theta} > 0$ and $\check{\beta} > 0$ satisfy the following LMI

$$\Xi^T + \Xi < -\tilde{h}I_2, \tag{4.28}$$

with $\tilde{h} > 0$ and

$$\Xi = \begin{bmatrix} \alpha\check{\beta} - \lambda_2(\mathcal{L})\tilde{\theta} + \frac{s}{N} & 1 \\ \frac{1}{N} & \alpha\check{\beta} - \frac{\rho_s}{N} \end{bmatrix}.$$

Besides, the weighted gain β_i in (4.24) is designed as

$$\beta_i = \frac{\rho_i\bar{\beta}}{\bar{\rho}} = \frac{\rho_i}{N\bar{\rho}\check{\beta}}. \tag{4.29}$$

Let $e(t) = \text{col}\{e_i(t)\}$, $i \in \mathcal{V}$ be the global estimation error, it can be derived from (4.23)–(4.24) that

$$\dot{e}(t) = [I_N \otimes A - \theta(\mathcal{L} \otimes I_n)]e(t) + \tilde{\Delta}_t\tilde{\beta}\tilde{L}\tilde{z}(t), \tag{4.30}$$

where $\tilde{\Delta}_t = \text{diag}\{\delta_i(t)\} \otimes I_n$, $\tilde{\beta} = \text{diag}\{\beta_i\} \otimes I_n$, $\tilde{L} = \text{Diag}\{C_i^\dagger\}$, and $\tilde{z}(t) = \text{col}\{\tilde{z}_i(t)\}$, $i \in \mathcal{V}$.

Motivated by the above design process, a distributed secure state monitoring algorithm against vulnerability-related sparse attacks is given in Algorithm 4.1.

Note that the structure of the proposed distributed SSM algorithm shares similarities with the resilient distributed state estimation strategies documented in [37,43]. Nevertheless, the methodology presented in this chapter introduces critical innovations in two fundamental operational components that substantially advance the field. The first distinction lies in the autonomous heterogeneous configuration of the feedback weights β_i, which assigns a different weight to each observation feedback by means of the vulnerability metric ρ_i of the sensor. The second innovation manifests in our novel threshold function $\gamma(t)$, which incorporates the vulnerability metric ρ for heterogeneous sensor networks.

The following variables are introduced to assist in analyzing the performance of state monitoring:

$$\bar{e}(t) = \frac{1}{N}(\mathbf{1}_N^T \otimes I_n)e(t),$$
$$\tilde{e}_i(t) = e_i(t) - \bar{e}(t), \ i \in \mathcal{V}.$$

Define matrix $\mathcal{I} = I_N - \frac{1}{N}\mathbf{1}_N\mathbf{1}_N^T$, it can be concluded that

$$\mathcal{L}^T\mathbf{1}_N = \mathcal{L}\mathbf{1}_N = \mathbf{0}_N, \quad \mathcal{I}\mathcal{L} = \mathcal{L}\mathcal{I} = \mathcal{L}.$$

Algorithm 4.1 Distributed SSM algorithm for heterogeneous sensors

---

Input: A, B, C_i, $\tilde{y}_i(t)$, $\hat{x}_i(0)$, η, s;

1: Assign unique vulnerability-related weight $\Lambda(i) = \rho_i$ to each node (sensor and observer), and compute to obtain the average weight $\bar{\rho}$;

2: $\alpha \geq \frac{1}{2}\lambda_{max}(A + A^T)$;

3: Solving LMI (4.28) to get feasible $\tilde{\theta}$ and $\check{\beta}$;

4: Compute the gains $\theta = \tilde{\theta}/\check{\beta}$, $\beta_i = \rho_i/(\bar{\rho}N\check{\beta})$;

5: $\gamma_1(0) = \sqrt{N}\eta$, $\gamma_2(0) = \eta$;

6: **while** $t > 0$ **do**

7:     Broadcasting its state monitoring $\hat{x}_i(t)$ to neighbors and receiving estimations $\hat{x}_j(t)$ from them;

8:     Updating the threshold function as follow:

$$\dot{\gamma}_1(t) = (\alpha - \theta\lambda_2(\mathcal{L}) + \bar{\beta}s)\gamma_1(t) + N\bar{\beta}\gamma_2(t),$$

$$\dot{\gamma}_2(t) = \left(\alpha - \rho_s\bar{\beta}\right)\gamma_2(t) + \bar{\beta}\gamma_1(t);$$

9:     $\gamma(t) = \gamma_1(t) + \gamma_2(t)$;

10:     $\tilde{z}_i(t) = \tilde{y}_i(t) - C_i\hat{x}_i(t)$;

11:     **if** $\|C_i^\dagger\tilde{z}_i(t)\| \leq \gamma(t)$ **then**

12:         $\delta_i(t) = 1$;

13:     **else**

14:         $\delta_i(t) = 0$;

15:     **end if**

16:     $\mu_i(t) = \sum_{j=1}^N a_{ij}(\hat{x}_j(t) - \hat{x}_i(t))$;

17:     Compute the state monitoring as follows:

$$\dot{\hat{x}}_i(t) = A\hat{x}_i(t) + Bu(t) + \theta\mu_i(t) + \delta_i(t)\beta_i L_i\tilde{z}_i(t);$$

18: **end while**

Output: State estimation $\hat{x}_i(t)$.

Then, one has $\tilde{e}(t) = \tilde{\mathcal{I}}e(t)$ with $\tilde{e}(t) = \text{col}\{\tilde{e}_i(t)\}$, $i \in \mathcal{V}$ and $\tilde{\mathcal{I}} = \mathcal{I} \otimes I_n$.

Denote $\tilde{\Delta}_\Omega = \text{diag}\{\delta_i^\Omega(t)\} \otimes I_n$, $i \in [N]$, $\forall\Omega \subseteq \mathcal{V}$, where $\delta_i^\Omega(t) = \delta_i(t)$ if $i \in \Omega$, and $\delta_i^\Omega(t) = 0$ otherwise. Then, the dynamics of $\tilde{e}(t)$ and $\bar{e}(t)$ can be written as

$$\dot{\tilde{e}}(t) = [I_N \otimes A - \theta(\mathcal{L} \otimes I_n)]\tilde{e}(t) + \tilde{\sigma}(t), \tag{4.31}$$

$$\dot{\bar{e}}(t) = A\bar{e}(t) + \frac{1}{N}\left(\sum_{i\in\mathcal{V}_{A,t}}\delta_i(t)\beta_i C_i^\dagger\tilde{z}_i(t) - \sum_{i\in\mathcal{V}_{N,t}}\beta_i C_i^\dagger C_i e_i(t)\right)$$

$$= \left(A - \bar{\beta}G(\rho)/N\right)\bar{e}(t) + \bar{\sigma}(t), \tag{4.32}$$

where $\tilde{\sigma}(t) = \tilde{\mathcal{I}}\tilde{\Delta}_{\mathcal{V}_{A,t}}\tilde{\beta}\tilde{L}\tilde{z}(t) - \tilde{\mathcal{I}}\tilde{\Delta}_{\mathcal{V}_{N,t}}\tilde{\beta}\tilde{L}\tilde{C}e(t)$, $\bar{\sigma}(t) = \frac{1}{N}\sum_{i\in\mathcal{V}_{A,t}}\beta_i C_i^\dagger(C_i\bar{e}(t) + \delta_i(t)\tilde{z}_i(t)) - \frac{1}{N}\sum_{i\in\mathcal{V}_{N,t}}\beta_i C_i^\dagger C_i\tilde{e}_i(t)$.

Theorem 4.3 Consider the NAS (4.23) under any vulnerability-related s-sparse attacks. Suppose that Assumption 4.4 holds. For any given initial value $\hat{x}_i(0)$ under Algorithm 4.1, the state monitoring $\hat{x}_i(t)$, $i \in \mathcal{V}$ can exponentially asymptotically converge to the system state $x(t)$ if the NAS (4.23) is $(\rho, 2s)$-weighted sparse redundant, i.e.,

$$\lim_{t \to \infty} \|e_i(t)\| = 0, \quad \forall i \in \mathcal{V}.$$

Proof 4.5 *The following Lyapunov functions are constructed:*

$$V_1(t) = \|\tilde{e}(t)\| = \left(\tilde{e}^T(t)\tilde{e}(t)\right)^{\frac{1}{2}},$$

$$V_2(t) = \|\bar{e}(t)\| = \left(\bar{e}^T(t)\bar{e}(t)\right)^{\frac{1}{2}}.$$

Then, take the differential of $V_1(t)$ along (4.31) as follows:

$$\dot{V}_1(t) = V_1^{-1}(t) \cdot \tilde{e}^T(t)\dot{\tilde{e}}(t)$$

$$= V_1^{-1}(t)\tilde{e}^T(t)\left[\frac{1}{2}I_N \otimes (A + A^T) - \theta(\mathcal{L} \otimes I_n)\right]\tilde{e}(t)$$

$$+ V_1^{-1}(t)\tilde{e}^T(t)\tilde{\sigma}(t)$$

$$\leq (\alpha - \theta\lambda_2(\mathcal{L}))\,V_1(t) + V_1^{-1}(t)\tilde{e}^T(t)\tilde{\sigma}(t), \tag{4.33}$$

where the inequality is deduced based on Lemma 2.10 and the fact that $(\mathbf{1}_N^T \otimes I_n)\tilde{e}(t) = \mathbf{0}_n$.

Based on (4.25) and (4.26), one has $\delta_i(t)\|C_i^\dagger\tilde{z}_i(t)\| \leq \gamma(t)$, $\forall i \in \mathcal{V}$ and $\tilde{\mathcal{I}}\tilde{e}(t) = \tilde{e}(t)$. Hence,

$$\tilde{e}^T(t)\tilde{\sigma}(t) \leq - \sum_{i \in \mathcal{V}_{N,t}} \delta_i(t)\beta_i\tilde{e}_i^T(t)C_i^\dagger C_i\tilde{e}_i(t)$$

$$+ \|\tilde{e}(t)\| \cdot \left\|\tilde{\beta}\left(\tilde{\Delta}_{\mathcal{V}_{N,t}}\tilde{L}\tilde{C}(\mathbf{1}_N \otimes \bar{e}(t)) + \tilde{\Delta}_{\mathcal{V}_{A,t}}\tilde{L}\tilde{z}(t)\right)\right\|$$

$$\leq \bar{\beta}V_1(t) \cdot \|\varrho(t)\|/\bar{\rho}$$

$$\overset{(a)}{\leq} \bar{\beta}V_1(t)\left(s^2\gamma^2(t) + (N^2 - s^2)\gamma_2^2(t)\right)^{\frac{1}{2}}$$

$$\leq \bar{\beta}V_1(t) \cdot (s\gamma_1(t) + N\gamma_2(t)),$$

where $\varrho(t) = \mathrm{col}_{i \in \mathcal{V}}\{\varrho_i(t)\}$ with $\varrho_i(t) = \rho_i\gamma_2(t)$ if $i \in \mathcal{V}_{N,t}$, and $\varrho_i(t) = \rho_i\gamma(t)$ otherwise. The inequality (a) is deduced by noting that $\frac{1}{\bar{\rho}^2}\sum_{i \in \mathcal{V}_{N,t}}\rho_i^2 \leq N^2 - s^2$.

Then, it can be further derived that

$$\dot{V}_1(t) \leq (\alpha - \theta\lambda_2(\mathcal{L}))V_1(t) + s\bar{\beta}\gamma_1(t) + N\bar{\beta}\gamma_2(t). \tag{4.34}$$

Combining (4.34) with (4.27a), and the fact that $V_1(0) = \|\tilde{e}(0)\| \leq \sqrt{N}\eta = \gamma_1(0)$, one can conclude that

$$V_1(t) \leq \gamma_1(t), \quad \forall t \geq 0. \tag{4.35}$$

Next, take the differential of $V_2(t)$ along (4.32) as follows:

$$\dot{V}_2(t) \leq \left[\alpha - \frac{\bar{\beta}}{N}\lambda_{min}(G(\rho))\right] V_2(t) + \|\bar{\sigma}(t)\|. \tag{4.36}$$

It can be derived from (4.26), (4.32), and (4.35) that

$$\|\bar{\sigma}(t)\| \leq \frac{1}{N} \sum_{i\in\mathcal{V}_{A,t}} \beta_i \|C_i^\dagger(C_i\bar{e}(t) + \delta_i(t)\tilde{z}_i(t))\| + \frac{1}{N} \sum_{i\in\mathcal{V}_{N,t}} \beta_i \|C_i^\dagger C_i\tilde{e}_i(t)\|$$

$$\leq \frac{\bar{\beta}}{N}(V_2(t) + \gamma(t)) \sum_{i\in\mathcal{V}_{A,t}} \frac{\rho_i}{\bar{\rho}} + \frac{\bar{\beta}}{N}\gamma_1(t) \sum_{i\in\mathcal{V}_{N,t}} \frac{\rho_i}{\bar{\rho}}$$

$$\leq s\frac{\bar{\beta}}{N}(V_2(t) + \gamma_2(t)) + \bar{\beta}\gamma_1(t). \tag{4.37}$$

Substituting (4.37) into (4.36), it follows that

$$\dot{V}_2(t) \leq \left[\alpha - \frac{\bar{\beta}}{N}(\lambda_{min}(G(\rho)) - s)\right] V_2(t) + \frac{\bar{\beta}}{N}s\gamma_2(t) + \bar{\beta}\gamma_1(t).$$

One can rewrite (4.27b) in the following form

$$\dot{\gamma}_2(t) = \left[\alpha - \frac{\bar{\beta}}{N}(\lambda_{min}(G(\rho)) - s)\right] \gamma_2(t) + \frac{\bar{\beta}}{N}s\gamma_2(t) + \bar{\beta}\gamma_1(t)$$

Then, by the same token, from $V_2(0) = \|\bar{e}(0)\| \leq \eta = \gamma_2(0)$, one has

$$V_2(t) \leq \gamma_2(t), \ \forall t \geq 0. \tag{4.38}$$

From (4.35) and (4.38), the convergence of the state monitoring translates into the asymptotic stability of $\gamma_1(t)$ and $\gamma_2(t)$. Then, rewrite (4.27a) and (4.27b) as the following compact form

$$\begin{bmatrix} \dot{\gamma}_1(t) \\ \dot{\gamma}_2(t) \end{bmatrix} = N\bar{\beta}\Xi \begin{bmatrix} \gamma_1(t) \\ \gamma_2(t) \end{bmatrix}. \tag{4.39}$$

As is known that the sensor network is connected and $(\rho, 2s)$-weighted sparse redundant, one has that

$$\theta\lambda_2(\mathcal{L}) > 0, \quad \rho_s = (\lambda_{min}(G(\rho)) - 2s)/N > 0,$$

which guarantees the existence of the solution to LMI (4.28).

Given the Lyapunov function $W(t) = (\gamma_1^2(t) + \gamma_2^2(t))^{1/2}$, it follows that

$$\dot{W}(t) \leq \frac{N}{2}\bar{\beta} \left\|(\Xi^T + \Xi) \begin{bmatrix} \gamma_1(t) \\ \gamma_2(t) \end{bmatrix}\right\| < -\frac{N}{2}\tilde{h}\bar{\beta}W(t),$$

where $\tilde{h}\bar{\beta} > 0$. Hence, one has $W(t)$ is exponentially asymptotically stable and

$$\lim_{t\to\infty} \gamma(t) \le \lim_{t\to\infty} \sqrt{2}W(t) = 0.$$

Further, since $e_i(t) = \tilde{e}_i(t) + \bar{e}(t)$, one can conclude from (4.35) and (4.38) that, for any $i \in \mathcal{V}$,

$$\|e_i(t)\| \le \|\tilde{e}(t)\| + \|\bar{e}(t)\| = V_1(t) + V_2(t) \le \gamma(t),$$

which implies that $\lim_{t\to\infty} \|e_i(t)\| \le \lim_{t\to\infty} \gamma(t) = 0$ holds for all $i \in \mathcal{V}$. This completes the proof.

Remark 4.10 *It can be obtained from (4.27a) and (4.34) that (4.35) holds only when the initial value of $\gamma_1(t)$ satisfies $\gamma_1(0) \ge V_1(0) = \|\tilde{e}(0)\|$. Therefore, if the initial states of all the observers $\hat{x}_i(0)$, $i \in \mathcal{V}$ are set as the same value, one has that $\tilde{e}_i(0) = \mathbf{0}_n$, $\forall i \in \mathcal{V}$ and thus $V_1(0) = 0$, then the initial value of $\gamma_1(t)$ can be simplified as $\gamma_1(0) = 0$.*

Remark 4.11 *In the attack-isolation-based SSM architecture, the threshold function $\gamma(t)$ provides accurate attack detection ultimately, while the $(\rho, 2s)$-weighted sparse redundant of the network makes the negative impact of the attacks be restrained until the attack detection is accomplished. The combination of $\gamma(t)$ and $(\rho, 2s)$-weighted sparse redundant ensures that the SSM can be achieved under arbitrary vulnerability-related s-sparse attacks.*

Remark 4.12 *For the case that any attack signal $a_i(t) \neq \mathbf{0}_{m_i}$, there must exist a time instant $t > T$ such that $\gamma(t) < \|C_i^\dagger \tilde{z}_i(t)\| = \|C_i^\dagger(C_i e_i(t) + a_i(t))\|$ holds due to $\lim_{t\to\infty} \gamma(t) = 0$, and thus, one has $\delta_i(t) = 0$ according to (4.25). This means that all the malicious attacks can be successfully detected and isolated as time evolves.*

Remark 4.13 *The $(\rho, 2s)$-weighted sparse redundant condition implies that $2s < \lambda_{min}(G(\rho)) \le \frac{1}{\bar{\rho}} \sum_{i=1}^N \rho_i \le N$, then one has the number s can be maximized as $\lceil \frac{N}{2} - 1 \rceil$. Note that this constraint form has been reported in [20] and is in agreement with it, namely, that the maximum tolerance limit against s-sparse attacks is $s \le \lceil \frac{N}{2} - 1 \rceil$. However, the difference is that for the network with a large number of vulnerable sensors, even if the attack damages more than $\lceil \frac{N}{2} - 1 \rceil$ vulnerable sensors, it may still satisfy $s \le \lceil \frac{N}{2} - 1 \rceil$ under the mechanism of vulnerability metrics. Furthermore, considering the extreme scenario where $C_i^\dagger C_i = I_n$ and $\rho_1^d \ge \frac{N}{2}\bar{\rho}$, the maximum tolerable number of attacks can reach its peak, i.e., $N - 1$.*

● *Distributed security monitoring with trusted sensors*

Intuitively, the security of state monitoring becomes increasingly assured with the incorporation of trusted sensors. Therefore, in this subsection, our focus shifts to the problem of identifying sensors under attack and simplification of SSM approach. To address this, the sensors can be divided into two subsets based on their trustworthiness: the trusted subset $\mathcal{T}$ and the un-trusted one $\bar{\mathcal{T}}$. Then, one has that $\bar{\mathcal{T}} = \mathcal{V}\backslash\mathcal{T}$ and $\mathcal{T} \cap \mathcal{V}_{A,t} = \emptyset$.

The following assumption on the trusted sensors is provided to facilitate the subsequent analysis.

Assumption 4.6 *The measurement matrix C_i of the trusted sensors satisfies*

$$\lambda_{min}\left(\sum_{i \in \mathcal{T}} C_i^{\dagger} C_i\right) \geq r_{\mathcal{T}} > 0. \tag{4.40}$$

Under Assumption 4.6, the measurements from all trusted sensors are sufficient to reconstruct the actual state of the system, however, since the trusted sensors are distributed all over the network, other un-trusted observers are still needed as communication relays. Therefore, in order to simplify the construction of the SSM algorithm by fully utilizing the measurements from trusted sensors and to quickly detect and locate the point of action of sparse attacks in the sensor network, the observation gain β_i is considered to be set as follows:

$$\beta_i = \begin{cases} \bar{\beta}, & \text{if } i \in \mathcal{T}, \\ 0, & \text{if } i \in \bar{\mathcal{T}}. \end{cases} \tag{4.41}$$

Besides, the dynamics of the threshold functions in (4.27a) and (4.27b) are further simplified as

$$\dot{\gamma}_1(t) = (\alpha - \theta\lambda_2(\mathcal{L}))\gamma_1(t) + \sqrt{N - \tau}\bar{\beta}\gamma_2(t), \tag{4.42a}$$

$$\dot{\gamma}_2(t) = \left(\alpha - \bar{\beta}\frac{r_{\mathcal{T}}}{N}\right)\gamma_2(t) + \bar{\beta}(1 - \frac{\tau}{N})\gamma_1(t), \tag{4.42b}$$

where $\tau = |\bar{\mathcal{T}}|$, the selection of the initial values is the same as in (4.27), while the variables $\theta > 0$ and $\bar{\beta} > 0$ should satisfy the following updated LMI:

$$\tilde{\Xi}^T + \tilde{\Xi} < -\tilde{h}I_2, \tag{4.43}$$

with $\tilde{h} > 0$ and

$$\tilde{\Xi} = \begin{bmatrix} \alpha - \lambda_2(\mathcal{L})\theta & \sqrt{N - \tau}\bar{\beta} \\ (1 - \tau/N)\bar{\beta} & \alpha - \bar{\beta}r_{\mathcal{T}}/N \end{bmatrix}.$$

Remark 4.14 *Importantly, only two supplementary assumptions are introduced in this context. First, each observer i is assumed capable of determining whether its associated sensor is reliable. Second, the intrinsic sensor trustworthiness parameters, $r_{\mathcal{T}}$ and $\tau = |\bar{\mathcal{T}}|$, are assumed to be available. As a result, it is not required for the full set of trusted sensors $\mathcal{T}$ to be known from a global perspective.*

Theorem 4.4 *Consider the continuous NAS (4.23) with trusted sensors subset $\mathcal{T}$. Suppose that Assumption 4.6 holds. By designing the observer gains and the threshold function as (4.41)–(4.43), the distributed SSM can be reached under any τ-sparse attacks and the identified malicious attack set satisfies $\lim_{t \to \infty} |\mathcal{V}_{A,t} \setminus \hat{\mathcal{V}}_{A,t}| = 0$ with $\hat{\mathcal{V}}_{A,t} = \{i \mid \delta_i(t) = 0, i \in \mathcal{V}\}$.*

Proof 4.6 Consider the extreme worst case where $\mathcal{V}_{A,t} = \bar{\mathcal{T}}$ when $t \geq 0$. Then, it follows that

$$\tilde{e}^T(t)\check{\sigma}(t) \leq -\sum_{i \in \mathcal{T}} \delta_i(t)\beta_i \tilde{e}_i^T(t) C_i^\dagger C_i \tilde{e}_i(t) + \|\tilde{e}(t)\| \cdot \left\|\bar{\beta}\tilde{\Delta}_{\mathcal{T}}\tilde{L}\tilde{C}(\mathbf{1}_N \otimes \bar{e}(t))\right\|$$

$$\leq \sqrt{N - \tau}\bar{\beta}V_1(t)\gamma_2(t).$$

Similar to (4.33) in the proof of Theorem 4.3, one has

$$\dot{V}_1(t) \leq (\alpha - \theta\lambda_2(\mathcal{L}))V_1(t) + \sqrt{N - \tau}\bar{\beta}\gamma_2(t).$$

Hence, it follows form $V_1(0) \leq \gamma_1(0)$ that $V_1(t) = \|\tilde{e}(t)\| \leq \gamma_1(t)$ holds.

Besides, the dynamics of the error $\bar{e}(t)$ can be rewritten as

$$\dot{\bar{e}}(t) = \left(A - \frac{\bar{\beta}}{N}\sum_{i \in \mathcal{T}} C_i^\dagger C_i\right)\bar{e}(t) + \check{\sigma}(t), \tag{4.44}$$

where $\check{\sigma}(t) = -\frac{1}{N}\bar{\beta}\sum_{i \in \mathcal{T}} C_i^\dagger C_i \tilde{e}_i(t)$.

Similar to the proof of Theorem 4.3, taking the differential of $V_2(t)$ along (4.44), one has

$$\dot{V}_2(t) \leq \left(\alpha - r_{\mathcal{T}}\bar{\beta}/N\right)V_2(t) + \|\check{\sigma}(t)\|, \tag{4.45}$$

where $\|\check{\sigma}(t)\| \leq \bar{\beta}(1 - \tau/N)\gamma_1(t)$. This suggests that

$$\dot{V}_2(t) \leq \left(\alpha - r_{\mathcal{T}}\bar{\beta}/N\right)V_2(t) + \bar{\beta}(1 - \tau/N)\gamma_1(t).$$

Thus, the same conclusion $V_2(t) \leq \gamma_2(t)$ can be derived. The subsequent analysis is similar to the proof of Theorem 4.3 and Remark 4.12 and is therefore omitted for brevity.

Remark 4.15 *The presence of condition (4.40) ensures that even if all un-trusted sensors are compromised, the measurements from the trusted sensors are still sufficient to complete the attack isolation and achieve a reliable SSM.*

4.3.4 Numerical simulations

To verify the validity of Theorems 4.3 and 4.4, the verification is continued by simulating the SSM of the IEEE 6-bus power system with three generators and six buses [108, 109]. Then, the Kron-reduced form of the power network [108] is obtained with the small-signal method, expressed as:

$$\begin{bmatrix} \dot{\zeta}(t) \\ M_g\dot{\omega}(t) \end{bmatrix} = \begin{bmatrix} 0 & I \\ L_{Ig}^T L_{II}^{-1} L_{Ig} - L_{gg} & -D_g \end{bmatrix} \begin{bmatrix} \zeta(t) \\ \omega(t) \end{bmatrix} + \begin{bmatrix} 0 & 0 \\ I & -L_{Ig}^T L_{II}^{-1} \end{bmatrix} \begin{bmatrix} p_\omega(t) \\ p_\theta(t) \end{bmatrix}, \tag{4.46}$$

where $\zeta(t) = [\zeta_1(t), \zeta_2(t), \zeta_3(t)]^T$ and $\omega(t) = [\omega_1(t), \omega_2(t), \omega_3(t)]^T$ represent the generator rotor angles and frequencies. The matrices M_g and D_g are diagonal matrices that denote the generator inertial and damping coefficients, respectively. The inputs $p_\omega(t) = [p_{1\omega}(t), \ p_{2\omega}(t), \ p_{3\omega}(t)]^T$ and $p_\theta(t) = [p_{1\theta}(t), \ p_{2\theta}(t), \ p_{3\theta}(t)]^T$ correspond to known changes in the mechanical input power and real power demand at the loads, respectively. To fit the system model given by (4.23), the system matrices can be expressed as follows:

$$A = \begin{bmatrix} 0 & I \\ M_g^{-1}\left(L_{Ig}^T L_{II}^{-1} L_{Ig} - L_{gg}\right) & -M_g^{-1}D_g \end{bmatrix},$$

$$B = \begin{bmatrix} 0 & 0 \\ M_g^{-1} & -M_g^{-1}L_{Ig}^T L_{II}^{-1} \end{bmatrix},$$

where the matrix parameters are set as $M_g = \text{diag}\{0.125, 0.034, 0.016\}$, $D_g = \text{diag}\{0.125, 0.068, 0.48\}$, $L_{gg} = \text{diag}\{0.058, 0.063, 0.059\}$, and

$$L_{Ig} = \begin{bmatrix} -0.058 & 0 & 0 & 0 & 0 & 0 \\ 0 & -0.063 & 0 & 0 & 0 & 0 \\ 0 & 0 & -0.059 & 0 & 0 & 0 \end{bmatrix}^T, \tag{4.47}$$

$$L_{II} = \begin{bmatrix} 0.235 & 0 & 0 & -0.085 & -0.092 & 0 \\ 0 & 0.296 & 0 & -0.161 & 0 & -0.072 \\ 0 & 0 & 0.330 & 0 & -0.170 & -0.101 \\ -0.085 & -0.161 & 0 & 0.246 & 0 & 0 \\ -0.092 & 0 & -0.170 & 0 & 0.262 & 0 \\ 0 & -0.072 & -0.101 & 0 & 0 & 0.173 \end{bmatrix}. \tag{4.48}$$

In the simulation, the initial state of the system is set as $x(0) = [0.2, \ -0.5, \ 0.7, \ 0, \ 0, \ 0]^T$, and the control input is set as $u(t) = 0.1sin(5t) \times \mathbf{1}_9$. In addition, the state of the power system is monitored by 7 sensors with measurement matrices as follows:

$$C_2 = \begin{bmatrix} 0 & 1 & -1 & 0 & 0 & 0 \\ -1 & 1 & 0 & 0 & 0 & 0 \\ 1 & 0 & 1 & 0 & 0 & 0 \\ 0 & 0 & 0 & 0 & 1 & -1 \\ 0 & 0 & 0 & 1 & 1 & 1 \end{bmatrix}, \quad C_3 = \begin{bmatrix} -1 & 0 & 1 & 0 & 0 & 0 \\ 0 & 1 & 0 & 0 & 0 & 0 \\ 0 & 0 & 0 & 0 & 1 & 1 \\ 0 & 0 & 0 & -1 & 1 & 0 \\ 0 & 0 & 0 & 1 & 1 & -1 \end{bmatrix},$$

$$C_4 = \begin{bmatrix} 0 & -1 & 0 & 0 & 0 & 0 \\ -1 & 1 & 0 & 0 & 0 & 0 \\ 0 & 0 & 0 & 0 & 0 & -1 \\ 0 & 0 & 0 & 0 & -1 & 0 \end{bmatrix}, \quad C_5 = \begin{bmatrix} 1 & 0 & -1 & 0 & 0 & 0 \\ 0 & 1 & 0 & 0 & 0 & 0 \\ 0 & 0 & 0 & 0 & 1 & 0 \\ 0 & 0 & 0 & 1 & 1 & 0 \end{bmatrix},$$

$$C_6 = \begin{bmatrix} 1 & 0 & -1 & 0 & 0 & 0 \\ 0 & 1 & -1 & 0 & 0 & 0 \\ 0 & 1 & 0 & 0 & 0 & 0 \\ 0 & 0 & 0 & -1 & 0 & -1 \\ 0 & 0 & 0 & 0 & -1 & 1 \end{bmatrix}, \quad C_1 = C_7 = I_6,$$

such that the collective observation matrix C satisfies the $(\rho, 5.5)$-weighted sparse

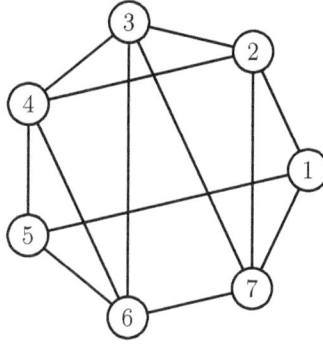

Figure 4.13   Communication topology of 7 heterogeneous sensors.

redundant with $\rho = [6, 1, 1.5, 0.5, 0.4, 0.6, 4]$. The communication topology of the sensors and observers is given in Fig. 4.13.

Assume that the attackers could collide with each other in the expectation of causing a uniform estimation deviation $\hat{x}^f(t)$ to the observers. Specifically, consider that the dynamics of $\hat{x}^f(t)$ follow the following form:

$$\dot{\hat{x}}^f(t) = A\hat{x}^f(t), \tag{4.49}$$

where the initial value is set as $\hat{x}^f(0) = [-1.2, -0.5, -1, 0.4, -1, 0.5]^T$. Thus, the malicious attack injection is set to be $a_i(t) = C_i\hat{x}^f(t)$, $i \in \mathcal{V}_{A,t}$ if i-th sensor is compromised. Furthermore, assume that the malicious attack is vulnerability-related 2.1-sparse and the attack target is randomly switched every 1 second.

It is easy to derive that $\alpha = 0.1696$, and then to solve the LMI in (4.28) by MATLAB robust toolbox, the observer gain $\theta = 297.21$, $\bar{\beta} = 1.49$, and $\tilde{h} = 0.19$ can be obtained subsequently.

Example 4.3: To illustrate the effectiveness of attack detection (4.25), a function $\tilde{\delta}_i(t)$ is introduced to characterize the detection result for sensor i. In detail,

$$\tilde{\delta}_i(t) = \begin{cases} 0, & \text{if } \delta_i(t) = 1, \\ i, & \text{if } \delta_i(t) = 0, \end{cases}$$

where $\tilde{\delta}_i(t) = 0$ suggests that the sensor i is normal, and $\tilde{\delta}_i(t) = i$ suggests that the sensor i is detected as being compromised. The attack detection results are depicted in Fig. 4.14, and it is obvious that for the inflicted vulnerability-related sparse attacks, the anomaly detector $\delta_i(t)$ constructed based on threshold $\gamma(t)$ designed in (4.27) can achieve precise detection and isolation of all attacks.

Fig. 4.15 depicts the actual state trajectory (solid lines) of the power system along with its state estimations (dashed lines) generated by each local observer. Additionally, the relationship between the state estimation error norm $\|e_i(t)\|$ and the threshold function $\gamma(t)$ is illustrated in Fig. 4.16. Obviously, it can be noted that all observers can asymptotically estimate the true state of the power system, and the estimation errors comply with the constraint (4.26) due to $\|C_i^\dagger C_i\| \leq 1$.

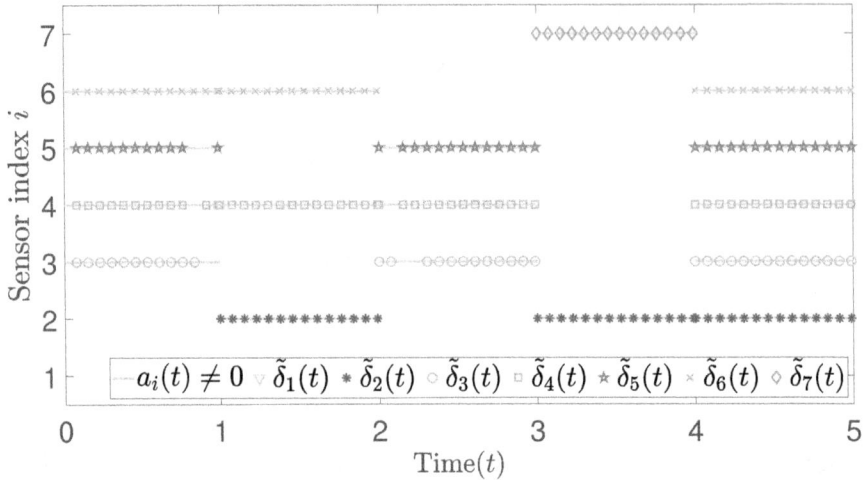

Figure 4.14 The compromised sensors and the attack detection result $\tilde{\delta}_i(t)$ in Example 4.3.

Overall, it can be observed that even when the number of attacks exceeds half of the sensor channels but in a vulnerability-related sparse attack environment, the proposed weight-based SSM method in this chapter remains effective. Therefore, the conclusion can be drawn that by assigning different vulnerability metric weights to different sensors, further enhancement of resilience against sparse attacks is achievable.

Example 4.4: In this example, in order to facilitate the demonstration of algorithmic effects, it is assumed that $C_i = I_n$ and satisfy is $(\rho, 6.5)$-weighted sparse redundant with $\rho = [6, 0.4, 0.5, 1.5, 1, 0.6, 4]$.

For clarity, consider a scenario where the adversaries launch a malicious attack by selecting a fixed subset of sensors. Specifically, for the power system under consideration, the adversaries choose fixed two ($\mathbf{A}^1 = \{2,5\}$), four ($\mathbf{A}^2 = \{2,3,4,5\}$), and five ($\mathbf{A}^3 = \{2,3,4,5,7\}$) sensors to inject a deliberate attack. The attack signals are also formulated as $a_i(t) = C_i\hat{x}^f(t)$, $i \in \mathbf{A}^j$, $j \in \{1,2,3\}$ with $\hat{x}^f(t)$ set by (4.49) and the initial value is chosen as $\hat{x}^f(0) = [-8, 1, -3, 2, -5, -6]^T$ in this example.

Here, the state estimation effects of the following three different methods are compared:

$m1$) Proposed Algorithm 4.1 with the weight factor ρ;

$m2$) Proposed SSM approach with trusted sensors $\{1,6\}$;

$m3$) Resilient SSM approach with SIU algorithm in [37].

Figs. 4.17–4.19 depict the state estimation results generated by observer 1 under the above three methods. Then, to evaluate the performance of each method in terms of estimation error, the performance function $\phi(t) = \frac{1}{7}\sum_{i=1}^{7}\|e_i(t)\|$ is introduced. The comparison results of function $\phi(t)$ under the three methods are shown in Fig. 4.20.

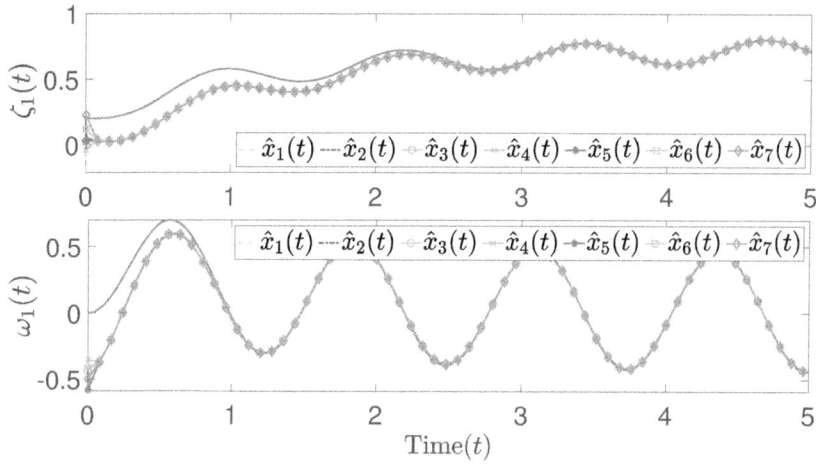

(a) $j = 1$

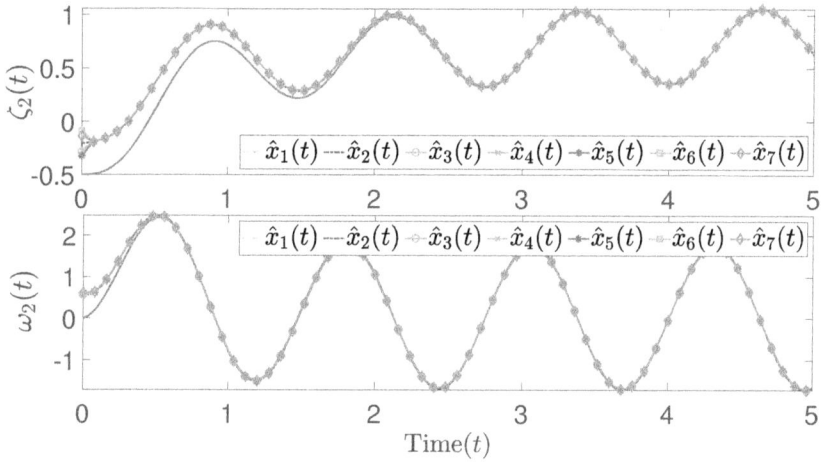

(b) $j = 2$

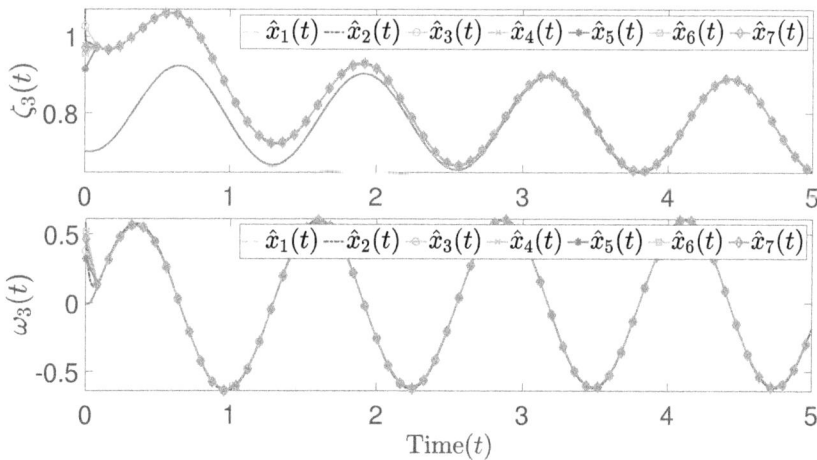

(c) $j = 3$

Figure 4.15   System state trajectories of $\zeta_j(t)$, $\omega_j(t)$, $j \in [3]$ and its estimation $\hat{x}_i(t)$ in Example 4.3.

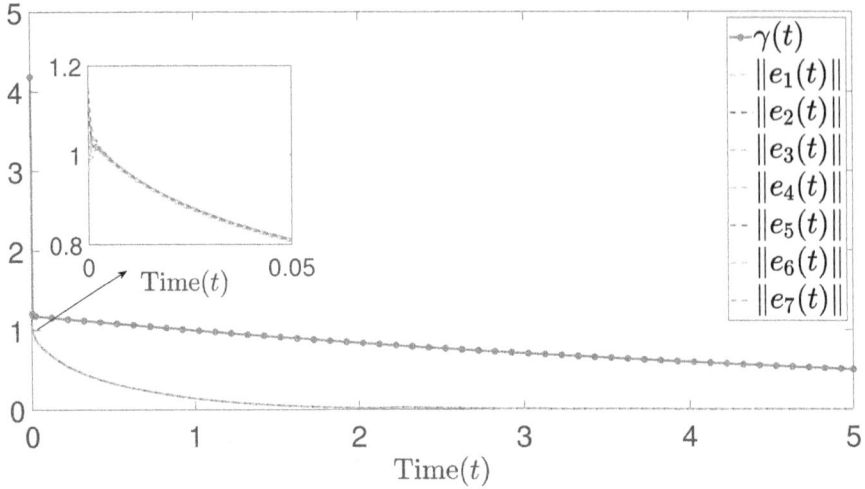

Figure 4.16   The relationship between the threshold $\gamma(t)$ and the error norm $\|e_i(t)\|$ in Example 4.3.

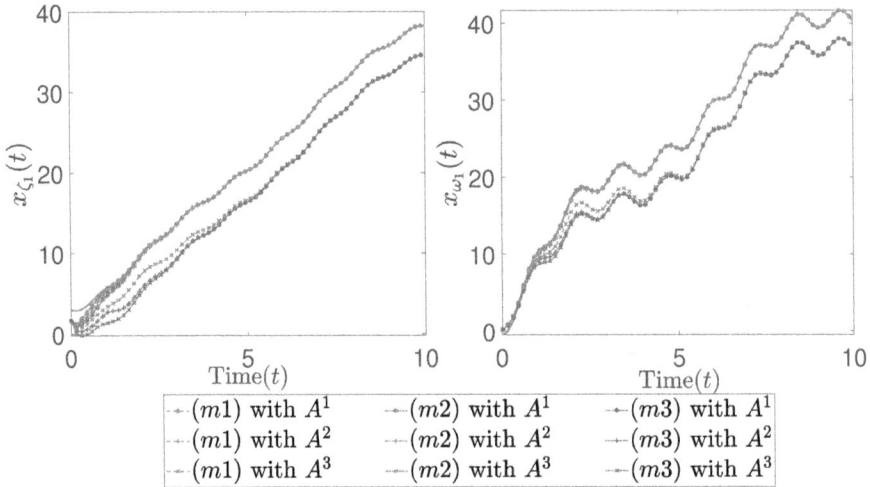

Figure 4.17   State monitoring $\hat{x}_1(t)$ generated by observer 1 under different attacks and approaches in Example 4.4.

It is evident that when the number of the attacks is smaller than half of the sensors (e.g., $\mathbf{A}^1$), the above three methods are able to effectively against these attacks. However, in the case that the number of the attacks exceeds half of the sensors but still meets the vulnerability-related s-sparse condition (e.g., $\mathbf{A}^2$), the two SSM approaches proposed in this chapter are able to function normally, while the algorithm presented in [37] fails. In particular, when all untrusted sensors are under attack (e.g., $\mathbf{A}^3$), only the algorithm in method ($m2$) can ensure SSM. Therefore, after weighing the sensors accordingly, the algorithm proposed in this chapter has a superior performance to resist more attacks, and of course, the presence of trusted sensors can further enhance the algorithm performance and simplify the design.

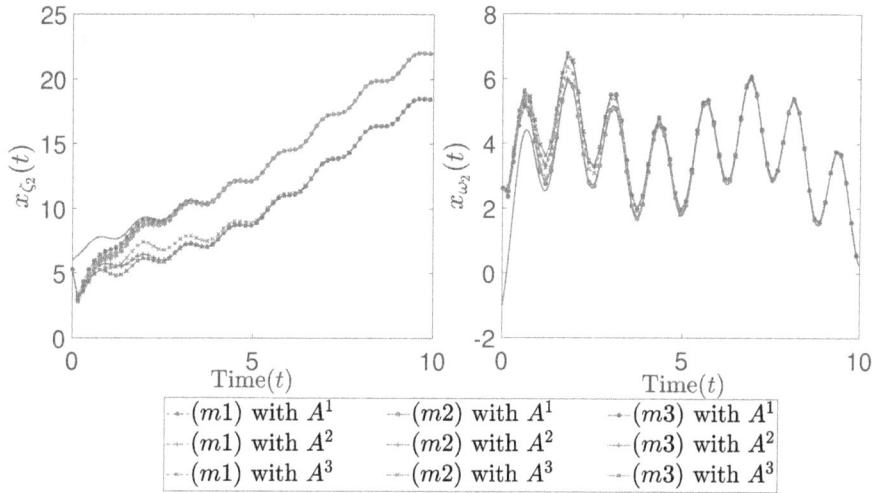

Figure 4.18 State monitoring $\hat{x}_2(t)$ generated by observer 1 under different attacks and approaches in Example 4.4.

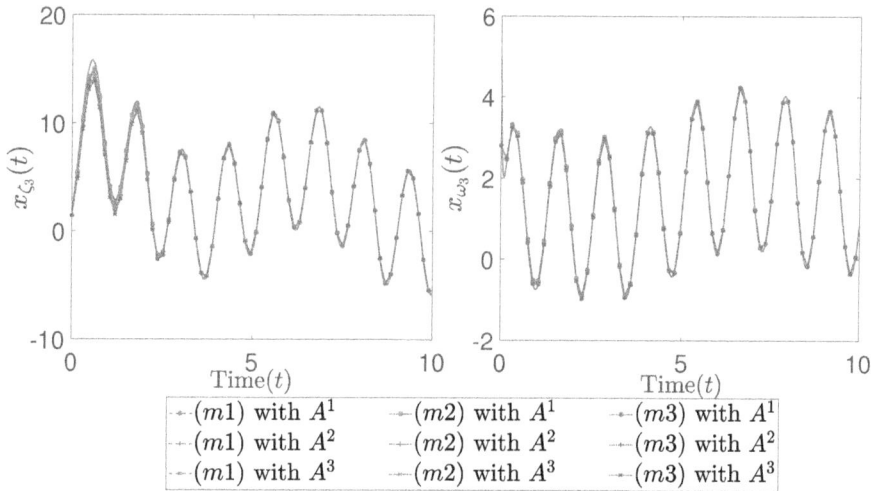

Figure 4.19 State monitoring $\hat{x}_3(t)$ generated by observer 1 under different attacks and approaches in Example 4.4.

4.4 CONCLUSIONS

This chapter systematically investigated defense mechanisms for heterogeneous sensor networks under sparse attack scenarios. Two principal advancements have been achieved: Firstly, a distributed security signal monitoring algorithm was developed, utilizing an adaptive sliding window mechanism designed to counteract sparse-varying attack patterns in heterogeneous sensor configurations. The proposed scheme demonstrated enhanced convergence characteristics through strategic incorporation of historical attack signatures, thereby optimizing both algorithmic architecture and estimation accuracy.

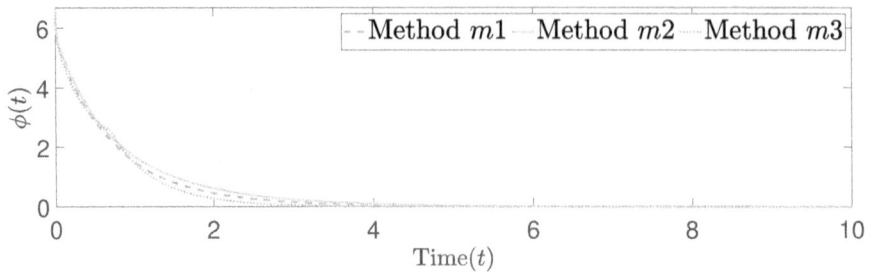

(a) With the attack set $\mathbf{A}^1$

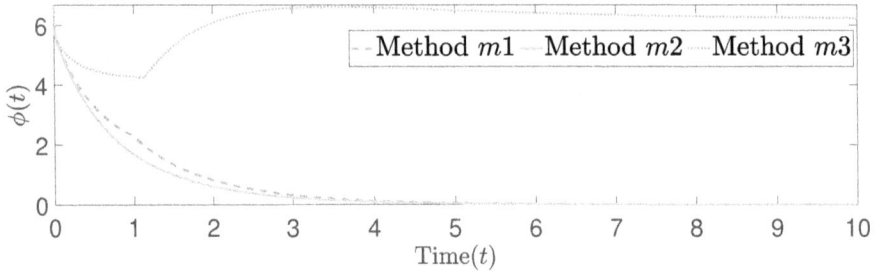

(b) With the attack set $\mathbf{A}^2$

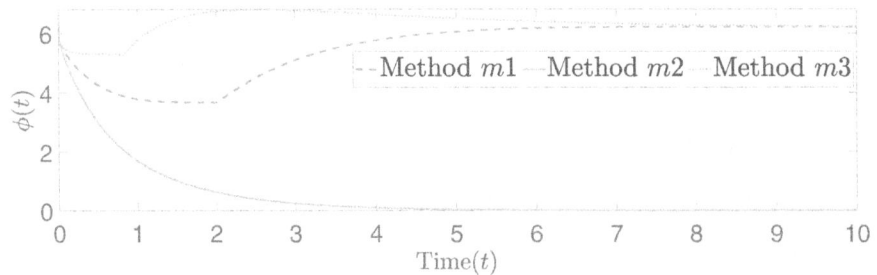

(c) With the attack set $\mathbf{A}^3$

Figure 4.20   Comparison of three different methods in estimation error.

Secondly, building upon the established vulnerability partitioning framework, we have developed an integrated security paradigm that encompassed: 1) a vulnerability-related s-sparse attack model quantifying differential susceptibilities across heterogeneous nodes, 2) a weighted sparse redundancy criterion establishing sufficient conditions for distributed SSM. The theoretical framework has been operationalized through dynamic detection thresholds that synergistically combined spatial-temporal attack characteristics, enabling precise malicious node identification and isolation. Comprehensive analytical validation has confirmed the methodology's efficacy in preserving network integrity against sophisticated sparse attack vectors.

CHAPTER 5

Distributed security monitoring against sparse attacks: From asymptotic to finite- and fixed-time convergence

This chapter investigates rapid distributed security monitoring for NASs under location-varying sparse sensor attacks. The exposition commences with an overview of existing detection methodologies, establishing the theoretical motivations through identified gaps in handling location-varying sparse attacks. Section 5.2 introduces an extra auxiliary compensation mechanism that is co-integrated with feedback monitoring gains to neutralize unknown dynamic attacks. In contrast to conventional anomaly detection paradigms, the proposed compensation framework eliminates unnecessary global detector redundancy while preserving non-smooth control properties, achieving enhanced computational efficiency without compromising robustness. Building upon this foundation, Section 5.3 establishes a co-design architecture for secure consensus protocols that enables distributed observers to reach security monitoring convergence from asymptotic to finite-/fixed-time. Finally, some simulation examples are presented to illustrate the effectiveness of theoretical results.

5.1   INTRODUCTION

Distributed security monitoring of NASs has garnered considerable attention in various fields such as smart grids, transportation networks, unmanned systems, and so on [118, 119]. Distributed SSM offers the advantage of reducing the workload on individual agents by distributing tasks among agents. Regrettably, fragile wireless communication networks inevitably render NASs powerless to prevent the effects of cyber-attacks or sudden faults, which poses a significant drawback to distributed

DOI: 10.1201/9781003669913-5

SSM [100, 120, 121]. Consequently, security analysis has become a crucial aspect in the field of distributed SSM for NASs, attracting extensive attention from various scientific communities [122–124].

From the perspective of resource-limited attackers, launching sparse attacks against local nodes in the network is a compromise approach. Such attacks are not limited in form and only the order of magnitude of attacks being constrained [125]. In response to such attacks, a class of SSM frameworks based on a static batch optimization technique was proposed in [17, 20]. The secure asymptotic monitoring under location-varying sparse attacks was implemented on the Luenberger observer employing the attack detection and isolation design in [37, 40, 93]. However, the design of the existing detection threshold is severely hampered by the presence of a nonlinear controller in carrying out the finite-/fixed-time control, and the results on finite-/fixed-time monitoring for large-scale NASs under sparse attacks are still insufficient.

On the other hand, finite-/fixed-time consensus control for NASs has attracted growing research interest owing to the advantages such as rapid convergence, superior tracking precision, and robustness against disturbances [126–130]. In [127, 128], fixed-time consensus tracking and leaderless consensus control protocols for nonlinear multi-agent systems were provided, respectively. For general linear multi-agent systems, some finite-time or fixed-time consensus results were given in [130–133] with the help of the sign function. In addition, [129] supplied a distributed finite-time consensus control protocol for nonlinear NASs with the network compromised by false data injection attacks. Regarding different multi-agent system models, [120] and [134] proposed appropriate design schemes based on distributed fixed-time estimation and control architectures, respectively, to mitigate the adverse effects of DoS attacks on cooperative target tracking. The research development and recent trends in fixed-time and prescribed-time cooperative control of NASs have been reviewed in [89]. It pointed out that the consensus control problem under unprescribed network-induced constraints is vital and yet challenging. It is important to note that when dealing with sparse attacks, the design of attack detection thresholds heavily depends on the error performance of NASs. However, the aforementioned finite-/fixed-time controllers significantly constrain precise performance analysis, making it difficult to seamlessly extend existing results [126–130] to address sparse attacks effectively. It remains a challenge for NASs to maintain the performance of distributed security consensus monitoring under sparse attacks.

Motivated by the aforementioned works on SSM and finite-/fixed-time control of NASs, this chapter focuses on the development of a new design framework of distributed SSM protocols for NASs against sparse attacks, aiming to achieve asymptotic or even finite-/fixed-time SSM. By analyzing the performance of consensus monitoring under worst-case attacks, a decaying auxiliary function is designed to counter sparse sensor attacks, allowing the security of asymptotic consensus monitoring. Then, the design framework of the original control gains is refined to accommodate consensus monitoring under different convergence rates. Particularly, a novel design approach for finite-/fixed-time SSM is proposed, which effectively avoids the analytical challenges caused by the utilization of the sign function in high-dimensional systems, as highlighted in [131–133].

5.2 PROBLEM FORMULATION

Consider a nonlinear NAS consisting of one monitored plant and N vulnerable sensor network with the system state $x(t) \in \mathbb{R}^n$ and the measurement output $\tilde{y}_i(t) \in \mathbb{R}^n$. The dynamics of NAS and corresponding distributed state monitors can be depicted as follows:

$$
\begin{cases}
\dot{x}(t) = f(t, x), \\
\tilde{y}_i(t) = b_i(t)(x(t) + a_i(t)), \\
\dot{\hat{x}}_i(t) = f(t, \hat{x}_i) + u_i(t) \quad i \in \mathcal{V} = [N],
\end{cases}
\tag{5.1}
$$

where $f(t, \cdot) : \mathbb{R}^+ \times \mathbb{R}^n \to \mathbb{R}^n$ is a nonlinear state transfer function satisfying that for any $x_a, x_b \in \mathbb{R}^n$, there exists a non-negative constant $\varrho > 0$ such that

$$
(x_a - x_b)^T (f(t, x_a) - f(t, x_b)) \leq \varrho \|x_a - x_b\|^2.
\tag{5.2}
$$

Particularly, $b_i(t) \in \{0, 1\}$ denote an index identifier to determine whether the current sensor i is capable of measuring the system, $a_i(t)$ is an unknown and abnormal attack injection on the sensor channel i, $\hat{x}_i(t)$ denote the estimation of $x(t)$ generated by the i-th observer, and $u_i(t)$ is the local control input of the observer i to be designed.

It is widely recognized that the security of sensor measurement information is pivotal in the state monitoring for NASs, making it highly susceptible to targeted attacks from adversaries. Hence, in light of potential cyber risks, a fraction of the communication data between system and sensors is assumed to undergo corruption, each of which may fulfill one of the following anomalies:

- Inevitably, some unknown artificial interventions or accidental mistakes occurs when sensors measure or capture the true state information of the system. It causes local sensors to exhibit perceptual bias of the system state without being aware of it, i.e., $a_i(t) \neq \mathbf{0}_n$ and $\tilde{y}_i(t) \neq b_i(t)x(t)$.

- In addition, during the process of sending measurement data from the sensor to the corresponding observer, some transmission channels are hijacked by an energy-limited attacker, resulting in part of the transmitted data being manipulated. Namely, some observers receive incorrect measurement information as $\tilde{y}_i(t) \neq b_i(t)x(t)$ at time t.

Note that here $\tilde{y}_i(t)$ denotes the false measurement obtained by the observer i. It is noteworthy that for simplicity and clarity, the above scenario is established in a noiseless environment. However, for noisy systems, imposing a bounded or statistical constraint is equally applicable.

Remark 5.1 *For the i-th compromised sensor, the abnormal data $\tilde{y}_i(t)$ can take any value except $x(t)$. Moreover, attackers are permitted to collaborate, fabricating* *falsified data capable of evading detection mechanisms. Thus, conventional methods of attack detection and compensation become ineffective, necessitating the development of advanced algorithms for attack suppression or isolation.*

The following general assumptions are listed for further advancement.

Assumption 5.1 *Suppose that the initial state error between observers and the system is within a bounded region of size σ, i.e., $\max_{i \in \mathcal{V}} \|\hat{x}_i(0) - x(0)\| \leq \sigma$.*

Assumption 5.2 *Assume that the topology $\mathcal{G}$ of observers is undirected and connected, the number of sensor measurements is satisfied with $|\mathcal{V}_{r,t}| = m \leq N$, where $\mathcal{V}_{r,t} \triangleq \{i | b_i(t) = 1, \ t \geq 0\}$.*

Assumption 5.3 *It is assumed that the attack $a_b(t) = \mathrm{col}\{b_i(t)a_i(t)\}$ is location-varying s-sparse, i.e., $|\mathcal{V}_{a,t}| \leq s$ with $\mathcal{V}_{a,t} = \{i \mid b_i(t)a_i(t) \neq \mathbf{0}_n\} \subseteq \mathcal{V}_{r,t}$. Besides, the data anomaly locations (i.e., $\mathcal{V}_{a,t}$) are unknown for all observers.*

In light of the above assumptions, the parameters m and s play a crucial role in determining the resistance of the NASs against sparse data anomalies. Specifically, the observers are unaware of which ones received biased data but are assumed that an upper bound on the number of worst-case data anomalies is learned through historical experience. This can also be interpreted as a level of performance of the NASs in terms of resistance to sparse data anomalies. Assumption 5.3 aligns with the common assumption of sparse sensor attacks addressed in works on SSM for NASs [17, 20, 33, 37, 40, 43]. Referring to previous chapters and Theorem 3.2 in [17], the necessary and sufficient condition for resisting sparse attacks and achieving secure estimation is that the number of attacks is less than half of the transmission channels, i.e., $s < m/2$.

It is noteworthy that, from the defender's perspective, $\mathcal{V}_{a,t}$ is modeled as a set with time-varying and unknown switching properties. This is primarily due to the fact that the attacker usually possesses knowledge of the current behavioral actions of a defender and could adaptively modify its attack strategy accordingly. Therefore, the objective of this chapter is to design an effective and security monitoring protocol framework for all observers, such that the NAS (5.1) achieves asymptotic to finite-/fixed-time SSM of the system in the presence of arbitrary location-varying sparse attacks.

5.3 DISTRIBUTED SECURITY MONITORING AGAINST SPARSE ATTACKS

This section focuses on developing a new framework of secure consensus controls against location-varying sparse sensor attacks. Specifically, a new framework for designing distributed SSM protocol is first presented. Subsequently, a comprehensive analysis is conducted under the worst-case scenario to ensure that all observers achieve consensus monitoring from asymptotic to finite/fixed-time convergence.

5.3.1 Distributed security state monitoring protocol

The core concept of distributed security state monitoring in NASs is for each observer to safeguard the resilience and security of its state through interactions with neighboring observers. Additionally, the sparsity of attacks is also leveraged, and the interaction magnitude of sensor measurement is homogenized to mitigate malicious effects and facilitate global consensus SSM.

Following this general idea, a new form of the SSM control protocol for the NAS (5.1) is proposed as follows:

$$u_i(t) = \sum_{j=1}^{N} a_{ij}\delta_{ij}(t)e_{ij}(t) + \gamma(t)\frac{z_i(t)}{\|z_i(t)\|}, \tag{5.3}$$

where $e_{ij}(t) = \hat{x}_j(t) - \hat{x}_i(t)$, the scalar functions $\delta_{ij}(t) \geq 0$ if $i \neq j$ and $\delta_{ii}(t) = \sum_{j=1,j\neq i}^{N} a_{ij}\delta_{ij}(t)/l_{ii}$, $\gamma(t) \geq 0$ are the control gain to be designed, and $z_i(t) = \tilde{y}_i(t) - \hat{x}_i(t)$ denote the innovation item with $\tilde{y}_i(t) \neq x(t)$ if $i \in \mathcal{V}_{a,t}$, $\tilde{y}_i(t) = \mathbf{0}_n$ if $i \notin \mathcal{V}_{r,t}$ otherwise $\tilde{y}_i(t) = x(t)$. Moreover, one has $z_i(t)/\|z_i(t)\| = \mathbf{0}_n$ if $\|z_i(t)\| = 0$. Incorporating the item $1/\|z_i(t)\|$ serves to homogenize the interaction magnitudes of all innovation items, thereby ensuring that the observers are resilient against the attempts of malicious adversaries to manipulate or fault-distort their behaviors.

Let $e_i(t) = x(t) - \hat{x}_i(t)$ be defined as the monitoring error of the observer i. Moreover, under Assumption 5.1, consider that the value σ is public for all observers in the set $\mathcal{V}$. Then, inspired by [37, 40, 42, 43, 93], the following auxiliary function is built to assist the gain $\gamma(t)$ design:

$$\dot{\tilde{\gamma}}(t) = \varrho\tilde{\gamma}(t) - \frac{m-2s}{N}\gamma(t), \tag{5.4}$$

where the initial value picked as $\tilde{\gamma}(0) = \sigma$, and remaining parameters ϱ, m, N are fixed global coefficients that can be obtained offline.

The auxiliary function $\tilde{\gamma}(t)$ is considered to fulfill the following criteria:

$$\frac{1}{N}\sum_{i=1}^{N}\|e_i(t)\| \leq \tilde{\gamma}(t), \qquad \lim_{t\to T_x}\tilde{\gamma}(t) = 0, \tag{5.5}$$

where $T_x \in \{+\infty, \ T(\sigma), \ T_{max}\}$ is required depending on the different monitoring performance. Intuitively, the first criterion ensures that the overall average monitoring error is governed by the auxiliary function $\tilde{\gamma}(t)$, facilitating the feedback control of the monitoring rate of observers. The second criterion assures the convergence rate of $\tilde{\gamma}(t)$, which indirectly affects the convergence rate of consensus monitoring.

In brief, the security performance of the state monitoring depends on two key factors: one is to homogenize data that may be compromised and resolve the abnormal effects by utilizing sparsity, which is mainly attributed to $z_i(t)/\|z_i(t)\|$; the other is to ensure that the monitoring rate of observers aligns with the prescribed convergence rate, which is governed by the design of the gain $\gamma(t)$ and the auxiliary function $\tilde{\gamma}(t)$.

The choice of $\gamma(t)$ not only affects the convergence rate of the SSM consensus but also involves the ability to suppress sparse attacks. Therefore, depending on different monitoring rate requirements, selecting an appropriate function $\gamma(t)$ will efficiently ensure security consensus monitoring.

5.3.2 Asymptotic consensus monitoring

The primary focus of this subsection is to design the gain $\gamma(t)$ that satisfies constraint (5.5) with $T_x = +\infty$, and determine the control gain $\delta_{ij}(t)$ in (5.3). Subsequently, the monitoring performance under the effect of sparse attacks will be analyzed.

It follows from (5.1), (5.3), and $e_{ij}(t) = e_i(t) - e_j(t)$ that

$$\dot{e}_i(t) = \tilde{f}_i(t) - l_{ii}\delta_{ii}(t)e_i(t) + \sum_{j=1,j\neq i}^{N} a_{ij}\delta_{ij}(t)e_j(t) - \gamma(t)\mu_i(t)$$

$$= \tilde{f}_i(t) - \sum_{j=1}^{N} l_{ij}\delta_{ij}(t)e_j(t) - \gamma(t)\mu_i(t), \tag{5.6}$$

where $\tilde{f}_i(t) = f(t,x) - f(t,\hat{x}_i)$ and $\mu_i(t) = \frac{z_i(t)}{\|z_i(t)\|}$.

Construct a group of auxiliary functions $g_i(t)$, $i \in \mathcal{V}$ with $\|e_i(0)\| \leq g_i(0) \leq \sigma$, and its dynamics given as

$$\dot{g}_i(t) = \varrho g_i(t) - \sum_{j=1}^{N} l_{ij}\delta_{ij}(t)g_j(t) + \rho_i(t)\gamma(t), \tag{5.7}$$

where

$$\rho_i(t) = \begin{cases} 0, & \text{if } i \notin \mathcal{V}_{r,t}, \\ 1, & \text{if } i \in \mathcal{V}_{a,t}, \\ -1, & \text{otherwise.} \end{cases}$$

Lemma 5.1 *For continuous dynamical systems as in (5.6)–(5.7), suppose that Assumption 5.1 holds. If the gain function $\gamma(t) \geq 0$ holds, then it can be concluded that $\|e_i(t)\| \leq g_i(t)$ holds for $\forall t \geq 0$.*

Proof 5.1 *From the fact that $x^T y \leq \|x\|\|y\|$ holds for $\forall x, y \in \mathbb{R}^n$ and $\mu_i(t) = e_i(t)/\|e_i(t)\|$ if $i \notin \mathcal{V}_{a,t}$, one conclude that*

$$\frac{e_i^T(t)\mu_i(t)}{\|e_i(t)\|} = \begin{cases} \frac{e_i^T(t)z_i(t)}{\|e_i(t)\|\|z_i(t)\|} \leq 1, & \text{if } i \in \mathcal{V}_{a,t}, \\ \frac{e_i^T(t)e_i(t)}{\|e_i(t)\|^2} = 1, & \text{otherwise.} \end{cases}$$

Define the Lyapunov function as $V_i(t) = \|e_i(t)\|$, $i \in \mathcal{V}$, one can then differentiate along (5.6) to obtain:

$$\dot{V}_i(t) = V_i^{-1}(t) \cdot e_i^T(t)\dot{e}_i(t)$$

$$\leq \frac{e_i^T(t)}{\|e_i(t)\|}\left[\tilde{f}_i(t) - \sum_{j=1}^{N} l_{ij}\delta_{ij}(t)e_j(t) - \mu_i(t)\gamma(t)\right]$$

$$\leq \varrho V_i(t) - \sum_{j=1}^{N} l_{ij}\delta_{ij}(t)V_j(t) + \rho_i(t)\gamma(t).$$

Combining the above equation with (5.7), and the fact that $V_i(0) \leq g_i(0)$ holds for any $i \in \mathcal{V}$, one obtain that

$$\|e_i(t)\| = V_i(t) \leq g_i(t), \quad \forall i \in \mathcal{V} \text{ and } t \geq 0.$$

Thus, the proof is completed.

Define the global form $g(t) = \text{col}\{g_i(t)\}$, $i \in \mathcal{V}$, then

$$\dot{g}(t) = (\varrho I_N - \mathcal{L}_\Delta(t))g(t) + \gamma(t)\tilde{\rho}_t,$$

where $\mathcal{L}_\Delta(t) = [l_{ij}\delta_{ij}(t)]_N \in \mathbb{R}^{N \times N}$ and $\tilde{\rho}_t = \text{col}\{\rho_i(t)\}_{i \in \mathcal{V}}$.

For the time-varying matrix $\mathcal{L}_\Delta(t)$, if $\delta_{ij}(t) = \delta_{ji}(t)$ holds for any $i, j \in \mathcal{V}$, it follows from $\delta_{ii}(t) = \sum_{j=1,j\neq i}^N a_{ij}\delta_{ij}(t)/l_{ii}$ that

$$\sum_{j=1}^N l_{ji}\delta_{ji}(t) = \sum_{j=1}^N l_{ij}\delta_{ij}(t) = \sum_{j=1,j\neq i}^N (a_{ij} + l_{ij})\,\delta_{ij}(t) = 0,$$

which implies that

$$1_N^T \mathcal{L}_\Delta(t) = 0_N^T, \quad \mathcal{L}_\Delta(t)1_N = 0_N, \quad \mathcal{I}\mathcal{L}_\Delta(t) = \mathcal{L}_\Delta(t)\mathcal{I} = \mathcal{L}_\Delta(t).$$

where $\mathcal{I} = I_N - \frac{1}{N}1_N 1_N^T$.

To continue the analysis, two intermediate variables $\bar{g}(t) = \frac{1}{N}1_N^T g(t)$ and $\tilde{g}_i(t) = g_i(t) - \bar{g}(t)$, $i \in \mathcal{V}$ are introduced. Hence, subject to $\delta_{ij}(t) = \delta_{ji}(t)$, the dynamics of $\tilde{g}(t)$ and $\bar{g}(t)$ can be written as:

$$\dot{\tilde{g}}(t) = (\varrho I_N - \mathcal{L}_\Delta(t))\tilde{g}(t) + \gamma(t)\mathcal{I}\tilde{\rho}_t, \tag{5.8a}$$

$$\dot{\bar{g}}(t) = \varrho\bar{g}(t) + \frac{1}{N}1_N^T\tilde{\rho}_t\gamma(t). \tag{5.8b}$$

where $\tilde{g}(t) = \mathcal{I}g(t)$ with $\tilde{g}(t) = \text{col}\{\tilde{g}_i(t)\}$, $i \in \mathcal{V}$.

The result of asymptotic consensus monitoring can now be provided.

Theorem 5.1 *For the NAS (5.1) under the controller (5.3) and the intermediate system (5.8) with the auxiliary system $\tilde{\gamma}(t)$ provided by (5.4), compatible with Assumptions 5.1–5.3. Under the condition $s < m/2$, the auxiliary systems $\tilde{\gamma}(t)$, $\tilde{g}(t)$, and $\bar{g}(t)$ are all asymptotically stable and the constraint (5.5) can be satisfied if the control gains are selected according to*

$$\delta_{ij}(t) = \delta_1 > \frac{\varrho}{\lambda_2(\mathcal{L})}, \quad \gamma(t) = \beta_1\tilde{\gamma}(t), \quad \beta_1 > \frac{N\varrho}{m - 2s}.$$

Proof 5.2 *It follows from $\gamma(t) = \beta_1\tilde{\gamma}(t)$ that*

$$\dot{\tilde{\gamma}}(t) = \left(\varrho - \frac{m - 2s}{N}\beta_1\right)\tilde{\gamma}(t),$$

which means that the system $\tilde{\gamma}(t)$ is asymptotically stable due to $\beta_1 > N\varrho/(m-2s)$, i.e., $\lim_{t\to\infty} \tilde{\gamma}(t) = 0$.

Since $\delta_{ij}(t) = \delta_1$ for $\forall i,j \in \mathcal{V}$, one can derive that $\delta_{ij}(t) = \delta_{ji}(t)$ and $\mathcal{L}_\Delta(t) = \delta_1 \mathcal{L}$. Furthermore, a Lyapunov function $V_g(t) = \|\tilde{g}(t)\|$ is considered. Then, taking its derivative, one has

$$\dot{V}_g(t) \leq \varrho V_g(t) - \delta_1 V_g^{-1}(t)\tilde{g}^T(t)\mathcal{L}\tilde{g}(t) + \|\mathcal{I}\|\|\tilde{\rho}_t\|\gamma(t)$$
$$\leq (\varrho - \delta_1\lambda_2(\mathcal{L}))V_g(t) + \sqrt{N}\beta_1\tilde{\gamma}(t),$$

where the inequality is derived based on Lemma 2.10 with $1_N^T\tilde{g}(t) = 0$ and $\|\mathcal{I}\| = 1$. Following that, from the fact that $\lim_{t\to\infty} \tilde{\gamma}(t) = 0$ and the condition $\delta_1 > \varrho/\lambda_2(\mathcal{L})$, it follows that $\lim_{t\to\infty} V_g(t) = \lim_{t\to\infty} \|\tilde{g}(t)\| = 0$. Therefore, asymptotic consensus is evidenced.

Moving on, according to Assumptions 5.2–5.3, it follows that $|\mathcal{V}_{a,t}| \leq s$ and $|\mathcal{V}_{n,t}| \geq N - s$ with $\mathcal{V}_{n,t} = \mathcal{V}_N\backslash\mathcal{V}_{a,t}$. Hence, one has

$$\dot{\bar{g}}(t) = \varrho\bar{g}(t) + \frac{|\mathcal{V}_{a,t}| - |\mathcal{V}_{n,t}|}{N}\gamma(t)$$
$$\leq \varrho\bar{g}(t) - \frac{m-2s}{N}\gamma(t).$$

Then, combining with $0 < \bar{g}(0) \leq \sigma = \tilde{\gamma}(0)$, one can get $\bar{g}(t) \leq \tilde{\gamma}(t)$ holds for any $t \geq 0$, i.e., $\lim_{t\to\infty} \bar{g}(t) \leq \lim_{t\to\infty} \tilde{\gamma}(t) = 0$.

Combined with Lemma 5.1, one can further obtain

$$\frac{1}{N}\sum_{i=1}^{N} \|e_i(t)\| \leq \frac{1}{N}\sum_{i=1}^{N} g_i(t) = \bar{g}(t) \leq \tilde{\gamma}(t),$$

which implies that (5.5) can be satisfied. The proof is completed.

Remark 5.2 *Synthesizing Lemma 5.1 and Theorem 5.1, it follows naturally that $\lim_{t\to\infty} \|\hat{x}_i(t) - x_0(t)\| = \lim_{t\to\infty} \|e_i(t)\| \leq \lim_{t\to\infty} g_i(t) \leq \lim_{t\to\infty}(\bar{g}(t) + \tilde{g}_i(t)) = 0$. Hence, the conclusion of Theorem 5.1 also means the realization of asymptotic consensus monitoring.*

In particular, inspired by the idea of attack isolation in [40, 43, 93], the controller (5.3) can be reorganized in the following form:

$$u_i(t) = \sum_{j=1}^{N} a_{ij}\delta_{ij}(t)e_{ij}(t) + \beta_1\theta_i(t)z_i(t), \tag{5.9}$$

where $\theta_i(t)$ is a dynamic attack-isolation factor which can be set as:

$$\theta_i(t) = \begin{cases} 1, & \text{if } \|z_i(t)\| \leq N\tilde{\gamma}(t), \\ 0, & \text{otherwise.} \end{cases}$$

Proposition 5.1 *Consider the NAS (5.1) under location-varying s-sparse attacks with $s < m/2$. Suppose that Assumptions 5.1–5.3 hold and the control protocol is picked as in (5.9); then the asymptotic consensus monitoring can be achieved if parameters $\delta_{ij}(t)$, $\gamma(t)$, and β_1 are the same as in Theorem 5.1.*

Proof 5.3 *From the criteria (5.5), one has $\sum_{i=1}^{N} \|e_i(t)\| \le N\tilde{\gamma}(t)$. For any $i \in \mathcal{V}_{n,t}$, it follows that $\|z_i(t)\| = \|e_i(t)\| \le \sum_{i=1}^{N} \|e_i(t)\| \le N\tilde{\gamma}(t)$.*

Hence, if $\|z_i(t)\| > N\tilde{\gamma}(t)$, one can conclude that $i \in \mathcal{V}_{a,t}$. Then, coupled with the fact that $\lim_{t\to\infty} \tilde{\gamma}(t) = 0$ and $\|z_i(t)\| > 0$ when $i \in \mathcal{V}_{a,t}$, means that all non-zero malicious attacks will eventually be isolated as time progresses. Following this process, the asymptotic consensus monitoring based on the remaining normal communication data will be achievable, and the proof is completed.

Remark 5.3 *From an intuitive perspective, control protocols (5.3) and (5.9) facilitate security consensus monitoring against arbitrarily location-varying s-sparse attacks. The key lies in normalizing the interaction strength of the innovation item $z_i(t)$ and leveraging consensus protocols to ensure that the negative impact of attacks/faults on entire NASs can be suppressed or isolated. Theorem 5.1 and Proposition 5.1 indicate that security monitoring can only be guaranteed when the number of anomalies is not dominant compared to the communication quantity between sensors and observers. This finding also aligns with the conclusions of previous works such as [17, 20, 33, 37] regarding the resistance against sparse attacks.*

5.3.3 Finite-time consensus monitoring

In this subsection, we will explore the finite-time consensus monitoring problem for the NAS (5.1) under the network vulnerable to sparse attacks. To this end, the gain function $\gamma(t)$ is replaced as follows:

$$\gamma(t) = \beta_1 \tilde{\gamma}(t) + \beta_2 \tilde{\gamma}^{p_1}(t), \tag{5.10}$$

where $0 < p_1 < 1$, $\beta_2 > 0$, and the selection of β_1 is consistent with Theorem 5.1. Hence, one has

$$\dot{\tilde{\gamma}}(t) = -\eta_1 \tilde{\gamma}(t) - \tilde{\beta}_2 \tilde{\gamma}^{p_1}(t), \tag{5.11}$$

where $\tilde{\beta}_2 = (m - 2s)\beta_2/N$ and $\eta_1 = (m - 2s)\beta_1/N - \varrho$.

Further, it follows from Lemma 2.15 that the system (5.11) is finite-time stable with the settling time $T_1 \le T(\tilde{\gamma}(0))$ and

$$T(\tilde{\gamma}(0)) = \frac{1}{\eta_1(1 - p_1)} \ln\left(1 + \frac{\eta_1}{\tilde{\beta}_2}\tilde{\gamma}^{1-p_1}(0)\right)$$

$$= \frac{\ln\left(\tilde{\beta}_2 + \eta_1\sigma^{1-p_1}\right) - \ln(\tilde{\beta}_2)}{\eta_1(1 - p_1)}.$$

Similarly, the following two auxiliary variables are introduced to support the analysis:

$$\bar{e}(t) = \frac{1}{N}\sum_{i=1}^{N} e_i(t), \quad \tilde{e}_i(t) = e_i(t) - \bar{e}(t), \ i \in \mathcal{V}.$$

Noting that if $\delta_{ij}(t)$ is designed to satisfy $\delta_{ij}(t) = \delta_{ji}(t)$ for any $i, j \in \mathcal{V}$, and adopting a similar analysis for $\bar{V}(t) = \|\bar{e}(t)\|$ following the previous subsection, it follows from Lemma 5.1 that

$$\dot{\bar{V}}(t) \le \frac{1}{N}\sum_{i=1}^{N} \frac{\partial\|e_i(t)\|}{\partial t} \le \dot{\bar{g}}(t) \le \varrho\bar{g}(t) - \frac{m-2s}{N}\gamma(t)$$

$$\le \varrho\tilde{\gamma}(t) - \frac{m-2s}{N}\gamma(t)$$

$$= -\eta_1\tilde{\gamma}(t) - \tilde{\beta}_2\tilde{\gamma}^{p_1}(t). \tag{5.12}$$

Similarly, it follows from $\bar{V}(0) \le \bar{g}(0) \le \sigma = \tilde{\gamma}(0)$ that $\|\bar{e}(t)\| \le \tilde{\gamma}(t)$ can be derived, i.e.,

$$\begin{cases} \lim_{t\to T_1} \|\bar{e}(t)\| \le \lim_{t\to T_1} \|\tilde{\gamma}(t)\| = 0, \\ \|\bar{e}(t)\| = \tilde{\gamma}(t) = 0, \quad \forall t \ge T_1. \end{cases}$$

Thereby, at time $t > T_1$, it follows from (5.6) that

$$\dot{\tilde{e}}_i(t) = \check{f}_i(t) - \sum_{j=1}^{N} a_{ij}\delta_{ij}(t)e_{ij}(t), \tag{5.13}$$

where $\check{f}_i(t) = \frac{1}{N}\left(\sum_{j=1}^{N} f(t,\hat{x}_j)\right) - f(t,\hat{x}_i)$.

Therefore, by the fact that $e_i(t) = \tilde{e}_i(t) + \bar{e}(t)$, one can effectively reformulate the finite-time consensus monitoring problem for the NAS (5.1) as a finite-time stability problem for the system (5.13). Following this, we proceed to analyze the conditions for designing control gains that guarantee finite-time monitoring for NAS (5.1) as outlined below.

Theorem 5.2 *For the NAS (5.1) under the control protocol (5.3) and the gain function $\gamma(t)$ designed in (5.10), suppose that Assumptions 5.1–5.3 hold. Then, with the premise that $s < N/2$, all observers can be achieved consensus monitoring in finite-time $T(\sigma) \le T_1 + T_2$ if the control gain $\delta_{ij}(t)$ is set to*

$$\delta_{ij}(t) = \begin{cases} \delta_1 + \delta_2\|e_{ij}(t)\|^{p_2}, & \text{if } i \ne j, \\ \sum_{j=1,j\ne i}^{N} \frac{a_{ij}\delta_{ij}(t)}{l_{ii}}, & \text{if } i = j, \end{cases} \tag{5.14}$$

and the parameters are selected to satisfy $\delta_1 > \varrho/\lambda_2(\mathcal{L})$, $\delta_2 > 0$, and $-2 < p_2 < 0$.

In addition, the settling-time $T_2 \le T(\tilde{e}(0))$ for the finite-time stable of the system (5.13) is given as

$$T(\tilde{e}(0)) = -\frac{\ln\left(1 + \frac{\delta_1\lambda_2(\mathcal{L})-\varrho}{\delta_2(\lambda_2(\mathcal{L}))^{1+\frac{p_2}{2}}}(2\sqrt{N}\sigma)^{-\frac{p_2}{2}}\right)}{p_2(\delta_1\lambda_2(\mathcal{L})-\varrho)}.$$

Proof 5.4 *Construct the Lyapunov candidate function as follows:*

$$W(t) = \frac{1}{2} \sum_{i=1}^{N} \tilde{e}_i^T(t)\tilde{e}_i(t).$$

Then, the derivative of $W(t)$ along the trajectory of (5.13) is given by

$$\dot{W}(t) = \sum_{i=1}^{N} \tilde{e}_i^T(t)\check{f}_i(t) - \sum_{i=1}^{N} \tilde{e}_i^T(t) \sum_{j=1}^{N} a_{ij}e_{ij}(t)\delta_{ij}(t)$$

$$= \sum_{i=1}^{N} \tilde{e}_i^T(t) \left(\frac{1}{N} \sum_{j=1}^{N} f(t, x_j) - f(t, \bar{x}) + f(t, \bar{x}) - f(t, \hat{x}_i) \right)$$

$$- \frac{1}{2} \sum_{i=1}^{N} \sum_{j=1}^{N} a_{ij}\tilde{e}_i^T(t)e_{ij}(t)\delta_{ij}(t) - \frac{1}{2} \sum_{i=1}^{N} \sum_{j=1}^{N} a_{ji}\tilde{e}_j^T(t)e_{ji}(t)\delta_{ji}(t),$$

where $\bar{x}(t) = \frac{1}{N} \sum_{i=1}^{N} \hat{x}_i(t)$.
Since $\tilde{e}_i(t) = \bar{x}(t) - \hat{x}_i(t)$, $e_{ij}(t) = \tilde{e}_i(t) - \tilde{e}_j(t)$, $\delta_{ij}(t) = \delta_{ji}(t)$, and $\sum_{i=1}^{N} \tilde{e}_i^T(t) \left(\frac{1}{N} \sum_{j=1}^{N} f(t, x_j) - f(t, \bar{x}) \right) = \mathbf{0}_n^T \left(\frac{1}{N} \sum_{j=1}^{N} f(t, \hat{x}_j) - f(t, \bar{x}) \right) = 0$, one gets

$$\dot{W}(t) \le 2\varrho W(t) - \frac{1}{2} \sum_{i=1}^{N} \sum_{j=1}^{N} a_{ij}\tilde{e}_i^T(t)e_{ij}(t)\delta_{ij}(t) + \frac{1}{2} \sum_{i=1}^{N} \sum_{j=1}^{N} a_{ij}\tilde{e}_j^T(t)e_{ij}(t)\delta_{ij}(t)$$

$$\le 2\varrho W(t) - \frac{1}{2} \sum_{i=1}^{N} \sum_{j=1}^{N} a_{ij}\|e_{ij}(t)\|^2 \delta_{ij}(t)$$

$$= 2\varrho W(t) - \frac{1}{2} \sum_{i=1}^{N} \sum_{j=1}^{N} a_{ij}\tilde{\delta}_{ij}(t),$$

where $\tilde{\delta}_{ij}(t) = \delta_1\|e_{ij}(t)\|^2 + \delta_2\|e_{ij}(t)\|^{p_2+2}$.
By invoking Lemmas 2.10 and 2.14, the following inequalities are obtained

$$\sum_{i=1}^{N} \sum_{j=1}^{N} a_{ij}\|e_{ij}(t)\|^2 \overset{(a)}{=} 2\tilde{e}^T(t)\tilde{\mathcal{L}}\tilde{e}(t) \ge 4\lambda_2(\mathcal{L})W(t),$$

$$\sum_{i=1}^{N} \sum_{j=1}^{N} a_{ij}\|e_{ij}(t)\|^{p_2+2} = \sum_{i=1}^{N} \sum_{j=1}^{N} (a_{ij}\|e_{ij}(t)\|^2)^{1+\frac{p_2}{2}}$$

$$\ge \left(\sum_{i=1}^{N} \sum_{j=1}^{N} a_{ij}\|e_{ij}(t)\|^2 \right)^{1+\frac{p_2}{2}}$$

$$\ge 2^{2+p_2} (\lambda_2(\mathcal{L})W(t))^{1+\frac{p_2}{2}},$$

where $\tilde{\mathcal{L}} = \mathcal{L} \otimes I_n$, and equation (a) is derived based on the fact that $\chi^T\mathcal{L}\chi =$

$\frac{1}{2} \sum_{i,j=1}^{N} a_{ij}(\chi_i - \chi_j)^2$ holds for any vector $\chi = \text{col}\{\chi_i\} \in \mathbb{R}^N$. Hence, it can be further deduced that

$$\dot{W}(t) \leq 2(\varrho - \delta_1 \lambda_2(\mathcal{L}))W(t) - 2^{1+p_2}\delta_2(\lambda_2(\mathcal{L}))^{1+\frac{p_2}{2}}W^{1+\frac{p_2}{2}}(t).$$

By denoting the notations $\bar{\delta}_2 = 2^{1+p_2}(\lambda_2(\mathcal{L}))^{1+\frac{p_2}{2}}\delta_2 > 0$ and $\tilde{p} = 1 + \frac{p_2}{2}$, one then has the following inequalities:

$$\dot{W}(t) \leq -2(\delta_1 \lambda_2(\mathcal{L}) - \varrho)W(t) - \bar{\delta}_2 W^{\tilde{p}}(t),$$
$$\delta_1 \lambda_2(\mathcal{L}) - \varrho > 0, \quad 0 < \tilde{p} < 1.$$

Similarly, it follows from Lemma 2.15 and $W(0) \leq \sqrt{N}\sigma/2$ that the error system (5.13) can stabilize to 0 within the given finite time T_2.

Finally, referring to $e_i(t) = \bar{e}(t) + \tilde{e}_i(t)$ and the fact that $\delta_{ij}(t) = \delta_{ji}(t)$, one can conclude that after time $T(\sigma) \leq T_1 + T_2$, the NAS (5.1) will necessarily reach a finite-time consensus monitoring under the control (5.3) with gains (5.10) and (5.14). This completes the proof.

Remark 5.4 Clearly, when $n = 1$, it should be noted that $\delta_{ij}(t)e_{ij}(t) = \delta_1 e_{ij}(t) + \delta_2 sig^{p_2+1}(e_{ij}(t))$. However, this equation is violated when $n \neq 1$. Consequently, in high-dimensional systems, the use of sig function for finite-time control design would present additional challenges in the scaling of some inequalities. The design approach proposed in this chapter addresses this issue explicitly, ensuring a seamless extension of analytical methods for finite-time stability of scalar systems. Furthermore, in the subsequent subsection, which focuses on the design of fixed-time monitoring control, a similar advantageous property is maintained.

5.3.4 Fixed-time consensus monitoring

In this subsection, to reach the objective of fixed-time consensus monitoring, the function $\gamma(t)$ is set to satisfy the following form:

$$\gamma(t) = \beta_3 \tilde{\gamma}^{p_1}(t) + \beta_4 \tilde{\gamma}^{q_1}(t), \tag{5.15}$$

with $0 < p_1 < 1 < q_1$ and $\beta_3 > N\varrho/(m - 2s)$, $\beta_4 > N\varrho/(m - 2s)$.

Based on $0 \leq \tilde{\gamma}(t) \leq \tilde{\gamma}^{p_1}(t) + \tilde{\gamma}^{q_1}(t)$, it follows from Lemma 2.16 that the system $\tilde{\gamma}(t)$ is fixed-time stable with the settling time

$$\tilde{T}_1 = \frac{1}{\eta_3(1 - p_1)} + \frac{1}{\eta_4(q_1 - 1)},$$

where $\eta_3 = (m - 2s)\beta_3/N - \varrho$ and $\eta_4 = (m - 2s)\beta_4/N - \varrho$.

Similarly, if $\delta_{ij}(t) = \delta_{ji}(t)$ holds for all $i, j \in \mathcal{V}$, one can still deduce that

$$\begin{cases} \lim_{t \to \tilde{T}_1} \|\bar{e}(t)\| \leq \lim_{t \to \tilde{T}_1} \|\tilde{\gamma}(t)\| = 0, \\ \|\bar{e}(t)\| = \tilde{\gamma}(t) = 0, \quad \forall t \geq \tilde{T}_1. \end{cases}$$

Subsequently, the following results about fixed-time consensus monitoring can be derived.

Theorem 5.3 *For the NAS (5.1) under the control protocol (5.3) with $\gamma(t)$ designed in (5.15), suppose that Assumptions 5.1–5.3 hold. Subject to $s < N/2$, if the control gain $\delta_{ij}(t)$ is set to*

$$\delta_{ij}(t) = \begin{cases} \tilde{\delta}_3\|e_{ij}(t)\|^{p_2} + \tilde{\delta}_4\|e_{ij}(t)\|^{q_2}, & \text{if } i \neq j, \\ \sum_{j=1,j\neq i}^{N} \frac{a_{ij}\delta_{ij}(t)}{l_{ii}}, & \text{if } i = j, \end{cases} \tag{5.16}$$

the parameters satisfy $-2 < p_2 < 0 < q_2$, and

$$\tilde{\delta}_3 = \delta_3 + \frac{\varrho}{2^{p_2}(\lambda_2(\mathcal{L}))^{1+\frac{p_2}{2}}}, \quad \tilde{\delta}_4 = \delta_4 + \frac{\varrho}{2^{q_2}N^{-\frac{q_2}{2}}(\lambda_2(\mathcal{L}))^{1+\frac{q_2}{2}}},$$

where $\delta_3, \delta_4 > 0$, then the NAS (5.1) solves the fixed-time consensus monitoring for $\forall i \in \mathcal{V}$, with the globally settling-time given as $T_{max} \leq \tilde{T}_1 + \tilde{T}_2$ and

$$\tilde{T}_2 = \frac{2}{\bar{\delta}_4 q_2} - \frac{2}{\bar{\delta}_3 p_2}, \tag{5.17}$$

where $\bar{\delta}_3 = 2^{1+p_2}(\lambda_2(\mathcal{L}))^{1+\frac{p_2}{2}}\delta_3$ and $\bar{\delta}_4 = 2^{1+q_2}N^{-\frac{q_2}{2}}(\lambda_2(\mathcal{L}))^{1+\frac{q_2}{2}}\delta_4$.

Proof 5.5 *By following a similar analysis as in the previous subsection, it is easy to show that the dynamics of $\tilde{e}_i(t)$ still conform to (5.13) with $\delta_{ij}(t)$ designed in (5.16) at time $t > \tilde{T}_1$.*

Therefore, for the same Lyapunov candidate function $W(t) = \frac{1}{2}\sum_{i=1}^{N} \tilde{e}_i^T(t)\tilde{e}_i(t)$, it can similarly be deduced that

$$\dot{W}(t) \leq 2\varrho W(t) - \frac{1}{2}\sum_{i=1}^{N}\sum_{j=1}^{N} a_{ij}\tilde{\delta}_{ij}(t),$$

where $\tilde{\delta}_{ij}(t) = \tilde{\delta}_3\|e_{ij}(t)\|^{p_2+2} + \tilde{\delta}_4\|e_{ij}(t)\|^{q_2+2}$.

According to Lemmas 2.10 and 2.14, one has

$$\sum_{i=1}^{N}\sum_{j=1}^{N} a_{ij}\|e_{ij}(t)\|^{q_2+2} \geq N^{-\frac{q_2}{2}}\left(2\tilde{e}^T(t)\tilde{\mathcal{L}}\tilde{e}(t)\right)^{1+\frac{q_2}{2}}$$

$$\geq 2^{2+q_2}N^{-\frac{q_2}{2}}(\lambda_2(\mathcal{L})W(t))^{1+\frac{q_2}{2}}.$$

Following this, an analogous result can be obtained:

$$\dot{W}(t) \leq 2\varrho W(t) - 2^{1+p_2}\tilde{\delta}_3(\lambda_2(\mathcal{L}))^{1+\frac{p_2}{2}}W^{1+\frac{p_2}{2}}(t)$$
$$- 2^{1+q_2}\tilde{\delta}_4 N^{-\frac{q_2}{2}}(\lambda_2(\mathcal{L}))^{1+\frac{q_2}{2}}W^{1+\frac{q_2}{2}}(t)$$
$$\leq -2^{1+p_2}\delta_3(\lambda_2(\mathcal{L}))^{1+\frac{p_2}{2}}W^{1+\frac{p_2}{2}}(t)$$
$$- 2^{1+q_2}\delta_4 N^{-\frac{q_2}{2}}(\lambda_2(\mathcal{L}))^{1+\frac{q_2}{2}}W^{1+\frac{q_2}{2}}(t).$$

Introducing the notations $\tilde{p} = 1 + \frac{p_2}{2}$, $\tilde{q} = 1 + \frac{q_2}{2}$, $\bar{\delta}_3$ and $\bar{\delta}_4$, one then has the following inequality

$$\dot{W}(t) \leq -\bar{\delta}_3 W^{\tilde{p}}(t) - \bar{\delta}_4 W^{\tilde{q}}(t),$$

Table 5.1: Comparison of the consensus SSM under different convergence rates

$u_i(t)$ in (5.3)	Asymptotic	Finite-time	Fixed-time
$\gamma(t)$ with $\tilde{\gamma}(t)$	$\beta_1\tilde{\gamma}(t)$	$\beta_1\tilde{\gamma}(t) + \beta_2\tilde{\gamma}^{p_1}(t)$	$\beta_3\tilde{\gamma}^{p_1}(t) + \beta_4\tilde{\gamma}^{q_1}(t)$
$\delta_{ij}(t)$ for $i \neq j$	δ_1	$\delta_1 + \delta_2\|e_{ij}(t)\|^{p_2}$	$\tilde{\delta}_3\|e_{ij}(t)\|^{p_2} + \tilde{\delta}_4\|e_{ij}(t)\|^{q_2}$
Settling-time	$+\infty$	$T(\sigma)$	T_{max}
Constant settling-time	✗	✗	✓
Bounded settling-time	✗	✓	✓
Low control load	✓	✓	✗

$$\bar{\delta}_3 > 0, \ \bar{\delta}_4 > 0, \ 0 < \tilde{p} < 1 < \tilde{q},$$

which by Lemma 2.16 immediately suggests that the system (5.13) reaches 0 within fixed time $\tilde{T}_2$ given in (5.17).

Further, due to $e_i(t) = \tilde{e}_i(t) + \bar{e}(t)$ and $\delta_{ij}(t) = \delta_{ji}(t)$, it follows that fixed-time consensus monitoring can be achieved with the globally settling-time $T_{max} \leq \tilde{T}_1 + \tilde{T}_2$. This completes the proof.

Based on the previous analyses, we summarize the design principles for security consensus monitoring under different convergence rate constraints considered in this chapter and provide a comparative analysis in Table 5.1.

Remark 5.5 *The settling times $T(\sigma) \leq T_1 + T_2$ and $T_{max} \leq \tilde{T}_1 + \tilde{T}_2$ depend only on the design parameters of the control protocol, the algebraic connectivity $\lambda_2(\mathcal{L})$, and the order N of the NASs. Additionally, from Table 5.1, both vertically and horizontally, it is revealed that with the deployment of parameters β_k and $\delta_k(t)$, $k \in [4]$, and coefficients p_l and q_l, $l \in [2]$, in gain functions $\gamma(t)$ and $\delta_{ij}(t)$, the effect of achieving security consensus monitoring becomes more productive. However, it also implies that the control amplitude is more excessive. Therefore, in practice, it is necessary to strike a trade-off between the convergence speed of monitoring control and associated control costs based on different task requirements.*

Corollary 5.1 *For a NAS (5.1) against location-varying s-sparse attacks with $s < m/2$, under the security monitoring control protocol (5.3) with different gains $\gamma(t)$ designed in Theorems 5.1 to 5.3. If the set $\mathcal{S}_t = \{i \mid \|z_i(t)\| > \tilde{\gamma}(t)\}$ is regarded as the recognized set for attacks, then it gives*

$$\lim_{t \to T_x} |(\mathcal{S}_t \setminus \mathcal{V}_{a,t}) \cap (\mathcal{V}_{a,t} \setminus \mathcal{S}_t)| = 0, \tag{5.18}$$

where $T_x \in \{+\infty, T(\sigma), T_{max}\}$ corresponds to different convergence rates, respectively, and the set $(\mathcal{S}_t \setminus \mathcal{V}_{a,t}) \cap (\mathcal{V}_{a,t} \setminus \mathcal{S}_t)$ denotes the set of disjoint elements of sets $\mathcal{S}_t$ and $\mathcal{V}_{a,t}$.

Proof 5.6 *Let $e_M(t) = \max\{\|e_i(t)\|\}_{i \in \mathcal{V}_{r,t}}$, one then has the following holds*

$$\dot{e}_M(t) \leq \varrho e_M(t) - \gamma(t).$$

It follows from (5.4) that once there exists some time $T' \leq T_x$ such that $e_M(t) \leq \tilde{\gamma}(t)$ holds for $t \geq T'$, one can conclude that $\|e_i(t)\| \leq \tilde{\gamma}(t)$ holds always for $i \in \mathcal{V}_{r,t}$ and $t \geq T'$. Hence, there must exist a time $T'' \leq T_x$ such that $\mathcal{S}_t \subseteq \mathcal{V}_{a,t}$ for $t \geq T''$.

In addition, according to $\lim_{t \to T_x} \tilde{\gamma}(t) = 0$ and $\|z_i(t)\| > 0$ if $i \in \mathcal{V}_{a,t}$, one has $\mathcal{V}_{a,t} \subseteq \mathcal{S}_t$. Therefore, in the light of the above analysis, it can be concluded (5.18) is proved.

Remark 5.6 *It is important to emphasize that the applicability of the control protocol (5.3) outlined in this chapter can be effectively and smoothly extended regardless of how frequently the location of attack/fault action changes, provided they adhere to the sparse property ($s < m/2$). However, it should be noted that the proposed algorithm may not be suitable in scenarios where attacks/faults occur densely. This is attributed to the potential for numerically superior attackers to collaborate, creating a convincing counterfeit system that entirely supplants the authentic one, relegating real information to sparse outliers. Consequently, in order to effectively study and counteract such potentially dense attack strategies, it becomes imperative to enforce stricter assumptions concerning the attackers, including, but not limited to, constraints on the frequency and magnitude of the attacks, thereby leveraging their inherent vulnerabilities to target. Otherwise, a thorough shutdown for overhaul is required.*

5.4   NUMERICAL SIMULATIONS

This section will present three simulation examples to validate the effectiveness and advancement of the proposed security monitoring protocol (5.3). Similarly, consider the nonlinear NASs consisting of one agent system and eight sensors with the agent consists of the IEEE 6-bus power system described by (4.46). Specifically, the nonlinear function $f(t, x)$ after transformation into the system depicted in (5.1) is expressed as follows:

$$f(t, x) = \begin{bmatrix} 0 & I \\ M_g^{-1} \left(L_{Ig}^T L_{II}^{-1} L_{Ig} - L_{gg} \right) & -M_g^{-1} D_g \end{bmatrix} x(t) \tag{5.19}$$

where the matrices M_g, L_{gg}, L_{Ig}, L_{II} are set same as in (4.47), $x(t) = \begin{bmatrix} \zeta(t) & \omega(t) \end{bmatrix}^T$.

Additionally, the communication topology of NASs is depicted in Fig. 5.1, where the black-directed dashed lines indicate fragile sensor transmission channels prone to attacks. It can be easily verified that the function satisfies the constraints of (5.2) with the constant $\varrho = 0.2$.

The initial state of the agent is selected as $x(0) = [0.2, -0.5, 0.7, 0.8, 1, -0.6]^T$, and the initial value of each observer is chosen as $\hat{x}_i(0) = \epsilon_i$ with $\epsilon_i \in \mathbb{R}^6$ is a random variable that obeys the uniform distribution on $[-1\ 1]$.

Assuming that the attacker randomly selects two transmission measurements from the sensor every 2 seconds among the 7 sensors and makes the following tampering operation:

$$\tilde{y}_i(t) = x(t) + (0.2\sin(\pi t) - e^{0.1t})1_n, \ i \in \mathcal{V}_{a,t}.$$

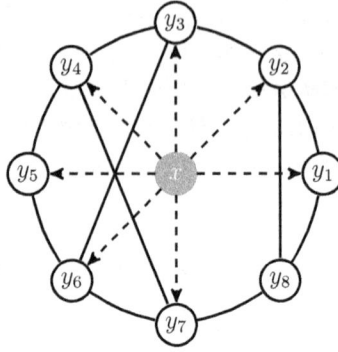

Figure 5.1 Communication topology of NASs.

Table 5.2: The locations of sparse sensor anomalies

Time range t	$[0,1)$	$[1,2)$	$[2,3)$	$[3,4)$	$[4,5)$
Anomalous sensors	$y_2(t)$ $y_7(t)$	$y_1(t)$ $y_5(t)$	$y_4(t)$ $y_6(t)$	$y_3(t)$ $y_7(t)$	$y_2(t)$ $y_5(t)$

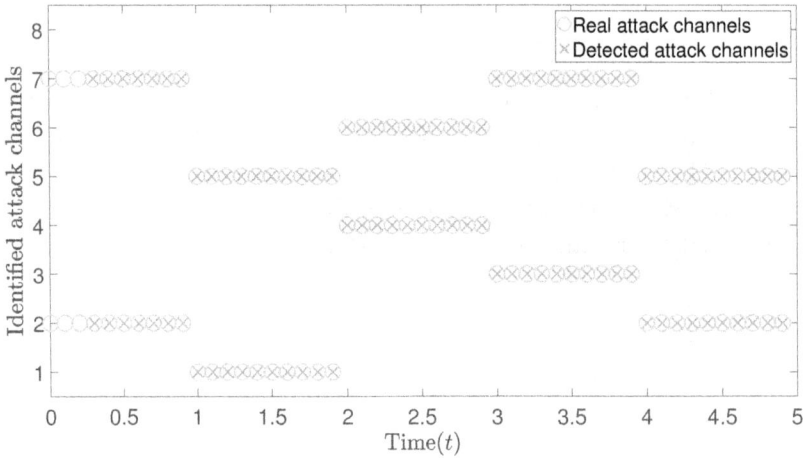

Figure 5.2 Abnormal sensor locations and identified attack channels.

Consequently, we depict the locations where attacks occurred on the sensors for each period in Table 5.2. It can be observed that the sensor attacks in this case exhibit sparsity, as indicated by the inequality $|\mathcal{V}_{a,t}| = 2 < 7/2 = |\mathcal{V}_{r,t}|/2$.

Example 1: The calculation shows that the values of parameters $\delta_1 = 9$ and $\beta_1 = 4.6$ are taken to satisfy the requirements of Theorem 5.1. Furthermore, asymptotic state monitoring simulations of the NAS (5.1) with $f(t,x)$ given in (5.19) are performed through control protocols (5.3) and (5.9), and the abnormal sensor locations and identified attack channels are illustrated in Fig. 5.2. Besides, the state trajectory monitoring results under the control protocol (5.3) are displayed in Fig. 5.3,

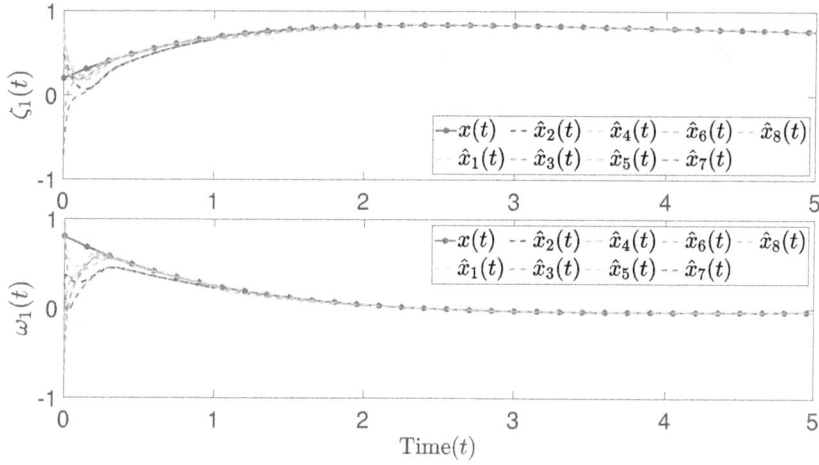

(a) $j = 1$

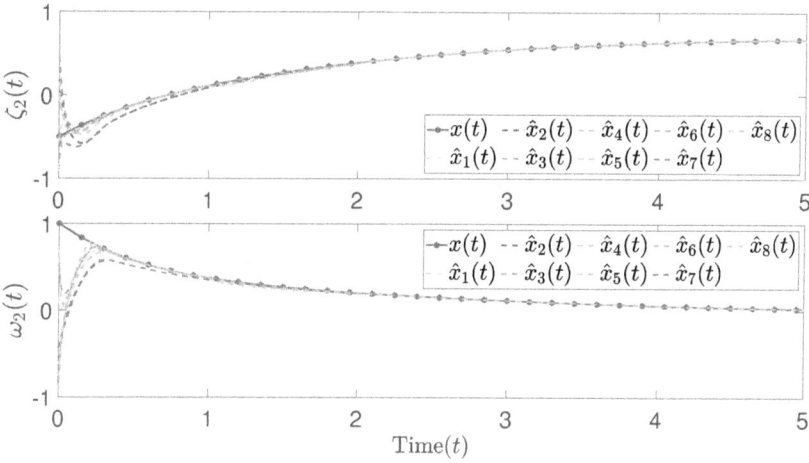

(b) $j = 2$

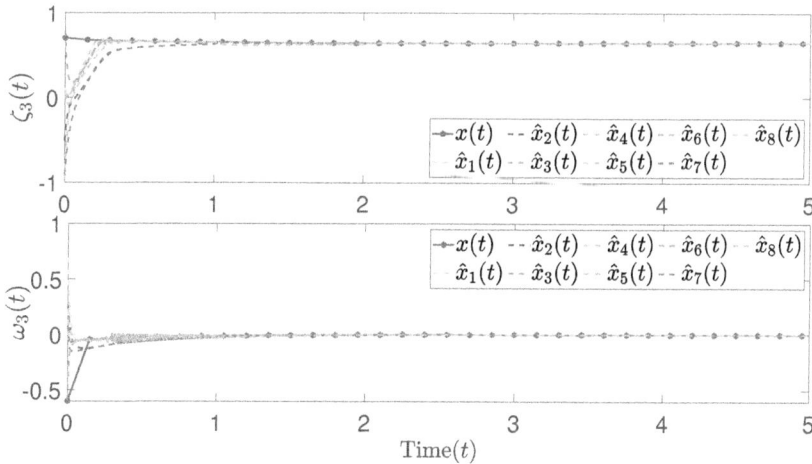

(c) $j = 3$

Figure 5.3   Asymptotic monitoring trajectories of $\zeta_j(t)$, $\omega_j(t)$, $j \in [3]$ under the protocol (5.3) with parameters given in Theorem 5.1.

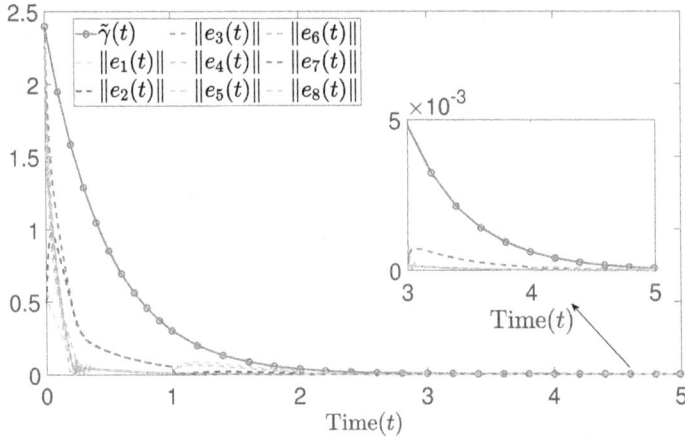

Figure 5.4   Profile curves of $\tilde{\gamma}(t)$ and each of $\|e_i(t)\|$, $i \in \mathcal{V}$.

and the profile curves of the auxiliary function $\tilde{\gamma}(t)$ and each of the norm $\|e_i(t)\|$ can be seen in Fig. 5.4. It can be seen that asymptotic state monitoring is reached, and the secure consensus control protocol (5.3) is robust against the above sparse sensor attacks.

For comparison, the monitoring performance is simulated under the following conventional resilience-free design of consensus control:

$$u_i(t) = \delta_1 \sum_{j=1}^{N} a_{ij}(x_j(t) - x_i(t)) + b_i \beta_1 z_i(t). \tag{5.20}$$

Then, under this resilience-free control protocol (5.20), the state monitoring error results are depicted in Fig. 5.5, which clearly demonstrate the successful biasing of the consensus monitoring under the sparse attack design. Subsequently, three simulation examples are executed for the NAS with (5.19) as mentioned above.

In addition, comparison results of the average monitoring error under two different control protocols are shown in Fig. 5.6, which shows that the control protocol (5.9) based on attack-isolation performs better compared to the control protocol (5.3) based on attack-suppression. This is mainly due to the fact that the impact of anomalous data on the monitoring performance can be completely eliminated once the attack is isolated, whereas the idea based on attack suppression only compresses the impact of the attack to a tolerable level. However, the control based on attack isolation has not yet broken through the demand for fast monitoring in finite/fixed time, and the work in this chapter addresses consensus monitoring with finite/fixed time convergence through the idea based on attack suppression.

Example 2: This example employs the finite-time and fixed-time control protocols to verify Theorems 5.2–5.3 with $\sigma = 5$, respectively. For intuition, the attack switching interval is set to $0.1s$, and the parameter values are designed as $p_1 = 0.2$, $q_1 = 2.0$, $p_2 = -0.8$, $q_2 = 1.5$, $\beta_2 = \beta_3 = \beta_4 = 4.5$, and $\delta_2 = \delta_3 = \delta_4 = 9$, which satisfy the conditions in Theorems 5.2–5.3. The simulation results of finite-time and fixed-time

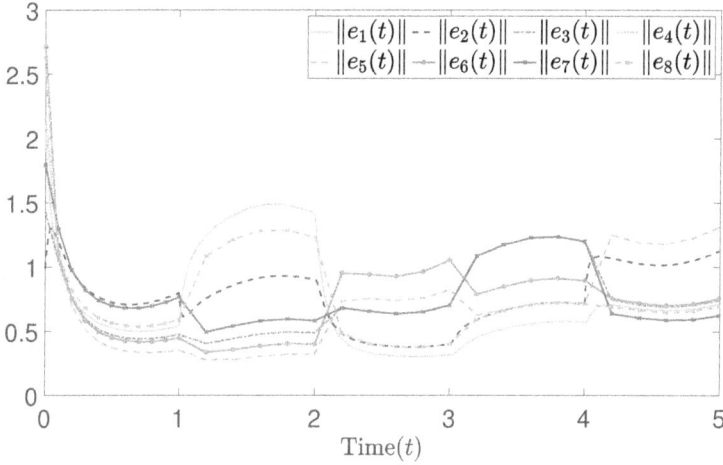

Figure 5.5   Monitoring errors of (5.19) under the resilience-free controller (5.20).

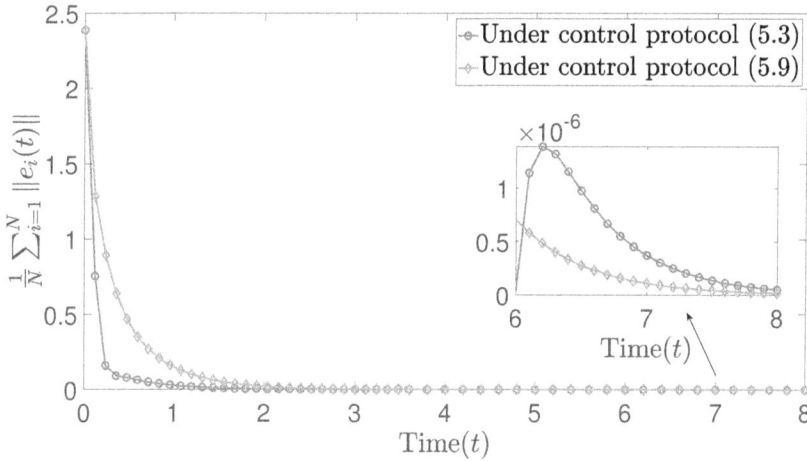

Figure 5.6   Comparison curves of average monitoring error for different control protocols (5.3) and (5.9).

consensus monitoring can be viewed in Figs. 5.7–5.8, respectively. The theoretical settling-time can be obtained as $T(\sigma) = 0.93s$ and $T_{max} = 1.61s$, while the real consensus monitoring time under the finite-time and fixed-time control protocols respectively is about $T_x = 0.5s$ and $T_x = 0.25s$. The proposed control protocols achieve security monitoring within a settling time far shorter than the theoretical estimation. In addition, the comparison of the mean control input effort $\bar{u}(t) = \frac{1}{N}\sum_{i\in\mathcal{V}}\|u_i(t)\|$ under different convergence rates is plotted in Fig. 5.9, which shows that the results are in line with the Table 5.1.

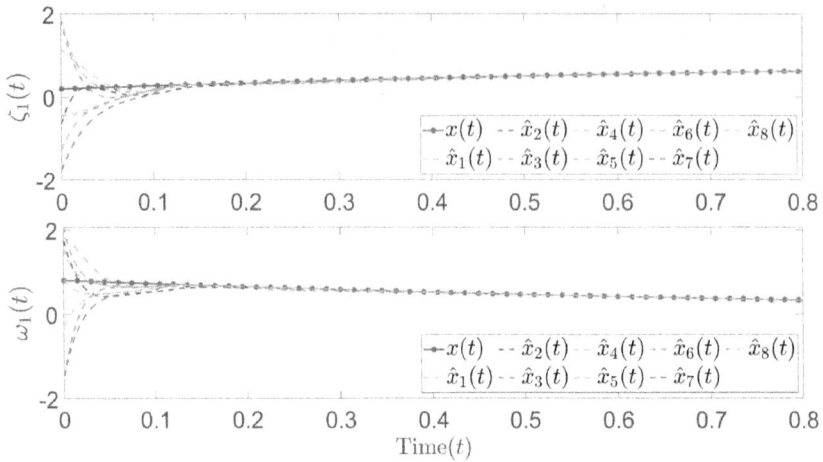

(a) $j = 1$

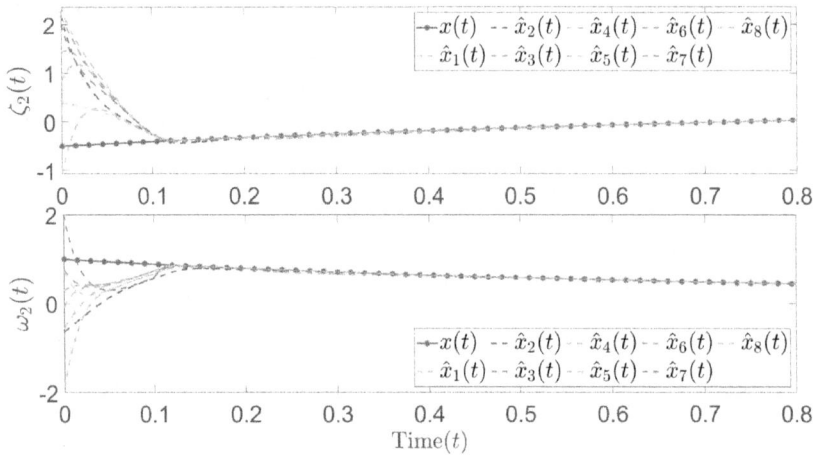

(b) $j = 2$

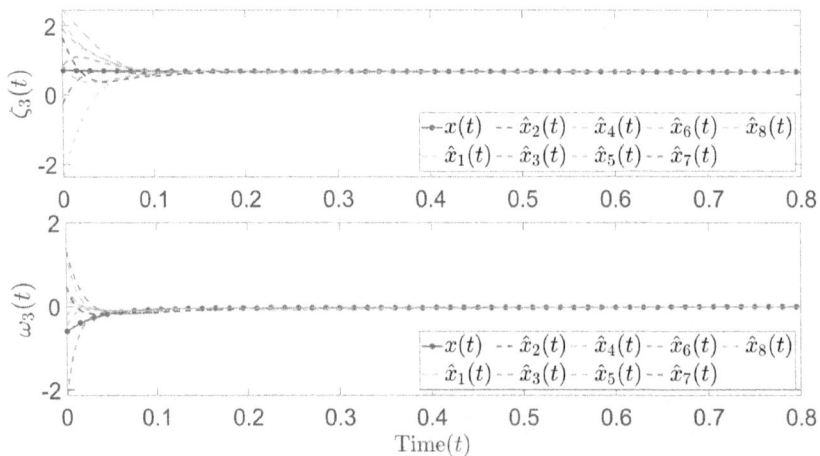

(c) $j = 3$

Figure 5.7   Finite-time monitoring trajectories of $\zeta_j(t)$, $\omega_j(t)$, $j \in [3]$ under the control given in Theorem 5.2.

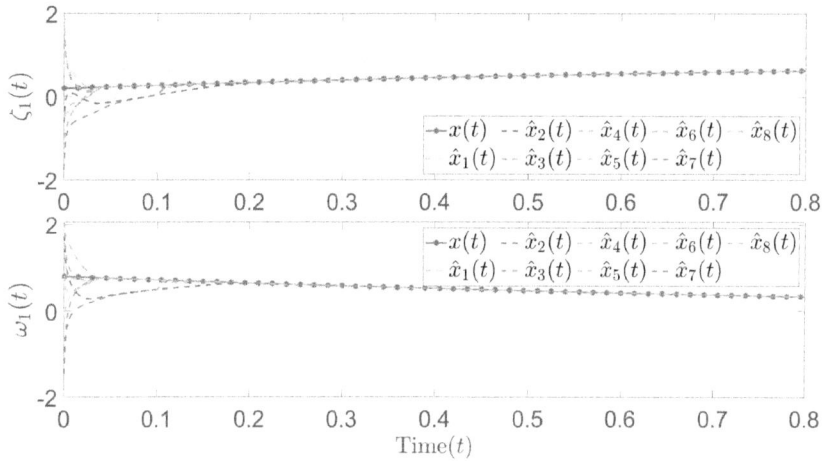

(a) $j = 1$

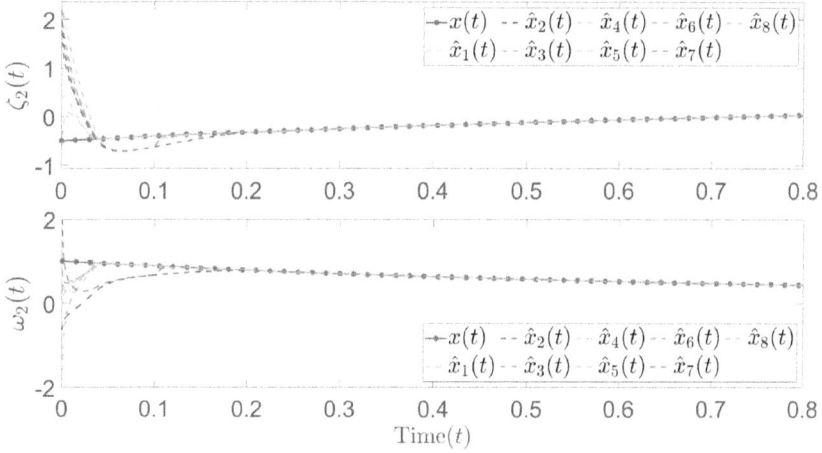

(b) $j = 2$

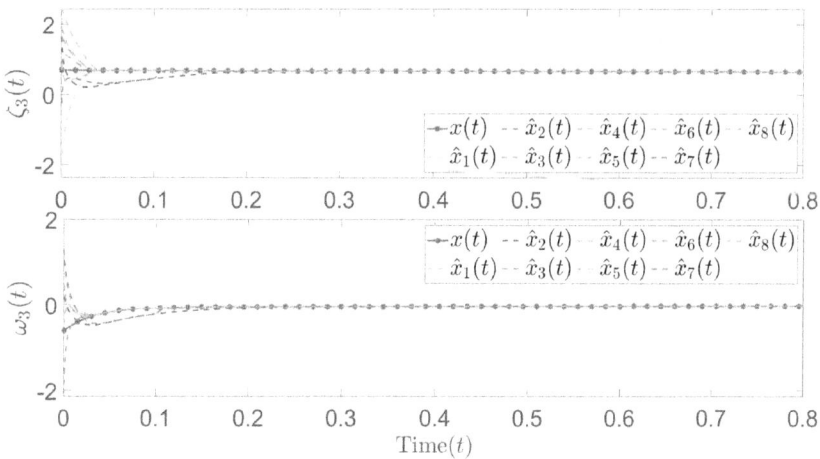

(c) $j = 3$

Figure 5.8 Fixed-time monitoring trajectories of $\zeta_j(t)$, $\omega_j(t)$, $j \in [3]$ under the control given in Theorem 5.3.

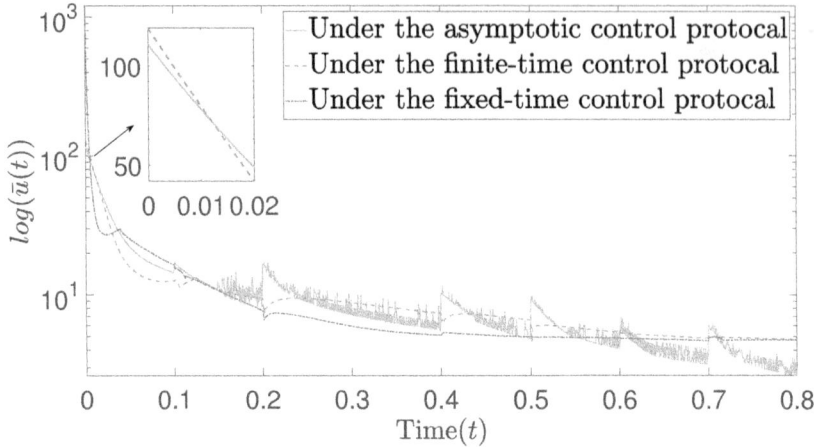

Figure 5.9   Comparison results of the mean control input $\bar{u}(t)$ under different convergence rates.

Example 3: To illustrate the efficacy of the proposed finite/fixed-time control protocol (5.3) compared to that in [131–133], the following comparative simulations are conducted. Consider adapting the control protocols in (5.3) to the following 4 forms for comparative simulations:

1) Set $\delta_{ij}(t)e_{ij}(t) = (\delta_1 + \delta_2\|e_{ij}(t)\|^{p_2})e_{ij}(t)$;

2) Set $\delta_{ij}(t)e_{ij}(t) = \delta_1 e_{ij}(t) + \delta_2\text{sig}^{1+p_2}(e_{ij}(t))$;

3) Set $\delta_{ij}(t)e_{ij}(t) = (\tilde{\delta}_3\|e_{ij}(t)\|^{p_2} + \tilde{\delta}_4\|e_{ij}(t)\|^{q_2})e_{ij}(t)$;

4) Set $\delta_{ij}(t)e_{ij}(t) = \tilde{\delta}_3\text{sig}^{1+p_2}(e_{ij}(t)) + \tilde{\delta}_4\text{sig}^{1+q_2}(e_{ij}(t))$.

To reflect the monitoring performance of each control protocol, define the average monitoring error as $\phi(t) = \frac{1}{N}\sum_{i=1}^{N}\|e_i(t)\|$, and present the comparative results in Fig. 5.10. It can be observed that the proposed control protocol in this chapter exhibits monitoring performance comparable to, if not superior to, the performance in [131–133] that relies on the sig function.

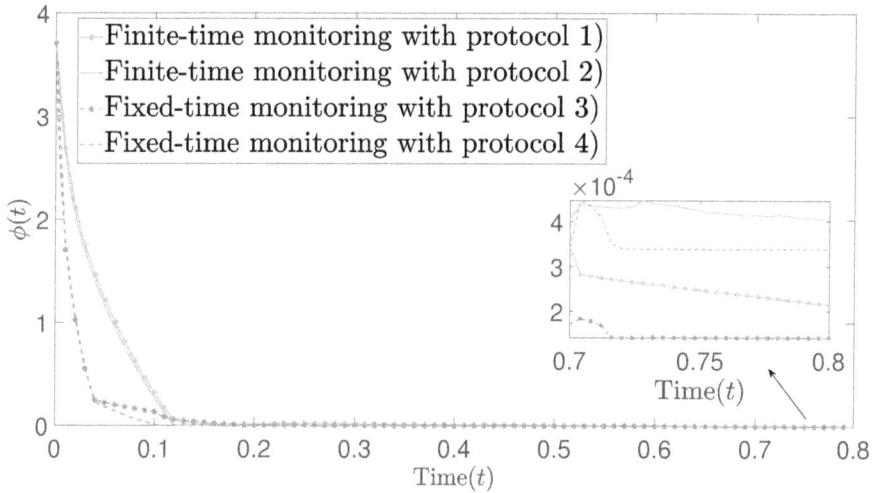

Figure 5.10   Comparison curves of the average monitoring error $\phi(t)$ under different control protocols.

5.5   CONCLUSIONS

This chapter has proposed a generalized, security consensus monitoring framework for NASs, designed to be resilient against location-varying sparse attacks and adaptable to a wide spectrum of convergence rates–including asymptotic, finite-time, and fixed-time monitoring scenarios. Within this framework, a dynamic auxiliary function has first been employed to mitigate the effects of sparse attacks, thereby enabling each observer to securely and asymptotically estimate the system states. Furthermore, two practical finite-/fixed-time control protocols have been seamlessly integrated with the security control protocol, resulting in finite-/fixed-time consensus monitoring.

CHAPTER 6

Distributed security monitoring of sensor networks against localized Byzantine attacks

This chapter investigates distributed SSM for sensor networks vulnerable to localized Byzantine attacks. The analysis commences with a systematic examination of existing defense strategies, highlighting critical limitations in current approaches for handling intelligent adversarial nodes with full network knowledge. Section 6.2 develops a novel *norm-based maximum-discard mitigation* (NMDM) framework within resilient consensus architectures. Addressing the challenge of Byzantine nodes capable of arbitrary protocol deviations and differential neighbor communications, the proposed mechanism implements a dual-component signal regulation architecture combining signum operators for directional control with absolute value modulation for magnitude confinement. This integrated approach ensures directional consensus and magnitude boundedness in the multi-agent data exchange while effectively containing malicious data propagation. Building upon this foundation, Section 6.3 is dedicated to countering Byzantine cyber link attacks characterized by flexible switching locations, which are more stealthy compared to previous fixed-location Byzantine node attacks. A dual-mode detection paradigm is established, integrating the NMDM core with topology robustness to achieve real-time identification of stealthy link-level intrusions. Finally, a distributed SSM algorithm against Byzantine attacks is summarized and constructed to ensure all observers achieve a consensus estimate of the system state, no matter how frequently the attack locations switch.

6.1 INTRODUCTION

In distributed SSM processes, the main challenges arise from two aspects. One is that each sensor node only has access to partial measurement information of the global system, which makes it necessary to seek asymptotic estimation of the full

DOI: 10.1201/9781003669913-6

state of the system through communication interaction among neighboring sensors [135, 136]. The other is that the cyber-attack environment imposes severe damages on the performance of state estimation, for example, local attacks may already be able to degrade the security of the whole network [15, 137, 138]. These challenges pose significant difficulties in the design of the distributed SSM and therefore have stimulated considerable research interest in the field.

As one of the significant forms of cyber-attacks in distributed SSM, sparse sensor attacks have been extensively investigated, generating a large number of attack detection and isolation algorithms [23, 26, 29, 33, 37, 40, 43, 138–140]. However, the above results mainly address sparse attacks on sensor-to-observer channels, which also inevitably receive malicious attacks when local observers communicate directly with each other. Based on security considerations on the interaction among observers, the distributed SSM problem against malicious or Byzantine node attacks was discussed in [58, 82, 84, 85, 141–143]. In [84], the impact of the number of adversarial nodes on the performance of the consensus-based distributed optimization protocol was analyzed. Combining the Jordan regular decomposition technology [136] and the *Weighted-MSR* (W-MSR) algorithm [49], a distributed SSM algorithm was proposed in [83] for networks with the strong $(3F + 1)$-robust topology against F-local Byzantine node attacks, where each node has at most F Byzantine neighbors. However, this algorithm enforces stringent limitations in terms of the redundancy of the network topology, thereby posing a significant challenge to the scalability of the distributed SSM algorithm. In response, this constraint was relaxed in [85] by adopting a diversity model and integrating trusted nodes into the network. Subsequently, a *local min-switching decision* (LMSD) approach was developed in [81] based on residual evaluation, in which an event-triggered control strategy was combined to achieve the distributed SSM over the $(2F + 1)$-robust graph. Under the interference of randomly bounded noise, a new robust topology condition was derived in [86], which satisfies the collective observability, utilizing projection operators and the W-MSR algorithm. Similarly, the event-triggered technology was incorporated to reduce the computational effort of the distributed SSM algorithm. However, the conservatism of the LMSD approach is easily exposed by targeted malicious attack injections. As a result, there is still a need to improve the distributed SSM algorithm utilizing the observer construction with resilience against more general Byzantine attacks.

Motivated by the above discussions, this chapter first considers the problem of security signal reconstruction under localized Byzantine attacks. In particular, under F-local Byzantine node attacks, a new algorithm is formed by decomposing the interaction data into two dimensions, symbol (direction) and norm (magnitude), which effectively improves the estimation efficiency of the W-MSR algorithm [83, 84, 142]. Secondly, an efficient distributed SSM algorithm is designed to counter F-local Byzantine edge-attacks. In this context, adversaries can arbitrarily and stealthily modify the attack patterns and change their locations at any time. To address potential challenges in acquiring sensor model knowledge amid cyber-attacks, a distributed SSM framework based on the observability decomposition technique is developed. The NMDM technique and the signum summation operation are then employed to mitigate the negative effects of the attacks in direction and magnitude dimensions, ensuring a robust consensus in the state estimation and achieving the distributed SSM.

6.2 DISTRIBUTED SECURITY MONITORING AGAINST LOCAL BYZANTINE NODE ATTACKS

6.2.1 Problem formulation

Consider a wireless sensor network consisting of N agents, where each agent generates a local data stream by measuring a static vector signal $\theta \in \mathbb{R}^n$. The measurement model of the i-th sensor is expressed as

$$y_i = C_i\theta, \quad i \in \mathcal{V} = [N], \tag{6.1}$$

where $y_i \in \mathbb{R}^{m_i}$ is the measurement output of the i-th sensor, C_i is the local measurement matrix, and denote the global measurement matrix as $C = \mathrm{col}\{C_i\}_{i\in\mathcal{V}}$. Thus, the global measurement vector is written as $y = [y_1^T, \ldots, y_N^T]^T = C\theta$.

In this study, let $\hat{\theta}_i(t)$ denote the state estimation of the signal θ maintained by sensor i at moment t. The objective of each sensor is to estimate the complete global state θ by receiving estimates from its neighboring sensors through the undirected network $\mathcal{G}$. To ensure the validity of the estimation and subsequent progress, the matrix C is considered to have global full column-rank. It should be noted that under this setting, the local matrix C_i may have a column rank deficiency, indicating that mutual communication interactions between some (or all) nodes are indispensable to achieve global state estimation.

Consider a vertices subset $\mathcal{V}_A \subset \mathcal{V}$ in sensor network to be adversarial, and assume that the adversaries have the ability to be fully aware of the network topology, the estimation results and the behavior taken by normal nodes. Such an assumption of an omniscient adversary is common in the literature with resilient distributed algorithms [81, 83, 84, 141, 142], and fully takes into account the 'worst-case' adversarial behavior. In terms of capabilities, an adversarial node can exploit the above information to deviate from its own estimation and communication behaviors at will while colluding with other adversaries. Additionally, according to the Byzantine attacks model [47], adversaries are permitted to send varying falsified data to different neighbors simultaneously. Referring to [58, 81, 83], we give the following assumption to characterize the above attack threat model.

Assumption 6.1 The graph $\mathcal{G}$ is subject to F-local Byzantine node attacks.

Throughout history, the actual number and location of Byzantine nodes infested by an adversary is usually unknown to the defenders in the network. Therefore, the parameter F here can be interpreted as the level at which the sensor network can withstand the maximum number of Byzantine attacks. Note that the updating rules of Byzantine nodes are not controlled by the algorithm, and the subsequent design rules and behaviors of the algorithm apply only to the regular nodes. Building on the above discussion, the problem explored in this section can be formally displayed as follows.

Problem 6.1. Given a sensor network (6.1) with the undirected communication topology $\mathcal{G}$, how does one go to design a set of rules for signal reconstruction updates and communication exchanges such that $\lim_{t \to \infty} \|\hat{\theta}_i(t) - \theta\| = 0$, $\forall i \in \mathcal{V}_R$, regardless of any acts of F-local Byzantine node attacks suffered by the topology $\mathcal{G}$.

6.2.2 Distributed security signal monitoring algorithm

The objective of this subsection is to present a basic framework of a distributed security signal monitoring algorithm against Byzantine node attacks.

Before proceeding, a simple consensus tracking control problem of the leader-following multi-agent system is first formulated, facilitating the evaluation for the consensus control performance after decoupling directions and magnitudes.

Consider a leader-following multi-agent system with a directed communication topology $\mathcal{G}_z = \{\mathcal{V}_z, \mathcal{E}_z\}$ and $\mathcal{V}_z = \{0\} \cup [N]$, for which the dynamics of each agent are described as

$$\begin{cases} \dot{z}_0(t) = g(t, z_0), \\ \dot{z}_i(t) = - \sum_{j \in \mathcal{N}_i} (\delta_{z,i}(t) \circ \phi(z_{ij}(t)) + \beta_i z_{ij}(t)), \end{cases} \tag{6.2}$$

where $z_0(t)$, $z_i(t) \in \mathbb{R}^n$ respectively are the state of leader 0 and the i-th ($i \in [N]$) follower, $\delta_{z,i}(t) = \sum_{j \in \mathcal{N}_i} \text{Sign}(z_{ij}(t))$, $z_{ij}(t) = z_i(t) - z_j(t)$, the parameter $\beta_i > 0$, and the system function $g(t, z_0)$ satisfies $g(t, \mathbf{0}_n) = \mathbf{0}_n$ which allows the equilibrium point $z_0 = \mathbf{0}_n$ for leader 0 to be asymptotically stable, i.e., $\lim_{t \to \infty} \|z_0(t)\| = 0$.

Lemma 6.1 *If there exists a spanning tree rooted at leader 0 in graph $\mathcal{G}_z$, then system (6.2) can achieve consensus asymptotically stable for all followers and the leader, in the sense that*

$$\lim_{t \to \infty} \|z_i(t)\| = \lim_{t \to \infty} \|z_0(t)\| = 0, \ \forall i \in [N].$$

Proof 6.1 *According to the definition of the consensus error $z_{ij}(t) = z_i(t) - z_j(t)$ between agents i and j, it can follows from (6.2) that the dynamics of $z_{i0}(t)$, $i \neq 0$, are described by*

$$\dot{z}_{i0}(t) = - g(t, z_0) - \sum_{j \in \mathcal{N}_i} (\delta_{z,i}(t) \circ \phi(z_{ij}(t)) + \beta_i z_{ij}(t)). \tag{6.3}$$

Since $\lim_{t \to \infty} z_0(t) = \mathbf{0}_n$, the stability of (6.3) is equivalent to that of the following modified dynamics:

$$\dot{z}_{i0}^l(t) = - \sum_{j \in \mathcal{N}_i} \left(\delta_{z,i}^l(t) \phi(z_{ij}^l(t)) + \beta_i z_{ij}^l(t) \right), \quad \forall l \in [n]. \tag{6.4}$$

For any $j \in \mathcal{N}_i$, one has

$$\delta_{z,i}^l(t) \phi(z_{ij}^l(t)) = \sum_{h \in \mathcal{N}_i} \text{sign}(z_{ih}^l(t)) |z_{ij}^l(t)| = \sum_{h \in \mathcal{N}_i} \kappa_{i,jh}^l(t) z_{ih}^l(t),$$

where

$$\kappa_{i,jh}^l(t) = \begin{cases} \left| \dfrac{z_{ij}^l(t)}{z_{ih}^l(t)} \right|, & \text{if } z_{ih}^l(t) \neq 0, \\ 0, & \text{otherwise.} \end{cases}$$

Thus, one has

$$\dot{z}_{i0}^l(t) = - \sum_{h \in \mathcal{N}_i} \left(\beta_i + \kappa_{ih}^l(t) \right) z_{ih}^l(t),$$

where $\kappa_{ih}^l(t) = \sum_{j \in \mathcal{N}_i} \kappa_{i,jh}^l(t) > 0$.

For the communication topology $\mathcal{G}_z$, reconsider allocating the following time-varying communication weights:

$$a_{ij}^*(t) = \begin{cases} \beta_i + \kappa_{ij}^l(t), & \text{if } (j,i) \in \mathcal{E}_z, \\ 0, & \text{otherwise.} \end{cases}$$

Denote the corresponding topology as $\mathcal{G}_z^*$ with the time-varying Laplacian matrix $\mathcal{L}_z^*(t) = [l_{ij}^*(t)]$, where $l_{ii}^*(t) = \sum_{j=1}^N a_{ij}^*(t)$ and $l_{ij}^*(t) = -a_{ij}^*(t)$ if $j \neq i$. Subsequently, the matrix $\mathcal{L}_z^*(t)$ can be written in block form as

$$\mathcal{L}_z^*(t) = \begin{bmatrix} 0 & \mathbf{0}_{1 \times N} \\ b_0^*(t) & \tilde{\mathcal{L}}_z^*(t) \end{bmatrix},$$

where $b_0^*(t) = \text{col}\{l_{i0}^*(t)\}$, $i \in [N]$, and $\tilde{\mathcal{L}}_z^*(t)$ is a time-varying non-singular M-matrix since there exists a directed spanning tree rooted at the leader 0 in the topology $\mathcal{G}_z^*$.

Let $\tilde{z}^l(t) = \text{col}\{z_{i0}^l(t)\}_{i \in [N]}$ be the global tracking error with respect to the l-th component. From $z_{ij}^l(t) = z_{i0}^l(t) - z_{j0}^l(t)$, one has

$$\dot{\tilde{z}}^l(t) = - \tilde{\mathcal{L}}_z^*(t) \tilde{z}^l(t). \tag{6.5}$$

Hence, due to $-\tilde{\mathcal{L}}_z^*(t)$ is Hurwitz for all $t \geq 0$, the zero equilibrium point of the system (6.5) is asymptotically stable, i.e., $\lim_{t \to \infty} z_{i0}^l(t) = 0$ holds for all $i \in \mathcal{V}$ and $l \in [n]$.

Finally, it follows from $z_{i0}(t) = z_i(t) - z_0(t)$ that

$$\lim_{t \to \infty} \|z_i(t)\| = \lim_{t \to \infty} \|z_0(t)\| = 0.$$

Remark 6.1 *Compared with the traditional cooperative control protocol designed in [13, 144, 145], where $\dot{z}_i(t) = \beta_i \sum_{j \in \mathcal{N}_i} z_{ij}(t)$, the extra item $\delta_{z,i}(t) \circ \phi(z_{ij}(t))$ in (6.2) is innovative. Intuitively, the function $\delta_{z,i}(t)$ makes the update direction of agent i be consistent with more than half of its neighbors, aiming to enhance the defensive ability against local cyber-attacks. Since the sum of the directions of z_{ij} may become zero before reaching consensus, the item $\beta_i z_{ij}(t)$ plays a supplementary role in achieving consensus. For the case of $\delta_{z,i}(t) \neq 0$, it helps improve the convergence speed of consensus. Conversely, for the case where $\delta_{z,i}(t) = 0$, but in fact some $z_{ij}(t) \neq 0$,*

Figure 6.1 A directed topology $\mathcal{G}_z$.

the presence of $\beta_i z_{ij}(t)$ will then inherit and dominate the function of consensus reaching. Here, a counter example is provided to illustrate the significance of $\beta_i z_{ij}(t)$ when $\delta_{z,i}(t) = 0$ and $z_{ij}(t) \neq 0$. Consider a multi-agent system (6.2) with the topology shown in Fig. 6.1 and the initial state set $z_0(0) = 0$, $z_1(0) = 1$, and $z_2(0) = 2$. In this setting, the dynamics (6.2) with $\beta_i = 0$ is reduced to

$$\dot{z}_0(t) = 0, \quad \dot{z}_1(t) = 0, \quad \dot{z}_2(t) = z_1(t) - z_2(t).$$

It can be verified that $\lim_{t\to\infty} z_2(t) = \lim_{t\to\infty} z_1(t) = 1 \neq z_0(t) = 0$, so that the presence of $\beta_i > 0$ is necessary.

To enable full-state signal reconstruction, drawing on the observer design in [144], the following form of the monitor is constructed for the sensor i:

$$\dot{\hat{\theta}}_i(t) = K_i u_i(t) + L_i[y_i - C_i \hat{\theta}_i(t)], \tag{6.6}$$

where $u_i(t)$ is the cooperative control to be designed. The matrices K_i and L_i are also the gains to be designed.

In the framework of distributed security signal monitoring, a trade-off typically needs to exist between topology connectivity (network redundancy) and state information contained in local measurements (information redundancy) [84]. In [81, 83], some necessary or sufficient conditions on graph robustness and system structure for the convergence of the estimation algorithm have been displayed. Similarly, we present the following assumption that characterizes a weak local redundancy.

Assumption 6.2 *Given the sensor network structure (6.1) and the communication topology $\mathcal{G}$, at least one of the following conditions holds for an arbitrary nonempty subset $\mathcal{S} \subset \mathcal{V}$ satisfaction:*

i) *The subset $\mathcal{S}$ is $2F + 1$-reachable;*

ii) *The matrix $C_{\mathcal{S} \backslash \mathcal{V}_A}$ is full column-rank.*

For the range space and the null space of the matrix C_i as $\mathcal{R}(C_i)$ and $\mathcal{N}(C_i)$, one can get two matrices $U_{1i} \in \mathbb{R}^{n_i \times n}$ and $U_{2i} \in \mathbb{R}^{(n-n_i) \times n}$ with $n_i = rank(C_i)$ such that their row vectors are the orthonormal basis of the $\mathcal{R}(C_i^T)$ and $\mathcal{N}(C_i)$, respectively.
Let matrices $U_i = [U_{1i}^T \quad U_{2i}^T]^T$ and $V_i = U_i^{-1} = U_i^T = [V_{1i} \quad V_{2i}]$, thus one has

$$U_{1i} V_{2i} = 0, \quad U_{2i} V_{1i} = 0, \quad C_i V_i = [\bar{C}_{1i}\ 0],$$

where the matrix $\bar{C}_{1i} \in \mathbb{R}^{m_i \times n_i}$ is full column-rank.
Consider that the message sent by the Byzantine node $j \in \mathcal{V}_A$ to its neighbor i has been maliciously falsified as $\tilde{\theta}_{j,i}(t) \neq \hat{\theta}_j(t)$, while the normal one is $\tilde{\theta}_{j,i}(t) = \hat{\theta}_j(t)$.

Upon receipt of all the neighborhood information $\tilde{\theta}_{j,i}(t)$ by the monitor i, it could compute $\gamma(\xi_{ij}^t) \triangleq |\xi_{ij}^t|$ separately with $\xi_{ij}^t = U_{2i}(\hat{\theta}_i(t) - \tilde{\theta}_{j,i}(t))$, and arrange the $\|\gamma(\xi_{ij}^t)\|$ in descending order as

$$\gamma_{i,1}^t \geq \gamma_{i,2}^t \geq \cdots \geq \gamma_{i,N_i}^t.$$

To block the impact of the malicious attack, each normal node $i \in \mathcal{V}_R$ performs the following action

$$\rho_{ij}(t) = \begin{cases} 1, & \|\gamma(\xi_{ij}^t)\| \leq \gamma_{i,F+1}^t, \\ 0, & \text{otherwise.} \end{cases} \tag{6.7}$$

where $\rho_{ij}(t)$ denotes the detection logic of node i for its neighbor $j \in \mathcal{N}_i$. That is, $\rho_{ij}(t) = 1$ means that the data $\tilde{\theta}_{j,i}(t)$ is normal and available for estimation updates, and conversely $\rho_{ij}(t) = 0$ means that the data $\tilde{\theta}_{j,i}(t)$ may have been maliciously falsified and is selected for discard and isolation.

Combining the rules constructed in (6.7), the cooperative control protocol $u_i(t)$ is designed as

$$u_i(t) = -\sum_{j \in \mathcal{N}_i} \rho_{ij}(t) \left(\delta_i(t) \circ \gamma(\xi_{ij}^t) + \beta_i \xi_{ij}^t \right), \tag{6.8}$$

where $\delta_i(t) = \sum_{j \in \mathcal{N}_i} \text{Sign}(\xi_{ij}^t)$ and $0 < \beta_i < 1$.

6.2.3 Monitoring performance against local Byzantine node attacks

Now, the main result of distributed security signal monitoring under Byzantine node attacks is presented.

Theorem 6.1 Given the sensor network (6.1) under local Byzantine node attacks, suppose that Assumptions 6.1–6.2 hold. Then the distributed estimator $\hat{\theta}_i(t)$, $i \in \mathcal{V}_R$ can asymptotically reconstruct the signal θ with the control protocol designed in (6.7)-(6.8) if, the gain matrices K_i and L_i are designed as follows:

$$K_i = \alpha_i V_{2i}, \quad L_i = \mu_i V_i \begin{bmatrix} \bar{C}_{1i}^\dagger \\ 0 \end{bmatrix} = \mu_i V_{1i} \bar{C}_{1i}^\dagger, \quad \forall i \in \mathcal{V}_R, \tag{6.9}$$

where $\alpha_i > 0$ and $\mu_i > 0$ are arbitrary real numbers.

Proof 6.2 *Defining $e_i(t) = \hat{\theta}_i(t) - \theta$ as the signal monitoring error, it follows from (6.1)-(6.6) that*

$$\dot{e}_i(t) = K_i u_i(t) - L_i C_i e_i(t), \quad \forall i \in \mathcal{V}_R.$$

Left-multiplying $e_i(t)$, $i \in \mathcal{V}_R$ by matrix U_i and denoting $\zeta_{1i}(t) = U_{1i} e_i(t)$ and $\zeta_{2i}(t) = U_{2i} e_i(t)$, one has

$$\dot{\zeta}_{1i}(t) = -\mu_i \zeta_{1i}(t), \quad \dot{\zeta}_{2i}(t) = \alpha_i u_i(t), \tag{6.10}$$

where the first equation is derived based on the full column-rank property of the matrix $\bar{C}_{1i}$, which satisfies $\bar{C}_{1i}^{\dagger}\bar{C}_{1i} = I_{n_i}$. Hence, one can conclude that $\lim_{t\to\infty}\|\zeta_{1i}(t)\| = 0$ holds for $\forall i \in \mathcal{V}_R$.

For $\forall i \in \mathcal{V}_R$, introduce the subset $\mathcal{S}_i(t) = \{i\}\cup\{j|\ j \in \mathcal{N}_i,\ \|\xi_{ij}(t)\| = 0\}$. Consider that the set $\mathcal{S}_i(t)$ is $2F + 1$-reachable, otherwise $\lim_{t\to\infty}\|e_i(t)\| = 0$, $i \in \mathcal{S}_i(t)$ can be obtained due to the full column-rank of matrix $C_{\mathcal{S}_i(t)\setminus\mathcal{V}_A}$.

There exists a node $g \in \mathcal{S}_i(t)$ that has at least $2F + 1$ neighbors outside $\mathcal{S}_i(t)$, and note that this neighborhood set is $\mathcal{H}_g^t$. For $\forall l \in [n - n_g]$ that makes $\mathrm{sign}(\xi_{gj}^{lt}) \neq 0$, where $j \in \mathcal{H}_g^t$ and ξ_{gj}^{lt} denotes the l-th element of vector ξ_{gj}^t. Hence, one has

$$\phi_g^l(t) = \sum_{j\in\mathcal{H}_g^t\cap\mathcal{V}_R}\mathrm{sign}(\xi_{gj}^{lt}) + \sum_{j\in\mathcal{H}_g^t\cap\mathcal{V}_a}\mathrm{sign}(\xi_{gj}^{lt}) = \sum_{j\in\mathcal{M}_g^t}\phi_{gj}^l(t)\mathrm{sign}(\xi_{gj}^{lt}),$$

where $\xi_{ij}^t = U_{2i}(\hat{\theta}_i(t) - x_j(t))$, $\phi_{gj}^l(t) \in \{1,2\}$, and the subset $\mathcal{M}_g^t \subseteq \mathcal{H}_g^t \cap \mathcal{V}_R$ with $|\mathcal{M}_g^t| \geq |\mathcal{H}_g^t| - 2F \geq 1$ due to $|\mathcal{H}_g \cap \mathcal{V}_A| \leq F$ and $\tilde{\mathcal{H}}_g^t \subseteq \mathcal{H}_g$.

Defining two subsets as $\mathcal{R}_g^t = \{j|\rho_{gj}(t) = 1,\ j \in \mathcal{V}_R\}$ and $\mathcal{A}_g^t = \{j|\rho_{gj}(t) = 1,\ j \in \mathcal{V}_A\}$, it follows that $|\mathcal{R}_g^t| \geq |\mathcal{H}_g^t| - 2F \geq 1$. Then, the control protocol (6.8) can be rewritten as

$$u_g^l(t) = -\sum_{k\in\mathcal{R}_g^t\cup\mathcal{A}_g^t}\left(\phi_g^l(t)\gamma(\xi_{gk}^{lt}) + \beta_g\xi_{gk}^{lt}\right)$$

$$= -\sum_{k\in\mathcal{R}_g^t\cup\mathcal{A}_g^t}\sum_{j\in\mathcal{M}_g^t}\phi_{gj}^l(t)\left|\frac{\xi_{gk}^{lt}}{\xi_{gj}^{lt}}\right|\xi_{gj}^{lt} - \sum_{k\in\mathcal{R}_g^t\cup\mathcal{A}_g^t}\beta_g\xi_{gk}^{lt}$$

$$= -\sum_{j\in\mathcal{M}_g^t}\psi_{gj}^l(t)\xi_{gj}^{lt} - \sum_{k\in\mathcal{R}_g^t\cup\mathcal{A}_g^t}\beta_g\xi_{gk}^{lt}$$

$$= -\tilde{\psi}_{gh}^l(t)\xi_{gh}^{lt} - \sum_{j\in\mathcal{R}_g^t}\beta_g\xi_{gj}^{lt} - \sum_{j\in\mathcal{M}_g^t\setminus\{h\}}\psi_{gj}^l(t)\xi_{gj}^{lt}, \qquad (6.11)$$

where $\psi_{gj}^l(t) = \phi_{gj}^l(t)\sum_{k\in\mathcal{R}_g^t\cup\mathcal{A}_g^t}\left(\left|\xi_{gk}^{lt}/\xi_{gj}^{lt}\right|\right)$, and there exists a node $h \in \mathcal{M}_g^t$ such that

$$\tilde{\psi}_{gh}^l(t) = \phi_{gh}^l(t)\sum_{k\in\mathcal{R}_g^t}\left|\frac{\xi_{gk}^{lt}}{\xi_{gh}^{lt}}\right| + \tilde{\phi}_{gh}^l(t)\sum_{k\in\mathcal{A}_g^t}\left|\frac{\xi_{gk}^{lt}}{\xi_{gh}^{lt}}\right|,$$

with $0 < \tilde{\phi}_{gh}^l(t) = \phi_{gh}^l(t) \pm \beta_g$, which due to $\left|\frac{\xi_{gk}^{lt}}{\xi_{gh}^{lt}}\right| \cdot |\xi_{gh}^{lt}| = \gamma(\xi_{gk}^{lt})$.

Let matrix $\tilde{W}_1 = \mathrm{col}[U_{1i}]_{i\in\mathcal{V}_R}$, and note that matrix $C_{\mathcal{V}_R}$ is full column-rank, means that $\tilde{W}_1$ is also full column-rank. Hence, for $\forall g \in \mathcal{V}_R$, one has

$$W_{2g} = W_{2g}\tilde{W}_1^{\dagger}\tilde{W}_1 = \sum_{i\in\mathcal{V}_R}\Gamma_{gi}U_{1i},$$

where $\Gamma_{gi} \in \mathbb{R}^{(n-n_g)\times n_i}$ is a parameter matrix.

Since for $\forall g, j \in \mathcal{V}_R$, one has $\zeta_{2g}(t) = \sum_{i\in\mathcal{V}_R}\Gamma_{gi}U_{1i}e_g(t)$ and

$$\xi_{gj}^t = U_{2i}(x_g(t) - \tilde{\theta}_{j,g}(t)) = \sum_{i\in\mathcal{V}_R}\Gamma_{gi}U_{1i}(e_g(t) - e_j(t)).$$

Combining (6.11) *with* $\lim_{t\to\infty} \Gamma_{gi} U_{1i} e_i(t) = \lim_{t\to\infty} \zeta_{gi} \xi_{1i}(t) = \mathbf{0}_{n-n_g}$, *one can get* $\lim_{t\to\infty} \zeta_{2g} = \lim_{t\to\infty} \sum_{i\in\mathcal{V}_R} \Gamma_{gi} U_{1i} e_i(t) = \mathbf{0}_{n-n_g}$, $g \in \mathcal{V}_R$.

Note that $\lim_{t\to\infty} \|\zeta_{1i}\| = 0$ *holds for* $\forall i \in \mathcal{V}_R$ *and* $e_i(t) = V_{1i}\zeta_{1i}(t) + V_{2i}\zeta_{2i}(t)$, *thus*

$$\lim_{t\to\infty} e_i(t) = V_{1i}\mathbf{0}_{n_i} + V_{2i}\mathbf{0}_{n-n_i} = \mathbf{0}_n.$$

This means that $\lim_{t\to\infty} \|\hat{\theta}_i(t) - \theta\| = 0$, $\forall i \in \mathcal{V}_R$.

Remark 6.2 *For distributed signal reconstruction in a deterministic attack-free environment, it is evident that the consensus control* $u_i(t)$ *can be simplified as* $u_i(t) = -U_{2i} \sum_{j\in\mathcal{N}_i} (\hat{\theta}_i(t) - x_j(t))$.

Remark 6.3 *Compared to the distributed estimation framework in [135, 136] that requires knowledge of the neighbor sensor models, the structure proposed in this chapter is more practical.*

6.3 DISTRIBUTED SECURITY MONITORING AGAINST LOCAL BYZANTINE LINK ATTACKS

6.3.1 Problem formulation

● *Sensor Networks*

Similarly, consider a sensor network under the disturbance of a noisy environment and with the communication topology $\mathcal{G}$ being a directed graph. The specific measurement flow is modeled as follows:

$$y_i(t) = C_i\theta + v_i(t), \quad \forall i \in \mathcal{V} = [N], \tag{6.12}$$

where $v_i(t)$ is the independent identically distributed Gaussian measurement noise with zero mean and variance being Σ_i, and other parameters are the same as in (6.1). The overall measurement matrix and the noise are respectively denoted by $C = \text{col}[C_i]_{i\in\mathcal{V}}$ and $v(t) = \text{col}[v_i(t)]_{i\in\mathcal{V}}$, and the overall measurement output is written as $y(t) = \text{col}[y_i(t)]_{i\in\mathcal{V}} = C\theta + v(t)$.

Assumption 6.3 *The measurement matrices* C_i *corresponding to the root node set* $R_\mathcal{G}$ *is jointly full column-rank, i.e.* $\text{rank}(C_{R_\mathcal{G}}) = n$.

● *Byzantine link attacks*

It is important to note that, under Assumption 6.3, the local matrix C_i may be column rank-deficient. Therefore, for global state estimation to be achievable, it is essential for some (if not all) nodes to be able to exchange information with their neighbors. Here, assume that some sensors are allowed to act as relay nodes that have only data processing and communication modules, i.e., there exist some i for which $y_i(t) \equiv 0$, $C_i = 0$ and $v_i(t) = 0$. Put all these relay sensors into a set $\mathcal{V}_R = \{i | y_i(t) \equiv 0\}$.

The communication links are inevitably susceptible to cyber-attacks for the openness of communication networks, and this section aims to improve the distributed state monitoring performance under the cyber-attack environment. In particular, focusing on malicious attacks occurring in communication links, the distributed SSM work is carried out under the condition that some of the links are compromised by Byzantine intrusions.

Without imposing any restriction on the attack injection, the link compromised by malicious attacks and the regular link set at time t are denoted by $\mathcal{E}_A^t$ and $\mathcal{E}_R^t$, respectively. Thus, one has $\mathcal{E}_R^t = \mathcal{E} \setminus \mathcal{E}_A^t$. The following assumption is introduced to quantify the number of compromised links in the network.

Assumption 6.4 *The sensor network is assumed to be F-local Byzantine link compromised at any time instant t, i.e., $|\mathcal{E}_A^t \cap \mathcal{K}_i| \leq F$ with $\mathcal{K}_i$ is incoming edges set of the node i defined in Chapter 2.*

Remark 6.4 *To ensure the universality of the distributed SSM algorithm, no constraints are imposed on the attacks other than the number of attacks. Therefore, the adversaries are assumed to be omniscient, knowing all the detailed information of the network structure and the system state. Besides, the adversaries are able to manipulate the data injected into the communication links arbitrarily and even collude with others to maintain covertness to cause serious damage. Since the actual number and locations of the compromised links are unknown to the sensor network, the parameter F in Assumption 6.4 can be assumed as a desired resilience level to resist the maximum number of malicious cyber-attacks. Note that this assumption is general and has been commonly adopted [81, 83, 85].*

Remark 6.5 *Compared with [81, 83–85, 142], where the locations of the attacks are fixed, the locations of the attacks considered in this section are allowed to be time-varying to avoid being detected, which makes the distributed SSM problem more realistic yet more challenging.*

• *Distributed observer architecture*

Consider the state monitoring framework discussed in [37,40,43,81,83,85], where each sensor is equipped with an observer, and each observer interacts with its neighbors over a communication network to make a full state estimation. This framework is shown in Fig. 6.2, where the solid lines represent the physical measurement channels that are immune to cyber-attacks, the dashed lines indicate the communication links that are vulnerable to cyber-attacks, and the gray ones indicate the compromised links at a certain time instant.

Inspired by the decentralized design of observers in [144], the following distributed observers are constructed in the cyber layer:

$$\dot{\hat{\theta}}_i(t) = K_i u_i(t) + L_i(\bar{y}_i(t) - C_i \hat{\theta}_i(t)), \tag{6.13a}$$

$$\dot{\bar{y}}_i(t) = \frac{1}{t + \chi_i} \left(y_i(t) - \bar{y}_i(t) \right), \tag{6.13b}$$

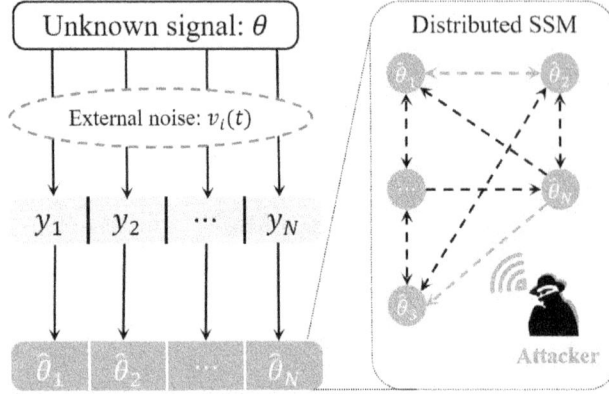

Figure 6.2 A distributed security signal monitoring framework under Byzantine link attacks.

where $\hat{\theta}_i(t)$ is the estimation of state θ by the i-th observer at the instant t, $u_i(t)$ is the consensus control input, which will be discussed in the next section, K_i and L_i are the decentralized consensus control and observation gain matrices to be designed, $\chi_i > 0$ is one optional parameter to avoid initial scaling, and $\bar{y}_i(t)$ denotes the running average of $y_i(t)$ with $\bar{y}_i(0) = y_i(0)$.

Remark 6.6 *It should be emphasized that the observer in (6.13a) exhibits a structural resemblance to the one introduced in [144]. Nevertheless, the critical distinction stems from its reliance on a generalized consensus control protocol, expressed as $u_i(t) = \beta \sum_{j \in \mathcal{N}_i} (\hat{\theta}_j(t) - \hat{\theta}_i(t))$, which inherently carries vulnerabilities to security threats and lacks robustness against cyber-attacks. As a result, this section prioritizes the development of attack-resilient distributed control strategies to mitigate such risks.*

Now, solving equation (6.13b) yields

$$\bar{y}_i(t) = \frac{\chi_i}{t + \chi_i} \bar{y}_i(0) + \frac{1}{t + \chi_i} \int_0^t y_i(s)ds = C_i \theta + \frac{\sigma_i}{t + \chi_i} v_i(0) + \bar{v}_i(t),$$

where $\bar{v}_i(t) = \frac{1}{t+\chi_i} \int_0^t v_i(s)ds$ is the time-averaged measurement noise. It is obvious that $\bar{v}_i(t)$ is also a Gaussian white noise with mean being 0 and variance being $\frac{t}{(t+\chi_i)^2}\Sigma_i$, i.e., $\lim_{t \to \infty} \bar{v}_i(t) \xrightarrow{a.s.} \mathbf{0}_{m_i}$. It follows that

$$\lim_{t \to \infty} \frac{\chi_i}{t + \chi_i} v_i(0) = \mathbf{0}_{m_i}, \quad \lim_{t \to \infty} \bar{y}_i(t) \xrightarrow{a.s.} C_i \theta.$$

In particular, $\bar{y}_i(t) = y_i(t) = C_i \theta$ holds constantly for the noise-free case.

Next, the problem to be studied in this section is formally formulated.

Problem 6.2. For a given sensor network (6.12) and a set of distributed observers (6.13) with a directed communication topology $\mathcal{G}$, design a distributed resilient control protocol $u_i(t)$ with the gain matrices K_i and L_i under any F-local Byzantine link attack, such that

$$\lim_{t \to \infty} \|\hat{\theta}_i(t) - \theta\| \xrightarrow{a.s.} 0, \quad \forall i \in \mathcal{V}. \tag{6.14}$$

6.3.2 Distributed security signal monitoring algorithm

In this subsection, a new decoupled design control protocol for distributed SSM will be proposed, and its consensus control performance will then be demonstrated in an attack-free environment. To that end, by combining the norm-based max-discard mechanism, an improved resilient control protocol will be developed against F-local Byzantine link attacks and achieve the distributed SSM.

Consider the information that observer i receives from the neighbor link $(j, i) \in \mathcal{K}_i$, modeled as

$$\tilde{\theta}_{j,i}(t) = \hat{\theta}_j(t) + w_{j,i}(t),$$

where $w_{j,i}(t)$ is the malicious data injected via the link (j, i), satisfying

$$w_{j,i}(t) \begin{cases} \neq \mathbf{0}_n, & \text{if } (j, i) \in \mathcal{E}_A^t, \\ = \mathbf{0}_n, & \text{if } (j, i) \in \mathcal{E}_R^t. \end{cases} \tag{6.15}$$

Denote $\hat{e}_{ij}(t) = \hat{\theta}_i(t) - \hat{\theta}_j(t)$ as the consensus error, and $e_i(t) = \hat{\theta}_i(t) - \theta$ as the signal monitoring error, for the observer i.

6.3.3 Monitoring performance against local Byzantine link attacks

In this subsection, a decoupling operation for the interaction information among observer neighbors is performed by distinguishing between their directions (i.e., Sign($\cdot$)) and magnitudes (i.e., $\phi(\cdot)$). It is demonstrated that the consensus control objective is not affected even if the direction and magnitude are processed separately.

Again for matrices U_i and V_i defined in the subsection 6.2.2, and denote $\tilde{U}_1 = \text{col}[U_{1h}]_{h \in R_{\mathcal{G}}}$, where each row vector in $\tilde{U}_1$ belongs to $\mathcal{R}(C_{R_{\mathcal{G}}}^T)$. Note that if $C_{R_{\mathcal{G}}}$ has full column-rank, then matrix $\tilde{U}_1$ also has full column-rank. Consequently, one has

$$U_{2i} = U_{2i}\tilde{U}_1^{\dagger}\tilde{U}_1 = \sum_{h \in R_{\mathcal{G}}} \Gamma_{ih} U_{1h}, \quad \forall i \in \mathcal{V}, \tag{6.16}$$

where $\Gamma_{ih} \in \mathbb{R}^{(n-n_i) \times n_h}$ is the parameter matrix.

- **Distributed signal monitoring under the attack-free environment**

Then, the control protocol for observer (6.13) under the attack-free environment is designed as

$$u_i(t) = - \sum_{j \in \mathcal{N}_i} (\delta_i(t) \circ \phi(\zeta_{ij}(t)) + \beta_i \zeta_{ij}(t)), \tag{6.17}$$

where $\delta_i(t) = \sum_{j \in \mathcal{N}_i} \text{Sign}(\zeta_{ij}(t))$, $\zeta_{ij}(t) = U_{2i}\hat{e}_{ij}(t)$, and $\beta_i > 0$.

The main result of distributed state estimation under the attack-free environment is presented.

Theorem 6.2 *Consider the sensor network (6.12) with its topology $\mathcal{G}$ containing spanning trees rooted at $R_{\mathcal{G}}$. Suppose that Assumption 6.3 holds. The distributed observers $\hat{\theta}_i(t)$, $i \in \mathcal{V}$, in (6.13) can achieve an asymptotic estimation of θ with the control protocol (6.17) if matrices K_i and L_i satisfy*

$$K_i = \mu_i V_{2i}, \quad L_i = \tau_i V_i \begin{bmatrix} \tilde{C}_{1i}^\dagger \\ 0 \end{bmatrix} = \tau_i V_{1i} \tilde{C}_{1i}^\dagger, \tag{6.18}$$

where $\mu_i > 0$ and $\tau_i > 0$ are arbitrary constants.

Proof 6.3 *It follows from (6.12)-(6.13) that the dynamics of $e_i(t)$ can be written as*

$$\dot{e}_i(t) = -L_i C_i e_i(t) + K_i u_i(t) + L_i \bar{v}_i(t).$$

Now, left-multiplying $e_i(t)$ by matrix U_i and setting $\xi_{1i} = U_{1i} e_i$ and $\xi_{2i} = U_{2i} e_i$, one has

$$\begin{cases} \dot{\xi}_{1i}(t) = -\tau_i \xi_{1i}(t) + \tau_i \tilde{C}_{1i}^\dagger \bar{v}_i(t), \\ \dot{\xi}_{2i}(t) = \mu_i u_i(t), \end{cases} \tag{6.19}$$

where K_i and L_i in (6.18) have been used. Since $\tau_i > 0$ and $\lim_{t\to\infty} \bar{v}_i(t) = \mathbf{0}_{m_i}$, one has

$$\lim_{t\to\infty} \|\xi_{1i}(t)\| \xrightarrow{a.s.} 0, \quad \forall i \in \mathcal{V}.$$

Combining (6.16)-(6.19) and Assumption 6.3, one gets

$$\begin{cases} \lim_{t\to\infty} \Gamma_{ih} U_{1h} e_h(t) = \lim_{t\to\infty} \Gamma_{ih} \xi_{1h}(t) = \mathbf{0}_{n-n_i}, \\ \dot{\xi}_{2i}(t) = -\mu_i \sum_{j \in \mathcal{N}_i} (\delta_i(t) \circ \phi(\zeta_{ij}(t)) + \beta_i \zeta_{ij}(t)), \ \forall i \in \mathcal{V}, \end{cases} \tag{6.20}$$

where $h \in R_{\mathcal{G}}$, $\zeta_{ij}(t) = \sum_{h \in R_{\mathcal{G}}} \Gamma_{ih} U_{1h}(e_i(t) - e_j(t))$, and $\xi_{2i} = \sum_{h \in R_{\mathcal{G}}} \Gamma_{ih} U_{1h} e_i(t)$. Then, it follows from Lemma 6.1 that system (6.20) can reach asymptotic consensus, i.e.,

$$\lim_{t\to\infty} \xi_{2i} = \lim_{t\to\infty} \sum_{h \in R_{\mathcal{G}}} \Gamma_{ih} \xi_{1h}(t) \xrightarrow{a.s.} \mathbf{0}_{n-n_i}.$$

Note that $\lim_{t\to\infty} \|\xi_{1i}\| \xrightarrow{a.s.} 0$ and $e_i(t) = V_{1i} \xi_{1i}(t) + V_{2i} \xi_{2i}(t)$, thereby

$$\lim_{t\to\infty} e_i(t) \xrightarrow{a.s.} V_{1i} \mathbf{0}_{n_i} + V_{2i} \mathbf{0}_{n-n_i} = \mathbf{0}_n.$$

This means that

$$\lim_{t\to\infty} \|\hat{\theta}_i(t) - \theta\| \xrightarrow{a.s.} 0, \quad \forall i \in \mathcal{V}.$$

The proof is completed.

Remark 6.7 *In [135], each observer i needs to know the measurement matrix C_j of its neighboring sensors $j \in \mathcal{N}_i$, and the one in [136] even resorts to the overall measurement matrix C to achieve state estimation. However, it follows from (6.13)-(6.18) that the i-th observer only has access to its own measurement matrix C_i. Thus, the proposed distributed state estimator decouples the measurement matrix information between neighbors and is more practical.*

- Distributed SSM against F-local Byzantine link attacks

Due to the presence of attacks, the information received by observer i from its neighbor j becomes $\tilde{x}_{j,i}(t)$, and the ζ_{ij} in (6.17) is changed to

$$\tilde{\zeta}_{ij}(t) = U_{2i}(\hat{\theta}_i(t) - \tilde{\theta}_{j,i}(t)).$$

For each observer i, the terms $\|\phi(\tilde{\zeta}_{ij}(t))\|$, $(j,i) \in \mathcal{K}_i$, are arranged in a descending order (simply denoted by $\phi_{i,1}(t), \ldots, \phi_{i,N_i}(t)$ with $N_i = |\mathcal{K}_i|$). Then, one has

$$\phi_{i,1}(t) \geq \phi_{i,2}(t) \geq \cdots \geq \phi_{i,N_i}(t).$$

The following norm-based maximum-discard mitigation (NMDM) framework is constructed to accomplish attack isolation:

$$\rho_{ij}(t) = \begin{cases} 1, & \text{if } \|\phi(\tilde{\zeta}_{ij}(t))\| \leq \phi_{i,F+1}(t), \\ 0, & \text{otherwise}, \end{cases} \tag{6.21}$$

where $\rho_{ij}(t)$ denotes the logical decision factor on link (j,i). Here, $\rho_{ij}(t) = 0$ means that observer i asserts that the link (j,i) may be compromised at instant t, and $\rho_{ij}(t) = 1$ indicates that the link (j,i) is normal.

Since the attack number satisfies $|\mathcal{E}_A^t \cap \mathcal{K}_i| \leq F$, one has $\phi_{i,F+1}(t) \leq \max_{(j,i)\in\mathcal{K}_i}\{\|\phi(\zeta_{ij}(t))\|\}$. As a result, the NMDM framework (6.21) implies that the attacks with extreme magnitudes can be effectively isolated, and their propagation impact can be blocked by setting $\rho_{ij}(t) = 0$, ensuring excellent state estimation performance.

Remark 6.8 *In the NMDM framework (6.21), the attack isolation factor $\rho_{ij}(t)$ is constructed by utilizing $\tilde{\zeta}_{ij}(t)$, which differs from the approach in [81], where the innovation item $r_{ij}(t) = y_i(t) - C_i\tilde{\theta}_{j,i}(t)$ is employed for this purpose. The main reason is that if a malicious attack $w_{j,i}(t)$ located in the null space $\mathcal{N}(C_i)$, one has $C_iw_{j,i}(t) = 0$ but $U_{2i}w_{j,i}(t) \neq 0$, which implies that the innovation term $r_{ij}(t)$ loses the information regarding the attack. In contrast, it remains present in $\tilde{\zeta}_{ij}(t)$. Hence, the isolation factor built on $r_{ij}(t)$ will be invalidated, while the one based on $\tilde{\zeta}_{ij}(t)$ can avoid this problem.*

Based on the above NMDM framework (6.21), the control protocol (6.17) is modified as

$$u_i(t) = -\sum_{j\in\mathcal{N}_i} \rho_{ij}(t)\left(\delta_i(t) \circ \phi(\tilde{\zeta}_{ij}(t)) + \beta_i\tilde{\zeta}_{ij}(t)\right), \tag{6.22}$$

where $0 < \beta_i < 1$.

Remark 6.9 *In the control protocol (6.22), the function $\delta_i(t)$ allows observer i to converge in the direction of over half of its neighbors, which contributes to ensuring the directional accuracy of observer i. Besides, the NMDM mechanism safeguards the normality of the updated magnitude of estimation by isolating extreme data. These two terms jointly resist the negative impact of attacks. It's worth noting that $\beta_i < 1$ here can ensure that the estimation is dominated by $\delta_i(t) \circ \phi(\tilde{\zeta}_{ij}(t))$, avoiding the defect of $\tilde{\zeta}_{ij}(t)$ in the convergence direction.*

Algorithm 6.1 Distributed SSM algorithm based on NMDM

Input: C_i, F, a_{ij}, $y_i(t)$, $\hat{\theta}_i(0)$, $\chi_i > 0$, $\mu_i > 0$, $\tau_i > 0$, $0 < \beta_i < 1$;
Output: State estimation $\hat{\theta}_i(t)$;
 1: Calculate $\mathcal{R}(C_i^T)$ and $\mathcal{N}(C_i)$ of the matrix C_i, and pick the matrices U_{1i}, U_{2i}, V_{1i} and V_{2i};
 2: $\tilde{C}_{1i} = C_i V_{1i}$, $K_i = \mu_i V_{2i}$, $L_i = \tau_i V_{1i} \tilde{C}_{1i}^\dagger$;
 3: **while** $t > 0$ **do**
 4:      Receive transmission data $\tilde{\theta}_{j,i}(t)$ from its in-neighbor and broadcast its estimation $\hat{\theta}_i(t)$ to its out-neighbor;
 5:      Compute $\tilde{\zeta}_{ij}(t) = U_{2i} \hat{e}_{ij}(t)$ and $\phi(\tilde{\zeta}_{ij}(t))$, and obtain the $(F+1)$-th maximum $\phi_{i,F+1}(t)$;
 6:      **if** $\|\phi(\tilde{\zeta}_{ij}(t))\| \le \phi_{i,F+1}(t)$ **then**
 7:          $\rho_{ij}(t) = 1$;
 8:      **else**
 9:          $\rho_{ij}(t) = 0$;
10:      **end if**
11:      $\delta_i(t) = \sum_{j \in \mathcal{N}_i} \text{Sign}(\zeta_{ij}(t))$;
12:      Calculate the control input

$$u_i(t) = -\sum_{j \in \mathcal{N}_i} \rho_{ij}(t) \Big(\delta_i(t) \circ \phi(\tilde{\zeta}_{ij}(t)) + \beta_i \tilde{\zeta}_{ij}(t) \Big);$$

13:      Update $\dot{\bar{y}}_i(t) = \frac{1}{t+\chi_i}(y_i(t) - \bar{y}_i(t))$;
14:      Update the state estimation:

$$\dot{\hat{\theta}}_i(t) = K_i u_i(t) + L_i(\bar{y}_i(t) - C_i \hat{\theta}_i(t));$$

15: **end while**
16: **return** $\hat{\theta}_i(t)$.

In summary of the distributed SSM analysis process presented above, the SSM processes of each observer can be operated in parallel according to the iterative steps delineated in Algorithm 6.1.

Remark 6.10 *In Algorithm 6.1, the higher-dimensional interaction term $\tilde{\zeta}_{ij}(t)$ between neighbors is processed as a whole vector without the need to split it into each dimension for individual processing as the W-MSR algorithm proposed in [49, 83, 86]. Specifically, for a single node i, the computational complexity of the W-MSR algorithm is $O(|\mathcal{N}_i|^2 \log |\mathcal{N}_i|)$, whereas that of the NMDM algorithm is $O(|\mathcal{N}_i| + |\mathcal{N}_i| \log |\mathcal{N}_i|)$. Therefore, compared to the W-MSR algorithm, Algorithm 6.1 greatly improves the attack isolation efficiency and reduces the complexity of the algorithm, making it more suitable for networked systems with limited resources.*

The main result for the distributed SSM is presented as follows.

Theorem 6.3 *For the sensor network* (6.12) *under F-local Byzantine link attacks, the distributed SSM* (6.14) *can be reached by the observer* (6.13) *under the control protocol* (6.22) *if, for any nonempty subset $\mathcal{S} \subseteq \mathcal{V}$, at least one of the followings holds:*

  i) *The matrix $C_{\mathcal{S}}$ has full column-rank;*

  ii) *The subset $\mathcal{S}$ is $(2F + 1)$-reachable.*

Proof 6.4 *Since each observer in the root node set $R_{\mathcal{G}}$ has no in-neighbors outside of $R_{\mathcal{G}}$, the subset $R_{\mathcal{G}}$ is necessarily not $(2F+1)$-reachable. It follows that the matrix $C_{R_{\mathcal{G}}}$ must have full column-rank, which is consistent with the Assumption 6.2.*

Assume that each observer retains only its $|N_i| - F$ incoming links. The graph $\mathcal{G}$ is divided into a series of connected components, $\mathcal{G}^q = \{\mathcal{V}^q, \mathcal{E}^q\}$ ($q \leq N$) with $R_{\mathcal{G}}^q$ being the set of root nodes in the q-th component. Then, each matrix $C_{R_{\mathcal{G}}^q}$ has full column-rank, due to the fact that each node in $R_{\mathcal{G}}^q$ has no in-neighbors outside of $R_{\mathcal{G}}^q$ in $\mathcal{G}^q$, implying that these nodes also have no more than F in-neighbors outside of $R_{\mathcal{G}}^q$ in $\mathcal{G}$.

Drawing on the proof of Theorem 6.2, it is needed to show that the system $\xi_{2i}(t)$ in (6.19) is asymptotically stable under the control protocol (6.22). To this end, for $i \in \mathcal{V}$, the following subset is introduced:

$$\mathcal{V}_{S,i}(t) = \{j|\ j \in \mathcal{N}_i,\ \|\zeta_{ij}(t)\| = 0\} \cup \{i\}.$$

Next, the sets $\mathcal{V}_{S,i}(t)$ are analyzed when condition i) or ii) is satisfied.

Case 1: If $\mathcal{V}_{S,i}(t)$ satisfies condition i), one has $\xi_{2g}(t) = \xi_{2j}(t)$ for $g, j \in \mathcal{V}_{S,i}(t)$ because $\|\zeta_{gj}(t)\| \leq \|\zeta_{ig}(t)\| + \|\zeta_{ij}(t)\| = 0$. Hence, it follows from Theorem 6.2 that all observers in $\mathcal{V}_{S,i}(t)$ can achieve the distributed SSM.

Case 2: If $\mathcal{V}_{S,i}(t)$ satisfies condition ii), the control (6.22) can be equivalently replaced by a form without attack. Then, the attack terms in $\delta_i(t)$ and $\tilde{\zeta}_{ij}(t)$ are analyzed, respectively.

It follows from condition ii) that there exists a node $g \in \mathcal{V}_{S,i}(t)$ with $(2F + 1)$ in-neighbors outside $\mathcal{V}_{S,i}(t)$. Denote the incoming link set of node g as $\tilde{\mathcal{K}}_g^t$. Thus, for any $(j, g) \in \tilde{\mathcal{K}}_g^t$, $l \in [n - n_g]$ and $\zeta_{gj}^l(t) \neq 0$, one has

$$\delta_g^l(t) = \sum_{(j,g)\in\mathcal{E}_R^t} \text{sign}(\zeta_{gj}^l(t)) + \sum_{(j,g)\in\mathcal{E}_A^t} \text{sign}(\tilde{\zeta}_{gj}^l(t))$$

$$= \sum_{(j,g)\in\mathcal{M}_g^t} \varphi_{gj}^l(t)\text{sign}(\zeta_{gj}^l(t)),$$

where $\varphi_{gj}^l(t) \in \{1, 2\}$, and the subset $\mathcal{M}_g^t \subseteq \tilde{\mathcal{K}}_g^t \cap \mathcal{E}_R^t$ with $|\mathcal{M}_g^t| \geq |\tilde{\mathcal{K}}_g^t| - 2F \geq 1$ because $|\mathcal{K}_g \cap \mathcal{E}_A^t| \leq F$ and $\tilde{\mathcal{K}}_g^t \subseteq \mathcal{K}_g$.

Let

$$\mathcal{R}_g^t = \{(j, g)|\rho_{gj}(t) = 1,\ (j, g) \in \mathcal{E}_R^t \cap \tilde{\mathcal{K}}_g^t\},$$
$$\mathcal{A}_g^t = \{(j, g)|\rho_{gj}(t) = 1,\ (j, g) \in \mathcal{E}_A^t \cap \tilde{\mathcal{K}}_g^t\}.$$

Similarly, it can be proved that $|\mathcal{R}_g^t| \geq |\tilde{\mathcal{K}}_g^t| - 2F \geq 1$.

For any $(k,g) \in \mathcal{A}_g^t$ *and* $(j,g) \in \mathcal{M}_g^t$, *one has*

$$\varphi_{gj}^l(t)\mathrm{sign}(\zeta_{gj}^l(t))|\tilde{\zeta}_{gk}^l(t)| + \beta_g\tilde{\zeta}_{gk}^l(t)$$

$$= \varphi_{gj}^l(t)\left|\frac{\tilde{\zeta}_{gk}^l(t)}{\zeta_{gj}^l(t)}\right|\zeta_{gj}^l(t) + \beta_g\tilde{\zeta}_{gk}^l(t)$$

$$= \left(\varphi_{gj}^l(t) + \beta_g\mathrm{sign}\left(\frac{\tilde{\zeta}_{gk}^l(t)}{\zeta_{gj}^l(t)}\right)\right)\left|\frac{\tilde{\zeta}_{gk}^l(t)}{\zeta_{gj}^l(t)}\right|\zeta_{gj}^l(t)$$

$$= \tilde{\varphi}_{gj}^l(t)\left|\frac{\tilde{\zeta}_{gk}^l(t)}{\zeta_{gj}^l(t)}\right|\zeta_{gj}^l(t),$$

where $0 < \tilde{\varphi}_{gj}^l(t) = \varphi_{gj}^l(t) + \beta_g\mathrm{sign}\left(\frac{\tilde{\zeta}_{gk}^l(t)}{\zeta_{gj}^l(t)}\right) \in \{\varphi_{gj}^l(t) - \beta_g, \varphi_{gh}^l(t) + \beta_g\}$.

Hence, the control protocol (6.22) can be converted to

$$u_g^l(t) = -\sum_{(k,g)\in\mathcal{R}_g^t\cup\mathcal{A}_g^t}\left(\delta_g^l(t)|\tilde{\zeta}_{gk}^l(t)| + \beta_g\tilde{\zeta}_{gk}^l(t)\right)$$

$$= -\sum_{(k,g)\in\mathcal{R}_g^t\cup\mathcal{A}_g^t}\sum_{(j,g)\in\mathcal{M}_g^t}\varphi_{gj}^l(t)\mathrm{sign}(\zeta_{gj}^l(t))|\tilde{\zeta}_{gk}^l(t)| - \sum_{(k,g)\in\mathcal{R}_g^t\cup\mathcal{A}_g^t}\beta_g\tilde{\zeta}_{gk}^l(t)$$

$$= -\sum_{(j,g)\in\mathcal{M}_g^t}\psi_{gj}^l(t)\zeta_{gj}^l(t) - \sum_{(j,g)\in\mathcal{R}_g^t\cup\mathcal{A}_g^t}\beta_g\tilde{\zeta}_{gj}^l(t)$$

$$= -\tilde{\psi}_{gh}^l(t)\zeta_{gh}^l(t) - \sum_{(j,g)\in\mathcal{M}_g^t,\ j\neq h}\psi_{gj}^l(t)\zeta_{gj}^l(t) - \sum_{(j,g)\in\mathcal{R}_g^t}\beta_g\zeta_{gj}^l(t), \qquad (6.23)$$

where $(h,g) \in \mathcal{M}_g^t$ *and*

$$0 < \psi_{gj}^l(t) = \varphi_{gj}^l(t)\sum_{(k,g)\in\mathcal{R}_g^t\cup\mathcal{A}_g^t}\left|\frac{\tilde{\zeta}_{gk}^l(t)}{\zeta_{gj}^l(t)}\right|,$$

$$0 < \tilde{\psi}_{gh}^l(t) = \varphi_{gh}^l(t)\sum_{(k,g)\in\mathcal{R}_g^t}\left|\frac{\zeta_{gk}^l(t)}{\zeta_{gh}^l(t)}\right| + \tilde{\varphi}_{gh}^l(t)\sum_{(k,g)\in\mathcal{A}_g^t}\left|\frac{\tilde{\zeta}_{gk}^l(t)}{\zeta_{gh}^l(t)}\right|.$$

It follows from $\mathcal{M}_g^t \cap \mathcal{E}_A^t = \varnothing$ *and* $\mathcal{R}_g^t \cap \mathcal{E}_A^t = \varnothing$ *that the control protocol* $u_g^l(t)$ *can be equivalently expressed as a combination of certain benign control terms. Similar to the analysis in Theorem 6.2, it is concluded that all nodes in* $\mathcal{V}_{S,i}(t)$ *can reach asymptotic consensus, which implies that*

$$\lim_{t\to\infty} e_i(t) = \mathbf{0}_n, \ \forall i \in \mathcal{V}_{S,i}(t) \subset \mathcal{V}.$$

Hence, it follows that all $\hat{\theta}_i(t)$ *can reach the SSM (6.14) with respect to* θ *under arbitrary F-local Byzantine link attacks, i.e.,* $\lim_{t\to\infty} \|\hat{\theta}_i(t) - \theta\| = 0, \ \forall i \in \mathcal{V}$.

Remark 6.11 *It follows from Theorem 6.3 that redundancy of the matrix C can serve as a supplementary for the robustness constraint on the topology in the distributed SSM problem. Therefore, this section establishes a connection between the strong $(2f+1)$-robust graph and the sensor measurement redundancy during the distributed SSM process.*

Corollary 6.1 *Consider the sensor network (6.12) that executes Algorithm 4.1, then for any non-zero F-local Byzantine link attack, it will be successfully isolated as time progresses, i.e., $\lim_{t\to\infty}\{(j,i)\in\mathcal{E}_A^t|\rho_{ij}(t)=1\}\xrightarrow{a.s.}\varnothing$.*

Proof 6.5 *For a non-zero attack $w_{j,i}(t)$, one can find a sufficiently small $\epsilon>0$ such that $\|U_{2i}w_{j,i}(t)\|>\epsilon$. It follows from Theorem 6.3 that $\lim_{t\to\infty}\|\hat\theta_i-\theta\|\xrightarrow{a.s.}0$ holds for any $i\in\mathcal{V}$, there must exits a time T such that $\|\zeta_{ij}(t)\|=\|U_{2i}\hat e_{ij}\|\leq\epsilon/2$ for $t\geq T$.*

Then, for any $(j,i)\in\mathcal{E}_A^t$, one has

$$\|\tilde\zeta_{ij}(t)\|\geq\|U_{2i}w_{j,i}(t)\|-\|\zeta_{ij}(t)\|>\frac{\epsilon}{2},\quad t\geq T.$$

According to $|\mathcal{E}_A^t\cap\mathcal{K}_i|\leq f$, one has $\phi_{i,f+1}(t)\leq\frac{\epsilon}{2}$. It follows that $\rho_{ij}(t)=0$ for all $(j,i)\in\mathcal{E}_A^t$ and $t\geq T$, i.e.,

$$\lim_{t\to\infty}\{(j,i)\in\mathcal{E}_A^t|\rho_{ij}(t)=1\}\xrightarrow{a.s.}\varnothing.$$

The proof is completed.

6.4 NUMERICAL SIMULATIONS

In this chapter, two examples will be given to validate some theoretical results given in sections 6.2 and 6.3, respectively. The performance of Theorem 6.1 and Algorithm 4.1 will be evaluated by performing a state monitoring of an unknown parameter $\theta^*=[-3,\ 12,\ 5,\ -8]^T\in\mathbb{R}^4$. The signal θ^* may denote, for example, the position of a target, the pixel value of an observed image, the concentration of environmental pollutants in a specific area, etc.

Example 1: In this example, to ensure the security of the signal monitoring with $y_i=C_i\theta^*, i\in[8]$, the redundancy of the sensors is ensured by selecting output measurement matrices as follows:

$$C_1=\begin{bmatrix}1&0&2&0\\0&-1&0&1\end{bmatrix},\quad C_2=\begin{bmatrix}1&0&0&0\\0&0&0&1\end{bmatrix},\quad C_3=\begin{bmatrix}2&0&1&0\\0&1&0&1\end{bmatrix},$$

$$C_4=\begin{bmatrix}0&0&1&0\\0&1&0&0\end{bmatrix},\quad C_5=\begin{bmatrix}1&0&1&0\\0&2&0&1\end{bmatrix},\quad C_6=\begin{bmatrix}1&0&-1&0\\0&1&0&-1\end{bmatrix},$$

$$C_7=C_8=\begin{bmatrix}0&0&0&0\end{bmatrix}.$$

Assume that 8 sensors are distributed in the network and exchange its state estimation with its neighbors in order to obtain a full state monitoring of the signal

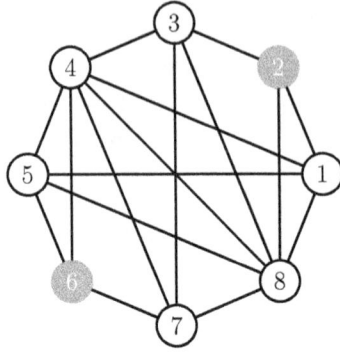

Figure 6.3   The distribution of 8 sensors with 2 Byzantine nodes.

θ^*. The corresponding communication topology can be seen in Fig. 6.3, where nodes $\{2, 6\}$ denote Byzantine nodes, and the rest are regular nodes.

It is seen that each local matrix C_i is column rank-deficient and that the whole topology $\mathcal{G}$ satisfies Assumption 6.2. In the simulation, consider the initial estimate values $\hat{\theta}_i(0)$, $i \in [8]$ randomly generated. Furthermore, the matrix C_i is spatially decomposed, and the gain matrix is obtained according to (6.18) as follows with selected $\alpha_i = 4$ and $\mu_i = 2$, $i \in \mathcal{V}_R$,

$$
K_1 = \begin{bmatrix} -1.6 & 0 \\ 0 & 2 \\ 0.8 & 0 \\ 0 & 2 \end{bmatrix}, \quad
K_2 = \begin{bmatrix} 0 & 0 \\ 4 & 0 \\ 0 & 4 \\ 0 & 0 \end{bmatrix}, \quad
K_3 = \begin{bmatrix} -1.6 & 0 \\ 0 & -2 \\ 3.2 & 0 \\ 0 & 2 \end{bmatrix},
$$

$$
K_4 = \begin{bmatrix} 4 & 0 \\ 0 & 0 \\ 0 & 0 \\ 0 & 4 \end{bmatrix}, \quad
K_5 = \begin{bmatrix} -2 & 0 \\ 0 & -1.6 \\ 2 & 0 \\ 0 & 3.2 \end{bmatrix}, \quad
K_6 = \begin{bmatrix} 2 & 0 \\ 0 & 2 \\ 2 & 0 \\ 0 & 2 \end{bmatrix},
$$

$$
L_1 = \begin{bmatrix} 0.4 & 0 \\ 0 & -1 \\ 0.8 & 0 \\ 0 & 1 \end{bmatrix}, \quad
L_2 = \begin{bmatrix} 2 & 0 \\ 0 & 0 \\ 0 & 0 \\ 0 & 2 \end{bmatrix}, \quad
L_3 = \begin{bmatrix} 0.8 & 0 \\ 0 & 1 \\ 0.4 & 0 \\ 0 & 1 \end{bmatrix},
$$

$$
L_4 = \begin{bmatrix} 0 & 0 \\ 0 & 2 \\ 2 & 0 \\ 0 & 0 \end{bmatrix}, \quad
L_5 = \begin{bmatrix} 1 & 0 \\ 0 & 0.8 \\ 1 & 0 \\ 0 & 0.4 \end{bmatrix}, \quad
L_6 = \begin{bmatrix} 1 & 0 \\ 0 & 1 \\ -1 & 0 \\ 0 & -1 \end{bmatrix},
$$

$$
K_7 = K_8 = 4I_4, \qquad L_7 = L_8 = \mathbf{0}_{4 \times 1}.
$$

Assume that the Byzantine node $\{2, 6\}$ performs its behavioral update with the following dynamics:

$$
\dot{\hat{\theta}}_i(t) = \theta^* - \hat{\theta}_i(t) + w_i(t).
$$

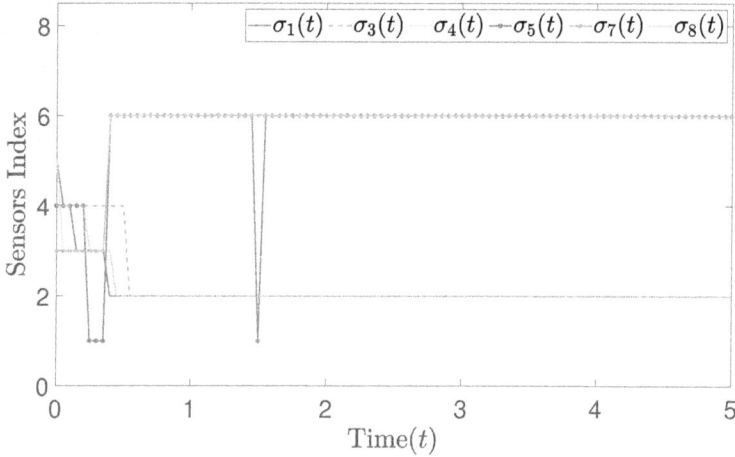

Figure 6.4 Switching signal $\sigma_i(t)$, $i \in \mathcal{V}_R$.

where $w_i(t) = i\sin(t\pi)[-2\ 1\ -1\ 2]^T$. We use the switching signal $\sigma_i(t)$, $i \in \mathcal{V}_R$ to describe the attack detection behaviors on the regular nodes, where $\sigma_i(t) = j$ if $\rho_{ij}(t) = 0$. Then, each regular node executes the estimation update as (6.6)–(5.3) to obtain the simulation results as shown in Figs. 6.4–6.5.

Fig. 6.4 depicts the switching behavior $\sigma_i(t)$ of each regular node regarding the attack isolation effect. It is evident that after a period of time, all the regular nodes can accurately identify the location of the Byzantine node and isolate it. The signal monitoring curves for each sensor can be seen in Fig. 6.5, where it can be seen that consensus signal monitoring is successfully achieved by the proposed algorithm in the section 6.2 under the effect of Byzantine node attacks.

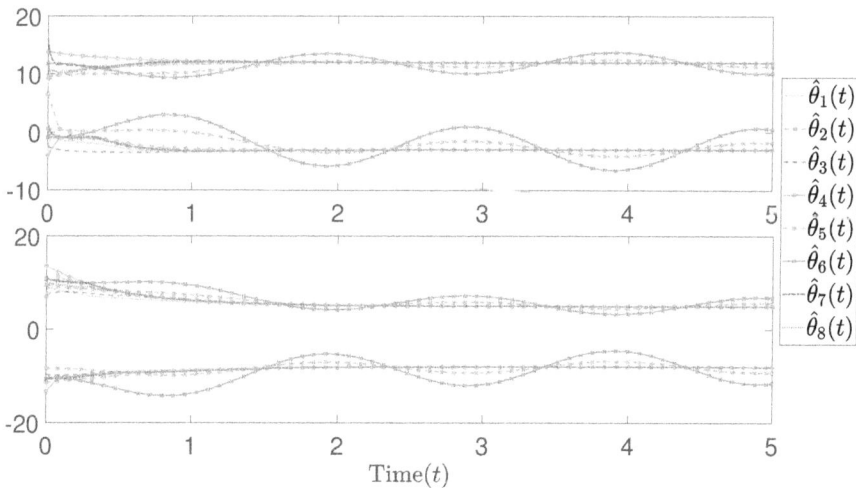

Figure 6.5 Signal monitoring curves for each sensor.

Example 2: In this example, the performance of Algorithm 6.1 will be evaluated. Assume that the unknown parameter $\theta^* \in \mathbb{R}^4$ is monitored through a sensor network consisting of 6 agents, in which the sensors $\{1, 2, 3, 4\}$ are normal measurement nodes and sensors 5 and 6 are relay nodes, i.e., $\mathcal{V}_R = \{5, 6\}$. The communication topology among the sensors is depicted in Fig. 6.6. Here, each sensor is equipped with a local observer, which is not shown. The dashed lines represent the communications among local observers, all of which may be hacked by malicious attackers, while the link attack model satisfies 1-local.

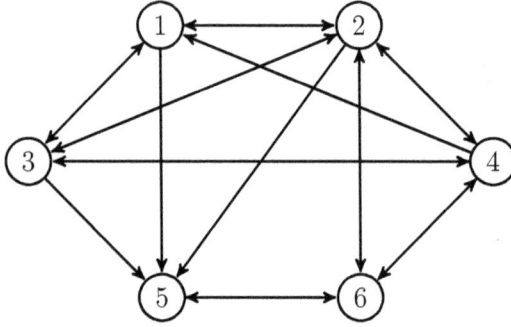

Figure 6.6   Communication topology of sensors network.

The measurement output of each sensor can be expressed as $y_i(t) = C_i \theta^* + v_i(t)$, $i \in [4]$, and $y_i(t) = 0$, $i \in \{5, 6\}$, with $v_i(t) \sim \mathcal{N}[0, 2I_2]$, and measurement matrices of these sensors are given by

$$C_1 = \begin{bmatrix} 1 & 0 & 0 & 0 \\ 0 & 1 & 0 & 0 \end{bmatrix}, \quad C_2 = \begin{bmatrix} 1 & 0 & -1 & 0 \\ 0 & 1 & 0 & -1 \end{bmatrix}, \quad C_3 = \begin{bmatrix} 1 & 0 & 1 & 0 \\ 0 & 1 & 0 & 1 \end{bmatrix},$$

$$C_4 = \begin{bmatrix} 0 & 0 & 1 & 0 \\ 0 & 0 & 0 & 1 \end{bmatrix}, \quad C_5 = C_6 = \mathbf{0}_4^T.$$

It should be noted that each local matrix C_i is also column rank-deficient, while the global matrix C possesses full column-rank. Obviously, the graph $\mathcal{G}$ in Fig. 6.6 together with matrix C meets the conditions of Theorem 6.3 with $F = 1$.

Furthermore, the initial estimation $\hat{\theta}_i(0)$, $i \in [6]$, is randomly generated. By the spatial decomposition of matrix C_i and picking $\mu_i = 8$ and $\tau_i = 3$, the following gain matrices are obtained:

$$K_1 = \begin{bmatrix} 0 & 0 \\ 0 & 0 \\ 8 & 0 \\ 0 & 8 \end{bmatrix}, \quad L_1 = \begin{bmatrix} 3 & 0 \\ 0 & 3 \\ 0 & 0 \\ 0 & 0 \end{bmatrix}, \quad K_2 = \begin{bmatrix} 4 & 0 \\ 0 & 4 \\ 4 & 0 \\ 0 & 4 \end{bmatrix}, \quad L_2 = \begin{bmatrix} 1.5 & 0 \\ 0 & 1.5 \\ -1.5 & 0 \\ 0 & -1.5 \end{bmatrix},$$

$$K_3 = \begin{bmatrix} -4 & 0 \\ 0 & -4 \\ 4 & 0 \\ 0 & 4 \end{bmatrix}, \quad L_3 = \begin{bmatrix} 1.5 & 0 \\ 0 & 1.5 \\ 1.5 & 0 \\ 0 & 1.5 \end{bmatrix}, \quad K_4 = \begin{bmatrix} 8 & 0 \\ 0 & 8 \\ 0 & 0 \\ 0 & 0 \end{bmatrix}, \quad L_4 = \begin{bmatrix} 0 & 0 \\ 0 & 0 \\ 3 & 0 \\ 0 & 3 \end{bmatrix},$$

$$K_5 = K_6 = 8I_4, \quad L_5 = L_6 = \mathbf{0}_4.$$

In this example, each observer is assumed to have one piece of data tampered with by the attacker among the data received from its neighbors at any instant t. It is assumed that the attack that performs data tampering in the form of $w_{j,i}(t) = \mu\hat{\theta}_j(t)$ with μ following the uniform distribution in $[-6, 6]$, and the compromised links are randomly switched every 1 second. Through the process, each local observer adopts Algorithm 6.1 and obtains the simulation results shown in Figs. 6.7–6.9.

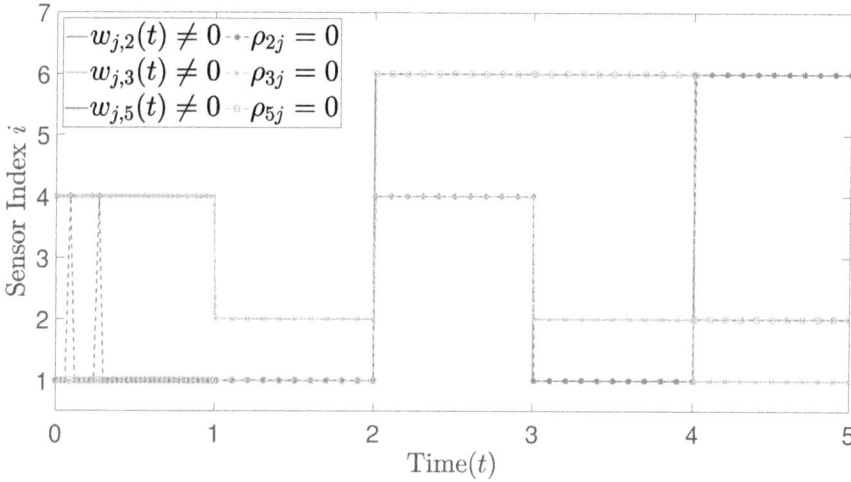

Figure 6.7   The actual attack injection channel and the identified attack channel for nodes $\{2, 3, 5\}$.

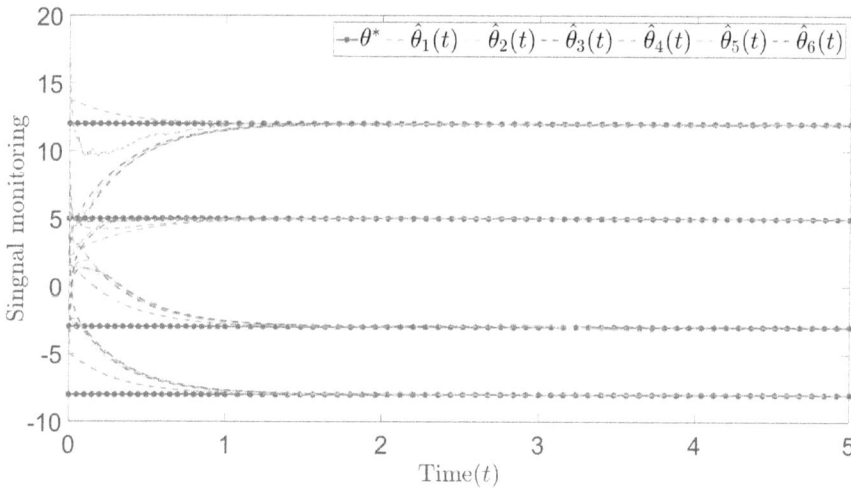

Figure 6.8   Unknown signal θ^* and its estimations under 1-local Byzantine link attacks.

Fig. 6.7 shows the neighbors associated with the actual attack injection links and the severed links at each instant on observers $\{2, 3, 5\}$. It can be seen that, after approximately 1 seconds, nodes 2, 3, and 5 can accurately identify the malicious

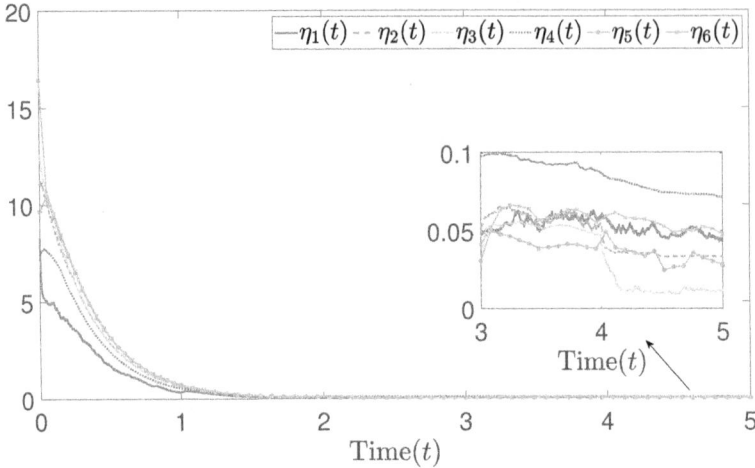

Figure 6.9 The profile of local estimation performance function $\eta_i(t)$ under Algorithm 6.1.

attack location and then isolate it. Similarly, the other nodes can also achieve the same effect.

Fig. 6.8 depicts the monitored parameter state trajectory and the estimation generated by each observer. In addition, the error performance function $\eta_i(t) = \|\hat{\theta}_i(t) - \chi^*\|$, $i \in [6]$, is introduced to measure the estimation performance and the profile of the function $\eta_i(t)$ is shown in Fig. 6.9. From the above, the validity of the distributed SSM Algorithm 4.1 under the F-local Byzantine link attacks is verified.

Example 3: The following simulation shows that the proposed Algorithm 6.1 outperforms the ones in [81, 83] in terms of resilience against F-local Byzantine link attacks under noise-free environment, i.e., $v_i(t) = 0$.

Assume that the malicious attacker chooses the channels $(2, 4)$ and $(6, 5)$ to inject the following false data:

$$w_{2,4}(t) = w_{6,5}(t) = [\sqrt{t}, \ 5\sin(t), \ 0, \ 0]^T.$$

Then, the following three distributed SSM algorithms are executed with respect to the parameter θ^*, respectively,

1) The NMDM-based SSM algorithm 6.1 proposed in this section;

2) The LMSD-based SSM algorithm proposed in [81];

3) The W-MSR-based SSM algorithm given in [83].

Fig. 6.10 shows the curve of the estimation performance function $\eta_i(t)$ with algorithms 1) and 2). It can be seen from Fig. 6.10 that, with the algorithm proposed in [81], the state estimation of nodes $\{4, 5, 6\}$ will be biased by the malicious attack, making local estimation ineffective.

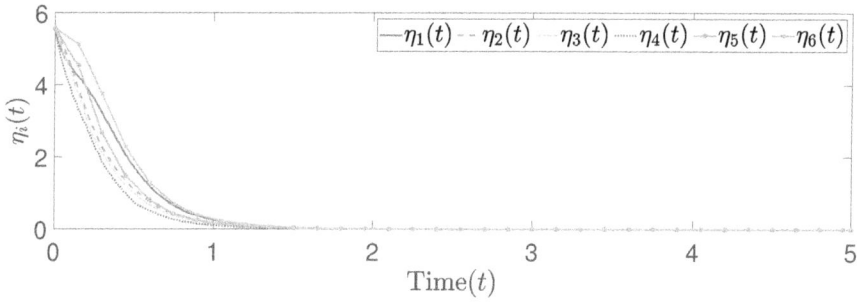

(a) $\eta_i(t)$ under NMDM-based method in 1)

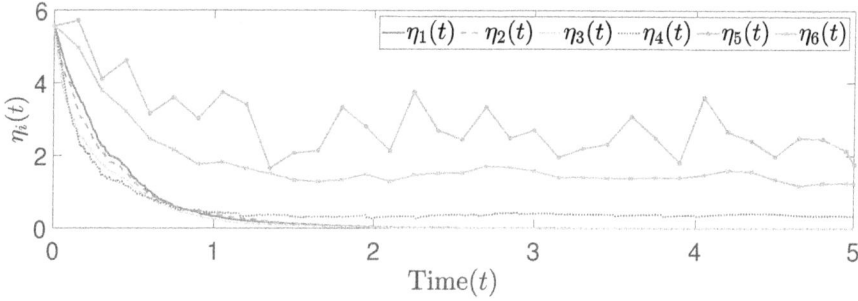

(b) $\eta_i(t)$ under LMSD-based method in 2)

Figure 6.10 The performance of function $\eta_i(t)$ under different distributed SSM algorithms.

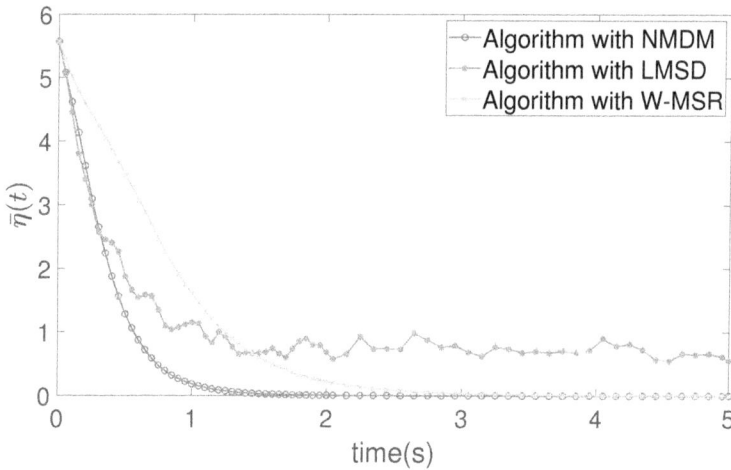

Figure 6.11 The average estimation error norms under the algorithms specified in $(1) - (3)$.

Furthermore, the manifestations of the average performance function $\bar{\eta}(t) = \frac{1}{6}\sum_{i=1}^{6}\eta_i(t)$ obtained by the above three algorithms are depicted in Fig. 6.11. It can be seen that the proposed NDMM algorithm has a higher estimation rate compared to the W-MSR algorithm [83] and is more resilient against attacks than the LMSD algorithm [81].

6.5 CONCLUSIONS

This chapter has solved the distributed security state monitoring problem for sensor networks suffering from F-local Byzantine node/link attacks. A norm-based max-discard mechanism has been proposed to accomplish the attack isolation. Based on this, a non-smooth decentralized control protocol has been constructed for each observer to achieve the distributed SSM over the whole network. A unique feature of the designed protocol is that the benign directions and magnitudes of consensus interactions are governed by the signum function and the absolute function. Finally, two simulation experiments validated the effectiveness and advantages of the proposed algorithms.

Resilient consensus of integrator-type NASs with switching topologies

This chapter studies the resilient consensus of first- and second-order integrator-type NASs. Section 7.1 briefly reviews the previous research findings of resilient consensus problem of NASs. Section 7.2 introduces the notions and properties of joint robustness of time-varying communication graphs. Sections 7.3 and 7.4 present the analysis of resilient consensus for first-order and second-order NASs, respectively. Sections 7.5 and 7.6 provide simulation examples and experiments to illustrate the validity of the theoretical results. Section 7.7 concludes this chapter.

7.1 INTRODUCTION

The distributed state consensus problem of NASs has been at the forefront of systems and control community for more than a decade [10,152,153]. However, existing studies within such a context mainly focus on the cases under various attack-free scenarios. Taking into account the cyber security, the resilient consensus problem of NASs with misbehaving individual agents has recently drawn increasing attention, where the control objective is to achieve state consensus among the normal individuals despite the misbehaving individuals trying to disrupt consensus by applying incorrect control signals or even sending inaccurate information to the normal ones [15,49,68,154,155].

The existing studies on resilient consensus can be roughly divided into two categories. One category is to detect and isolate all misbehaving agents by designing appropriate observers, and then to execute the consensus protocols designed based on the normal neighbors' information (see [15, 68, 154, 155] and references therein). The other is to construct consensus protocols based upon the Mean-Subsequence-Reduced (MSR) algorithms [49, 156, 157] so as to realize resilient consensus. In this chapter, the focus will be on the latter method, as it will yield a more straightforward protocol structure compared to the former.

DOI: 10.1201/9781003669913-7

An analysis framework for resilient consensus of NASs with misbehaving agents was built in [49, 158], where the graph robustness was defined to derive some conditions for realizing resilient consensus among the normal agents under the Weighted-MSR (W-MSR) algorithm. In particular, with the condition of having no more than F malicious agents in the NASs, it was revealed in [49] that resilient consensus of first-order NASs equipped with the W-MSR algorithm can be achieved if and only if the fixed communication graph is $(F + 1, F + 1)$-robust. Based on the notion of graph robustness proposed in [49], resilient consensus of first-order NASs was further investigated in [159], where the inherent dynamics of agents are allowed to switch between continuous-time and discrete-time systems. In [50], resilient consensus for a class of second-order NASs under F-local malicious model was studied by designing a Double-integrator Position-based MSR (DP-MSR) algorithm. Within the framework of the DP-MSR algorithm, it was further manifested in [51] that the fixed $(F+1, F+1)$-robust communication graph is also a necessary and sufficient condition for the achievement of resilient consensus in second-order NASs with no more than F malicious agents. Note that resilient consensus of high-order NASs was considered in [53] while resilient quantized consensus was explored in [160].

The aforementioned results have advanced our knowledge on how to achieve resilient consensus in NASs with malicious agents under fixed communication network. However, it should be pointed out that the underlying communication networks of many engineering NASs need generally be modelled as time-varying graphs [161]. Moreover, although the above-mentioned results employ the interesting notion of graph robustness to seek the criteria of resilient consensus, it also puts forward a high demand for the connectivity of the fixed communication graph. The vast amount of information required to be transmitted via communication channels at every time instant may limit the applications of the MSR algorithm and its variants in practical situations. Within the context of consensus in NASs under time-varying communication graphs and attack-free scenarios, the work in [153] demonstrated that a necessary and sufficient condition for consensus of first-order NASs is that the communication graph jointly has a directed spanning tree.

Considering the existence of misbehaving agents, it is much more difficult to derive necessary and sufficient conditions for resilient consensus of NASs subject to time-varying communication graphs. A few attempts have been made to tackle this issue [49, 53, 55–57, 162, 163]. It was revealed in [49, 53] that resilient consensus can be guaranteed if there exists an infinite subsequence of switching time instants such that the time-varying network satisfies certain robust conditions. The resilience consensus of high-order multi-agent systems under time-varying random graphs was investigated in [162], with the concept of $(2F + 1)$-excess robustness presented to provide sufficient conditions for the realization of the resilient consensus. The resilient tracking consensus of linear multi-agent systems under random graphs was further studied in [163]. In [55], an efficient Sliding-Weighted-MSR (SW-MSR) algorithm was suggested for achieving resilient consensus, where each agent stored and utilized its neighbors' values within T steps, and the notion of the $(T, 2F + 1)$-robust time-varying communication graph was introduced to ensure the resilient consensus under the SW-MSR algorithm. Later on, the results derived in [55] were further extended to resilient consensus of second-order NASs [56] under time-varying communication graphs, and resilient leader-follower consensus under strongly $(T, 2F+1)$-robust time-varying graph [57].

It is worth mentioning that the $(T, 2F + 1)$-robustness only requires that the union of graphs in T steps should be $(2F + 1)$-robust, indicating that the graph at any time instant may not be $(2F + 1)$-robust. Consequently, the shortcoming of the heavy communication burden is overcome by the SW-MSR algorithm. However, to implement the SW-MSR algorithm, it is required that all the agents should gain access to the value of T, which is the global information. Furthermore, since the relaxation of the communication burden comes at the price of the storage of the T-step information, it is desired to use a lower bound of T for a time-varying network, which however is not easy to calculate. Additionally, it is preferable to derive necessary and sufficient criteria for achieving resilient consensus over time-varying graph with misbehaving agents, while only sufficient conditions were reported in the literature.

These observations motivate the study of this chapter, which aims to develop some necessary and sufficient conditions for resilient consensus of NASs with time-varying communication graphs and misbehaving agents under the W-MSR and DP-MSR algorithms. Particularly, the notions of joint reachability and joint robustness are respectively introduced. Based upon the proposed notions, several necessary and sufficient criteria are respectively developed for guaranteeing resilient consensus in first-order and second-order NASs under time-varying communication graphs with misbehaving individuals. In particular, it is proven that resilient consensus under F-total malicious model is realized if and only if the time-varying communication graphs are jointly $(F + 1, F + 1)$-robust; while resilient consensus under F-total or F-local Byzantine model is guaranteed if and only if the time-varying subgraphs consisting of the normal agents are jointly $(F + 1)$-robust.

The main contributions of this study are at least twofold. Firstly, a novel notion of *joint robustness* is proposed to characterize the graph robustness properties for achieving resilient consensus of NASs with time-varying communication graphs. Different from [55] introducing $(T, 2F + 1)$-robustness to derive some sufficient conditions for resilient consensus of NASs with time-varying communication graphs under SW-MSR algorithm, the joint $(F + 1, F + 1)$-robustness herein is a necessary and sufficient condition for the resilient consensus of the F-total malicious time-varying communication graphs under the W-MSR or DP-MSR algorithms. Moreover, the relationship between the $(T, 2F + 1)$-robustness and joint $(2F + 1)$-robustness is clearly discussed. Secondly, compared with [49] focusing on the robustness of fixed communication graphs, the joint robustness of time-varying communication graphs presented in this chapter makes the communication burden substantially declined, thus providing possible opportunities for the applications of distributed resilient consensus in practice. Besides, the proposed notion of *joint robustness* of the time-varying communication graph also makes a contribution to the graph theory as it connects the network robustness of the fixed graph [49] and the time-varying communication graph jointly containing a directed spanning tree [153]. Note that a very preliminary version of this work appeared in [205] and was awarded the Zadeh Best Paper Award.

7.2   JOINT GRAPH ROBUSTNESS

In this section, several definitions of joint robust graph are introduced by extending those on robust network topologies presented in [49]. In the following, time instants are all in the set of natural numbers.

   The definition of the jointly r-reachable set is first given, which is inspired by the concept of the r-reachable set.

Definition 7.1 (Jointly r-reachable set) *For the time-varying network $\mathcal{G}[t] = (\mathcal{V}, \mathcal{E}[t])$, a nonempty subset $\mathcal{D} \subseteq \mathcal{V}$ is called a jointly r-reachable set, if there exists an infinite sequence of bounded time interval (ISBTI) $[t_l, t_{l+1})$ such that in every time interval $[t_l, t_{l+1})$, there exist $T_j \in [t_l, t_{l+1})$ and $i_j \in \mathcal{D}$ such that $|\mathcal{N}_{i_j}[T_j] \backslash \mathcal{D}| \geq r$.*

Remark 7.1 *Recall the definition of r-reachability that a nonempty set $\mathcal{D}$ is r-reachable if there exists one node in $\mathcal{D}$ which has at least r neighbors outside the set $\mathcal{D}$, i.e., $\exists i \in \mathcal{D}, |\mathcal{N}_i \backslash \mathcal{D}| \geq r$. Intuitively, the r-reachability of a set means that there exists at least one node inside the set which can be influenced by enough nodes (the number is no smaller than r) outside the set; while the joint r-reachability only requires that at some time instant, certain node inside the set satisfies such condition. Note that in Definition 7.1, the node i_j can be different in each time interval.*

Note that the subset $\mathcal{D}$ is a finite set, but the number of time intervals is infinite. According to the well-known Pigeonhole Principle [164], one has the following equivalent condition for the joint r-reachability.

Lemma 7.1 *For the time-varying network $\mathcal{G}[t] = (\mathcal{V}, \mathcal{E}[t])$, a nonempty subset $\mathcal{D} \subseteq \mathcal{V}$ is jointly r-reachable, if and only if there exist a node $i \in \mathcal{D}$ and an ISBTI $[t_l, t_{l+1})$ such that in each time interval $[t_l, t_{l+1})$, there exists $T_j \in [t_l, t_{l+1})$ such that $|\mathcal{N}_i[T_j] \backslash \mathcal{D}| \geq r$.*

Proof 7.1 *Only the necessity is shown as the sufficiency is obvious. For a jointly r-reachable subset $\mathcal{D} = \{i_1, \cdots, i_q\}$, by Definition 7.1, there exists an ISBTI $[t'_l, t'_{l+1})$ such that in each time interval $[t'_l, t'_{l+1})$, there exist $T'_j \in [t'_l, t'_{l+1})$ and $i_j \in \mathcal{D}$ such that $|\mathcal{N}_{i_j}[T'_j] \backslash \mathcal{D}| \geq r$. Let*

$$\mathcal{K} = \{[t'_1, t'_2), [t'_2, t'_3), \cdots, [t'_l, t'_{l+1}), \cdots\}.$$

Construct the following q sets $\mathcal{T}_h, h = 1, \cdots, q$ such that

$$\mathcal{T}_h = \{[t^l_h, t^{l'}_h) \in \mathcal{K} \mid \exists T^l_h \in [t^l_h, t^{l'}_h), \; |\mathcal{N}_{i_h}[T^l_h] \backslash \mathcal{D}| \geq r\}.$$

According to the Pigeonhole Principle [164], at least one set contains infinite intervals. Without loss of generality, it is assumed that $\mathcal{T}_1$ contains infinite intervals. Thus, one can conclude that for node $i_1 \in \mathcal{D}$, there exists an ISBTI $[t^l_1, t^{l+1}_1)$ such that in each time interval $[t^l_1, t^{l+1}_1)$, there exists $T^l_1 \in [t^l_1, t^{l+1}_1)$ such that $|\mathcal{N}_{i_1}[T^l_1] \backslash \mathcal{D}| \geq r$. This completes the proof.

Note that Lemma 7.1 indicates that the joint r-reachability requires a node getting influenced by no less than r nodes outside the set for infinite times.

Based on the definition of the jointly r-reachable set, one can present the following definition capturing the redundancy of the time-varying network.

Definition 7.2 (Joint r-robustness) *The time-varying network $\mathcal{G}[t] = (\mathcal{V}, \mathcal{E}[t])$ is said to be jointly r-robust if, for every pair of nonempty disjoint subsets of $\mathcal{V}$, at least one of the subsets is jointly r-reachable.*

Remark 7.2 *Intuitively, to realize the resilient consensus, for any pair of nonempty disjoint node subsets, at least one subset must contain some nodes having enough neighbors outside the set for infinite times. Consequently, a necessary condition for the joint r-robustness is that each node must possess at least r neighbors for infinite times.*

Similar to the discussion of the joint r-reachability, one has the following equivalent condition for the joint r-robustness.

Lemma 7.2 *The time-varying network $\mathcal{G}[t] = (\mathcal{V}, \mathcal{E}[t])$ is jointly r-robust, if and only if for every pair of nonempty disjoint subsets of $\mathcal{V}$, there exists an ISBTI $[t_l, t_{l+1})$ such that in each time interval $[t_l, t_{l+1})$, at least at one time instant, at least one node in these two subsets has at least r neighbors outside the subset which it belongs to.*

Proof 7.2 *Only the sufficiency is shown, since the necessity is straightforward. For each pair of nonempty disjoint subsets $\mathcal{D}_1, \mathcal{D}_2 \subseteq \mathcal{V}$, define two sets $\mathcal{T}_1$ and $\mathcal{T}_2$ by*

$$\mathcal{T}_h = \left\{ [t_h^l, t_h^{l'}) \in \mathcal{K} \mid \exists T_h^l \in [t_h^l, t_h^{l'}), \; \exists i_h^l \in \mathcal{D}_h, \; |\mathcal{N}_{i_h^l}[T_h^l] \backslash \mathcal{D}_h| \geq r \right\}, \; h = 1, 2,$$

where $\mathcal{K} = \{[t_1, t_2), [t_2, t_3), \cdots, [t_l, t_{l+1}), \cdots\}$. According to the Pigeonhole Principle [164], at least one set contains infinite intervals. Without loss of generality, assume that $\mathcal{T}_1$ contains infinite intervals. One then obtains that there exists an ISBTI $[t_1^l, t_1^{l+1})$ such that in each time interval, there exist $T_1^l \in [t_1^l, t_1^{l+1})$ and $i_1^l \in \mathcal{D}_1$ such that $|\mathcal{N}_{i_1^l}[T_1^l] \backslash \mathcal{D}_1| \geq r$. That is, $\mathcal{D}_1$ is jointly r-reachable. The proof is complete.

Another equivalent condition for the joint r-robustness can be formulated in the following result.

Lemma 7.3 *The time-varying network $\mathcal{G}[t] = (\mathcal{V}, \mathcal{E}[t])$ is jointly r-robust, if and only if there exists an ISBTI $[t_l, t_{l+1})$ such that for every pair of nonempty disjoint subsets of $\mathcal{V}$, at least at one time instant of each time interval $[t_l, t_{l+1})$, at least one node in these two subsets has at least r neighbors outside the subset which it belongs to.*

Proof 7.3 *The sufficiency is straightforward in light of Lemma 7.2. So only the necessity is shown. It is not difficult to calculate that the total number of different pairs of nonempty disjoint subsets of $\mathcal{V}$ is $\frac{3^N+1}{2} - 2^N$. Since the time-varying network is jointly r-robust, for each pair of nonempty disjoint subsets $\mathcal{D}_1^q, \mathcal{D}_2^q, q = 1, \cdots, \frac{3^N+1}{2} -$*

2^N, there exists an ISBTI $[t_l^q, t_{l+1}^q)$ such that at least at one time instant of each time interval $[t_l^q, t_{l+1}^q)$, at least one node in $\mathcal{D}_1^q \cup \mathcal{D}_2^q$ has at least r neighbors outside the subset which it belongs to. Then one can construct an ISBTI $[t_l, t_{l+1})$ such that for all $q = 1, \cdots, \frac{3^N+1}{2} - 2^N$ and each time interval $[t_l, t_{l+1})$, there exists q_j with $[t_{q_j}^q, t_{q_j+1}^q) \in [t_l, t_{l+1})$.

Remark 7.3 *Lemma 7.3 reveals that for a jointly r-robust network, the common ISBTI can be found for every pair of nonempty disjoint subsets.*

Then, lets recall the definition of (T, r)-robustness in [55].

Definition 7.3 ((T, r)-robustness [55]) *The time-varying network $\mathcal{G}[t]$ is said to be (T, r)-robust if $\bigcup_{\tau=0}^{T} \mathcal{G}[t - \tau]$ is r-robust for $t \geq T$.*

The following result is provided to clarify the relationship between the joint r-robustness and the (T, r)-robustness.

Lemma 7.4 *If the time-varying network $\mathcal{G}[t]$ is jointly r-robust, there exists T such that $\mathcal{G}[t]$ is (T, r)-robust.*

Remark 7.4 *Lemma 7.4 indicates that the joint r-robustness is a sufficient condition for the (T, r)-robustness with sufficiently large T. Specifically, T can be chosen by $T \geq 2\bar{T}$ with $\bar{T}$ being the maximum length of time intervals $[t_l, t_{l+1})$ in the definition of the joint r-robustness.*

Now, the joint (r, s)-reachability is introduced to describe the minimum number of nodes that can be influenced by no less than r nodes outside the set with infinite times.

Definition 7.4 (Jointly (r, s)-reachable set) *For the time-varying network $\mathcal{G}[t] = (\mathcal{V}, \mathcal{E}[t])$, a nonempty subset $\mathcal{D} \subseteq \mathcal{V}$ is called a jointly (r, s)-reachable set if there exists an ISBTI $[t_l, t_{l+1})$ such that in each time interval $[t_l, t_{l+1})$, there holds $|\Phi_{\mathcal{D}}^r[t_l, t_{l+1})| \geq s$ with*

$$\Phi_{\mathcal{D}}^r[t_l, t_{l+1}) = \{i \in \mathcal{D} \mid \exists T_{i_j} \in [t_l, t_{l+1}), \ s.t. \ |\mathcal{N}_i[T_{i_j}] \backslash \mathcal{D}| \geq r\}.$$

Remark 7.5 *The joint (r, s)-reachability means that there exist infinite time intervals such that in each interval, at least s nodes inside the set can be influenced by no less than r nodes outside the set at least at one time instant. Note that for each time interval $[t_l, t_{l+1})$, the subset $\Phi_{\mathcal{D}}^r[t_l, t_{l+1})$ can be distinct, and for each node i, the time instant T_{i_j} can be different.*

Since the number of time intervals is infinite while the subset number of $\mathcal{D}$ is finite, one has the following equivalent condition for the joint (r, s)-reachability.

Lemma 7.5 *For the time-varying network $\mathcal{G}[t] = (\mathcal{V}, \mathcal{E}[t])$, the nonempty subset $\mathcal{D} \subseteq \mathcal{V}$ is jointly (r, s)-reachable, if there exists a subset $\Phi_{\mathcal{D}}^r$ of $\mathcal{D}$ with no less than s nodes such that there exists an ISBTI $[t_l, t_{l+1})$, for each node $i \in \Phi_{\mathcal{D}}^r$, there exists $T_{i_j} \in [t_l, t_{l+1})$ such that $|\mathcal{N}_i[T_{i_j}] \backslash \mathcal{D}| \geq r$.*

Proof 7.4 *Only the necessity is shown below since the sufficiency is obvious. Let Ψ be the total number of subsets of $\mathcal{D}$ with no less than s nodes, one has*

$$\Psi = \sum_{k=s}^{|\mathcal{D}|} \frac{|\mathcal{D}|!}{k!(|\mathcal{D}| - k)!}.$$

Let these Ψ subsets be denoted by $\Phi_1, \cdots, \Phi_\Psi$. Define

$$\mathcal{T}_h = \{[t_h^l, t_h^{l'}) \in \mathcal{K} \mid \forall i \in \Phi_h, \ \exists T_i^l \in [t_h^l, t_h^{l'}),$$
$$s.t. \ |\mathcal{N}_i[T_i^l]\backslash\mathcal{D}| \geq r\}, \quad h = 1, \cdots, \Psi,$$

with $\mathcal{K} = \{[t_1, t_2), \cdots, [t_l, t_{l+1}), \cdots\}$ being the set of the infinite time intervals. It is clear that at least one set among $\mathcal{T}_1, \cdots, \mathcal{T}_\Psi$ contains infinite time intervals. Without loss of generality, assume that $\mathcal{T}_1$ contains infinite time intervals, i.e., there is an ISBTI $[t_1^l, t_1^{l+1})$ such that for each node $i \in \Phi_1$, there exists $T_i^l \in [t_1^l, t_1^{l+1})$ such that $|\mathcal{N}_i[T_i^l]\backslash\mathcal{D}| \geq r$. This completes the proof by noticing that $|\Phi_1| \geq s$.

Remark 7.6 *Lemma 7.5 implies that for a jointly (r, s)-reachable set $\mathcal{D}$ of Definition 7.4, there is at least one subset that appears infinite times in the sequence of subsets $\Phi_\mathcal{D}^r[t_l, t_{l+1}), l = 1, 2, \cdots$.*

The following lemma is a straightforward property of the joint (r, s)-reachability.

Lemma 7.6 *For the time-varying network $\mathcal{G}[t] = (\mathcal{V}, \mathcal{E}[t])$, the nonempty subset $\mathcal{D} \subseteq \mathcal{V}$ is jointly (r', s')-reachable for $1 \leq r' \leq r, 1 \leq s' \leq s$, if it is jointly (r, s)-reachable. Moreover, the joint r-reachability is equal to the joint $(r, 1)$-reachability.*

Now, it is ready to introduce the notion of joint (r, s)-robustness, which plays a vital role in our main results.

Definition 7.5 (Joint (r, s)-robustness) *The time-varying network $\mathcal{G}[t] = (\mathcal{V}, \mathcal{E}[t])$ is said to be jointly (r, s)-robust if, for every pair of nonempty disjoint subsets $\mathcal{D}_1, \mathcal{D}_2 \subseteq \mathcal{V}$, there exists an ISBTI $[t_l, t_{l+1})$ such that in each time interval, at least one of the following conditions holds:*

1) $|\Phi_{\mathcal{D}_1}^r[t_l, t_{l+1})| = |\mathcal{D}_1|$;

2) $|\Phi_{\mathcal{D}_2}^r[t_l, t_{l+1})| = |\mathcal{D}_2|$;

3) $|\Phi_{\mathcal{D}_1}^r[t_l, t_{l+1})| + |\Phi_{\mathcal{D}_2}^r[t_l, t_{l+1})| \geq s$.

Remark 7.7 *Intuitively, the joint (r, s)-robustness means that there are enough nodes in every pair of nonempty disjoint subsets having at least r neighbors outside their own subsets for infinite times. Clearly, the (r, s)-robustness proposed in [49] is a special case of the joint (r, s)-robustness.*

For the joint (r, s)-robustness, the following equivalence conditions are available, which indicates that for a jointly (r, s)-robust network, a common ISBTI can be found for every pair of nonempty disjoint subsets.

Lemma 7.7 *The time-varying network $\mathcal{G}[t] = (\mathcal{V}, \mathcal{E}[t])$ is jointly (r, s)-robust, if there exists an ISBTI $[t_l, t_{l+1})$ such that for every pair of nonempty disjoint subsets $\mathcal{D}_1$, $\mathcal{D}_2 \subseteq \mathcal{V}$, at least one of the following conditions holds:*

1) $|\Phi_{\mathcal{D}_1}^r[t_l, t_{l+1})| = |\mathcal{D}_1|$;

2) $|\Phi_{\mathcal{D}_2}^r[t_l, t_{l+1})| = |\mathcal{D}_2|$;

3) $|\Phi_{\mathcal{D}_1}^r[t_l, t_{l+1})| + |\Phi_{\mathcal{D}_2}^r[t_l, t_{l+1})| \geq s$.

Proof 7.5 *Only the necessity is shown below as the sufficiency is obvious. Similar to the proof in Lemma 7.3, for each pair of nonempty disjoint subsets $\mathcal{D}_1^q, \mathcal{D}_2^q, q = 1, \cdots, \frac{3^N+1}{2} - 2^N$, there exists an ISBTI $[t_l^q, t_{l+1}^q)$ such that at least one of the following three conditions holds:*

1) $|\Phi_{\mathcal{D}_1^q}^r[t_l^q, t_{l+1}^q)| = |\mathcal{D}_1^q|$;

2) $|\Phi_{\mathcal{D}_2^q}^r[t_l^q, t_{l+1}^q)| = |\mathcal{D}_2^q|$;

3) $|\Phi_{\mathcal{D}_1^q}^r[t_l^q, t_{l+1}^q)| + |\Phi_{\mathcal{D}_2^q}^r[t_l^q, t_{l+1}^q)| \geq s$.

Since each time interval $[t_l^q, t_{l+1}^q)$ is bounded, one can construct an ISBTI $[t_l, t_{l+1})$ such that, $\forall q = 1, \cdots, \frac{3^N+1}{2} - 2^N$, there exists q_l such that $[t_{q_l}^q, t_{q_l+1}^q) \in [t_l, t_{l+1})$. This completes the proof.

The following lemma is straightforward by noting that $\Phi_{\mathcal{D}}^{r'}[t_l, t_{l+1}) \subset \Phi_{\mathcal{D}}^r[t_l, t_{l+1})$, $\forall r' \leq r$.

Lemma 7.8 *The time-varying network $\mathcal{G}[t] = (\mathcal{V}, \mathcal{E}[t])$ is jointly (r', s')-robust for $1 \leq r' \leq r, 1 \leq s' \leq s$, if it is jointly (r, s)-robust. Moreover, the joint r-robustness is equal to the joint $(r, 1)$-robustness.*

The following lemma characterizes the relationship between the joint (r, s)-robustness and the joint spanning tree.

Lemma 7.9 *The time-varying network $\mathcal{G}[t] = (\mathcal{V}, \mathcal{E}[t])$ jointly contains a directed spanning tree, if and only if $\mathcal{G}[t]$ is jointly $(1, 1)$-robust.*

Proof 7.6 *The proof is divided into the following two parts.*

Necessity. According to [153], $\mathcal{G}[t]$ jointly contains a directed spanning tree if and only if there exists an ISBTI $[t_l, t_{l+1})$ such that in each time interval, the union of the graphs contains a directed spanning tree. Then for every pair of nonempty disjoint subsets $\mathcal{D}_1, \mathcal{D}_2 \subseteq \mathcal{V}$, there must be a node $i \in \mathcal{D}_1$ ($i \in \mathcal{D}_2$) having at least one neighbor outside $\mathcal{D}_1$ ($\mathcal{D}_2$) for at least one time in each time interval $[t_l, t_{l+1})$; otherwise, the union of the graphs cannot contain a directed spanning tree.

Sufficiency. If $\mathcal{G}[t]$ does not jointly contain a directed spanning tree, then for $t \geq \bar{t}$, the node set $\mathcal{V}$ can be divided into two nonempty disjoint subsets, $\mathcal{D}_1, \mathcal{D}_2 \subseteq \mathcal{V}$ such that, for each node $i \in \mathcal{D}_1$ ($i \in \mathcal{D}_2$), there holds $\mathcal{N}_i \cap \mathcal{D}_2 = \emptyset$ ($\mathcal{N}_i \cap \mathcal{D}_1 = \emptyset$). This implies that the network $\mathcal{G}[t]$ is not jointly $(1, 1)$-robust, which is a contradiction.

Remark 7.8 *It was proven in [153] that the consensus of the first-order NASs can be achieved if and only if the time-varying network jointly contains a directed spanning tree. This means that for the first-order 0-total malicious NAS, the resilient consensus can be realized under the W-MSR algorithm if and only if the time-varying network is jointly $(1, 1)$-robust. In the following section, some necessary and sufficient criteria for resilient consensus of first-order NASs with time-varying communication graphs will be derived under the F-total malicious model.*

7.3 RESILIENT CONSENSUS OF FIRST-ORDER NASS WITH DIRECTED SWITCHING TOPOLOGIES

In this section, the resilient consensus of first-order NASs with time-varying communication graphs will be studied.

7.3.1 Problem formulation

The dynamics of the i-th agent within the first-order NAS are described as

$$p_i[t + 1] = p_i[t] + u_i[t], \ i = 1, \cdots, N, \tag{7.1}$$

in which $p_i[t]$ and $u_i[t]$ are respectively the state and the control input of agent i at time instant t.

Before moving forward, the W-MSR algorithm is recalled below [49]:

1) At every time step t, each normal agent i receives the values of its neighbors $p_j[t]$, $j \in \mathcal{N}_i$, and sorts the relative values $p_j[t] - p_i[t]$ in an increasing order.

2) Each normal agent i removes the smallest F negative values $p_j[t] - p_i[t]$ and the largest F positive values $p_j[t] - p_i[t]$. Otherwise, it removes all the negative (positive) values if there are less than F negative (positive) ones. Denote by $\mathcal{S}_i[t]$ the removed neighboring set of agent i.

3) Each normal agent i generates its control protocol as

$$u_i[t] = \sum_{j=1}^{N} a_{ij}[t](p_j[t] - p_i[t]), \tag{7.2}$$

where $a_{ij}[t] > 0$ if $j \in \mathcal{N}_i[t]\backslash\mathcal{S}_i[t]$; otherwise, $a_{ij}[t] = 0$. Particularly, $\sum_{j=1}^{N} a_{ij}[t] < 1$.

Remark 7.9 *The prominent characteristic of the W-MSR algorithm is that each normal agent disregards enough extreme values deviated from its own value to alleviate the influence from misbehaving agents. For the 0-total malicious model, i.e., the case that all the agents are normal, no information of neighboring agents is removed, and the controller (7.2) is exactly the protocol given in [153].*

In what follows, the resilient consensus problem to be handled in this section is formulated.

Definition 7.6 (Resilient consensus [51]) *The NAS is said to realize resilient consensus if the following two statements hold for all initial values of nodes and any possible misbehaving node set.*

i) (Consensus) All the normal nodes satisfy

$$\lim_{t \to \infty} (p_i[t] - p_j[t]) = 0, \; \forall i, j \in \mathcal{N},$$

where $\mathcal{N} = \mathcal{V} \backslash \mathcal{V}_A$ denotes the set of normal agent nodes with $\mathcal{V}_A$ is the set of compromised agent nodes.

ii) (Safety) The values of the normal nodes stay within a bounded, invariant interval Υ for all time instants.

7.3.2 Resilient consensus for F-total malicious model

Let

$$\underline{p}[t] = \min(p_i[t]), \; \overline{p}[t] = \max(p_i[t]), \; \forall i \in \mathcal{N}.$$

The following result is provided for the resilient consensus of first-order time-varying NASs with the F-total malicious model under the W-MSR algorithm.

Theorem 7.1 *For the F-total malicious model, the resilient consensus of the first-order NAS with time-varying communication graph $\mathcal{G}[t]$ under the W-MSR algorithm is realized if and only if $\mathcal{G}[t]$ is jointly $(F+1, F+1)$-robust. Moreover, the safety interval is given by $\Upsilon = [\underline{p}[0], \overline{p}[0]]$.*

Proof 7.7 Necessity. *Prove it by contradiction. If $\mathcal{G}[t]$ is not jointly $(F+1, F+1)$-robust, then there are nonempty disjoint subsets $\mathcal{D}_1, \mathcal{D}_2 \subseteq \mathcal{V}$ such that for $t \geq \hat{t}$, the following three conditions hold:*

1) $|\Phi_{\mathcal{D}_1}^{F+1}[\hat{t}, \infty)| < |\mathcal{D}_1|$;

2) $|\Phi_{\mathcal{D}_2}^{F+1}[\hat{t}, \infty)| < |\mathcal{D}_2|$;

3) $|\Phi_{\mathcal{D}_1}^{F+1}[\hat{t}, \infty)| + |\Phi_{\mathcal{D}_2}^{F+1}[\hat{t}, \infty)| \leq F$.

Suppose that $p_i[0] = a, \forall i \in \mathcal{D}_1$ and $p_j[0] = b, \forall j \in \mathcal{D}_2$ with $a < b$. By Conditions 3), assume that all the agents in $\Phi_{\mathcal{D}_1}^{F+1}[\hat{t}, \infty) \cup \Phi_{\mathcal{D}_2}^{F+1}[\hat{t}, \infty)$ are malicious, with their values remaining unchanged. Besides, assume that the agents in $\mathcal{D}_1 \backslash \Phi_{\mathcal{D}_1}^{F+1}[\hat{t}, \infty)$ and $\mathcal{D}_2 \backslash \Phi_{\mathcal{D}_2}^{F+1}[\hat{t}, \infty)$ are normal. By Conditions 1) and 2), one has that

$$\mathcal{D}_1 \backslash \Phi_{\mathcal{D}_1}^{F+1}[\hat{t}, \infty) \neq \emptyset, \; \mathcal{D}_2 \backslash \Phi_{\mathcal{D}_2}^{F+1}[\hat{t}, \infty) \neq \emptyset.$$

There are at most F neighbors with values different from a for any node $i \in \mathcal{D}_1 \cap \mathcal{N}$, which implies that $p_j[t] = a, \forall j \in \{i\} \cup \mathcal{N}_i[t] \backslash \mathcal{S}_i[t], \forall t \geq \hat{t}$. Thus, one has that $p_i[t] = a, \forall i \in \mathcal{D}_1 \cap \mathcal{N}, \forall t \geq \hat{t}$. Similarly, one has that $p_j[t] = b, \forall j \in \mathcal{D}_2 \cap \mathcal{N}, \forall t \geq \hat{t}$. This implies that the resilient consensus cannot be realized.

Sufficiency. *The first provides safety conditions. Substituting (7.2) into (7.1) yields*

$$p_i[t+1] = \sum_{j=1}^{N} \bar{a}_{ij}[t]p_j[t] \tag{7.3}$$

where $\bar{a}_{ij}[t] = a_{ij}[t]$ if $i \neq j$ and $\bar{a}_{ii}[t] = 1 - \sum_{j=1}^{N} a_{ij}[t]$ if $i = j$. Note that $\bar{a}_{ij}[t] \geq 0$ means that for each normal agent i, $p_i[t+1]$ lies in the convex combination of the values of itself and its neighbors in the set $\mathcal{N}_i[t]\backslash\mathcal{S}_i[t]$. Since $p_j[t] \in [\underline{p}[t], \overline{p}[t]]$, $j \in \{i\} \cup \mathcal{N}_i[t]\backslash\mathcal{S}_i[t]$, $i \in \mathcal{N}$, one has that $p_i[t+1] \in [\underline{p}[t], \overline{p}[t]]$, $i \in \mathcal{N}$, implying that

$$[\underline{p}[0], \overline{p}[0]] \supset [\underline{p}[1], \overline{p}[1]] \supset \cdots \supset [\underline{p}[t], \overline{p}[t]] \supset \cdots .$$

Then, one can conclude that for each normal agent i, $p_i[t] \in \Upsilon$ with $\Upsilon = [\underline{p}[0], \overline{p}[0]]$.

Next, the achievement of the resilient consensus is presented. Since both $\overline{p}[t]$ and $\underline{p}[t]$ are monotonic and bounded, their limits are denoted by $\overline{P}$ and $\underline{P}$, respectively. Clearly, the resilient consensus is realized if and only if $\overline{P} = \underline{P}$. Next, a contradiction is used to prove the point.

Suppose that $\overline{P} > \underline{P}$. Then there is a constant $\varepsilon_0 > 0$ such that $\overline{P} - \varepsilon_0 > \underline{P} + \varepsilon_0$. Since $\mathcal{G}[t]$ is jointly $(F+1, F+1)$-robust, there exists an ISBTI $[t_l, t_{l+1})$ such that for every pair of nonempty disjoint subsets $\mathcal{D}_1$, $\mathcal{D}_2 \subseteq \mathcal{V}$, at least one of the following conditions holds:

1) $|\Phi_{\mathcal{D}_1}^{F+1}[t_l, t_{l+1})| = |\mathcal{D}_1|$;

2) $|\Phi_{\mathcal{D}_2}^{F+1}[t_l, t_{l+1})| = |\mathcal{D}_2|$;

3) $|\Phi_{\mathcal{D}_1}^{F+1}[t_l, t_{l+1})| + |\Phi_{\mathcal{D}_2}^{F+1}[t_l, t_{l+1})| \geq F + 1$.

Let $\delta \in (0, \frac{1}{2})$ be the constant satisfying

$$\bar{a}_{ij}[t] \geq \delta, \ \forall t \geq 0, \forall i, \forall j \in \{i\} \cup \mathcal{N}_i[t]\backslash\mathcal{S}_i[t].$$

Denote by T the maximum length of time intervals $[t_l, t_{l+1})$, and choose

$$\varepsilon = \frac{\delta^{N_0 T + 1}}{1 - \delta^{N_0 T + 1}} \varepsilon_0 < \varepsilon_0,$$

where $N_0 \geq N - F$ is the number of normal nodes. Let t_q be the finite time instant such that $\overline{p}[t] < \overline{P} + \varepsilon$ and $\underline{p}[t] > \underline{P} - \varepsilon, \forall t \geq t_q$. It is not difficult to verify

$$0 < \varepsilon < \frac{\delta^{t_{N_0+q} - t_q}}{1 - \delta^{t_{N_0+q} - t_q}} \varepsilon_0 < \varepsilon_0.$$

In the following, a strictly monotone decreasing sequence $\{\varepsilon_h\}$ and two node subsets sequences $\{\mathcal{Z}_1(t_q + h, \varepsilon_h)\}, \{\mathcal{Z}_2(t_q + h, \varepsilon_h)\}$ will be firstly constructed, and then demonstrate that the total number of normal nodes in $\mathcal{Z}_1(t_l, \varepsilon_{t_l - t_q}) \cup \mathcal{Z}_2(t_l, \varepsilon_{t_l - t_q})$ is strictly larger than that of normal nodes in $\mathcal{Z}_1(t_{l+1}, \varepsilon_{t_{l+1} - t_q}) \cup \mathcal{Z}_2(t_{l+1}, \varepsilon_{t_{l+1} - t_q})$, which will further lead to the contradiction.

The strictly monotone decreasing sequence $\{\varepsilon_h\}$ is chosen as

$$\varepsilon_{h+1} = \delta\varepsilon_h - (1-\delta)\varepsilon, \quad h = 0, \cdots, t_{N_0+q} - t_q - 1.$$

Noting that

$$\varepsilon_{t_{N_0+q}-t_q} = \delta^{t_{N_0+q}-t_q}\varepsilon_0 - \sum_{h=0}^{t_{N_0+q}-t_q-1} \delta^h(1-\delta)\varepsilon$$

$$= \delta^{t_{N_0+q}-t_q}\varepsilon_0 - (1 - \delta^{t_{N_0+q}-t_q})\varepsilon$$

$$> 0,$$

one has $\varepsilon_h > 0$ and $\overline{P} - \varepsilon_h > \underline{P} + \varepsilon_h$ for all h.

Let $\mathcal{Z}_1(t_q + h, \varepsilon_h)$, $\mathcal{Z}_2(t_q + h, \varepsilon_h) \subseteq \mathcal{V}$ with nodes of value larger than $\overline{P} - \varepsilon_h$ and smaller than $\underline{P} + \varepsilon_h$ at time instant $t_q + h$, i.e.,

$$
\begin{aligned}
\mathcal{Z}_1(t_q + h, \varepsilon_h) &= \{i \in \mathcal{V} | p_i[t_q + h] > \overline{P} - \varepsilon_h\}, \\
\mathcal{Z}_2(t_q + h, \varepsilon_h) &= \{i \in \mathcal{V} | p_i[t_q + h] < \underline{P} + \varepsilon_h\}.
\end{aligned}
\tag{7.4}
$$

It is clear that

$$\mathcal{Z}_1(t_q + h, \varepsilon_h) \cap \mathcal{Z}_2(t_q + h, \varepsilon_h) = \emptyset, \quad \forall h.$$

Then, it is shown that

$$
\begin{aligned}
&\left| \left(\mathcal{Z}_1(t_l, \varepsilon_{t_l-t_q}) \cup \mathcal{Z}_2(t_l, \varepsilon_{t_l-t_q})\right) \cap \mathcal{N} \right| \\
&> \left| \left(\mathcal{Z}_1(t_{l+1}, \varepsilon_{t_{l+1}-t_q}) \cup \mathcal{Z}_2(t_{l+1}, \varepsilon_{t_{l+1}-t_q})\right) \cap \mathcal{N} \right|.
\end{aligned}
\tag{7.5}
$$

First, it is shown that for all h,

$$
\begin{aligned}
\{\mathcal{Z}_1(t_q + h, \varepsilon_h) \cap \mathcal{N}\} &\supset \{\mathcal{Z}_1(t_q + h + 1, \varepsilon_{h+1}) \cap \mathcal{N}\}, \\
\{\mathcal{Z}_2(t_q + h, \varepsilon_h) \cap \mathcal{N}\} &\supset \{\mathcal{Z}_2(t_q + h + 1, \varepsilon_{h+1}) \cap \mathcal{N}\}.
\end{aligned}
$$

The normal agents can be divided into five disjoint subsets at time instant $t_q + h$:

$$
\begin{aligned}
\mathcal{Y}_1(t_q + h, \varepsilon_h) &= \{i \in \mathcal{Z}_1(t_q + h, \varepsilon_h) \cap \mathcal{N} \mid |\mathcal{N}_i[t_q + h] \backslash \mathcal{Z}_1(t_q + h, \varepsilon_h)| \geq F + 1\}, \\
\mathcal{Y}_2(t_q + h, \varepsilon_h) &= \{\mathcal{Z}_1(t_q + h, \varepsilon_h) \cap \mathcal{N}\} \backslash \mathcal{Y}_1(t_q + h, \varepsilon_h), \\
\mathcal{Y}_3(t_q + h, \varepsilon_h) &= \{i \in \mathcal{Z}_2(t_q + h, \varepsilon_h) \cap \mathcal{N} \mid |\mathcal{N}_i[t_q + h] \backslash \mathcal{Z}_2(t_q + h, \varepsilon_h)| \geq F + 1\}, \\
\mathcal{Y}_4(t_q + h, \varepsilon_h) &= \{\mathcal{Z}_2(t_q + h, \varepsilon_h) \cap \mathcal{N}\} \backslash \mathcal{Y}_3(t_q + h, \varepsilon_h), \\
\mathcal{Y}_5(t_q + h, \varepsilon_h) &= \mathcal{N} \backslash \{(\mathcal{Z}_1(t_q + h, \varepsilon_h) \cup \mathcal{Z}_2(t_q + h, \varepsilon_h)) \cap \mathcal{N}\}.
\end{aligned}
$$

For any agent $i \in \mathcal{Y}_1(t_q + h, \varepsilon_h)$, there is at least one neighbor satisfying

$$p_j[t_q + h] \leq \overline{P} - \varepsilon_h, \quad j \in \mathcal{N}_i[t_q + h] \setminus \mathcal{R}_i[t_q + h]$$

and one has

$$
\begin{aligned}
p_i[t_q + h + 1] &\leq \delta(\overline{P} - \varepsilon_h) + (1-\delta)\overline{p}[t_q + h] \\
&< \delta(\overline{P} - \varepsilon_h) + (1-\delta)(\overline{P} + \varepsilon) \\
&= \overline{P} - \varepsilon_{h+1},
\end{aligned}
$$

which implies that

$$\mathcal{Y}_1(t_q + h, \varepsilon_h) \cap \mathcal{Z}_1(t_q + h + 1, \varepsilon_{h+1}) = \emptyset.$$

Besides, since $p_j[t_q + h] \geq \underline{p}[t_q + h], j \in \mathcal{N}_i[t_q + h] \backslash \mathcal{S}_i[t_q + h],$ *one obtains*

$$\begin{aligned} p_i[t_q + h + 1] &> \delta(\overline{P} - \varepsilon_h) + (1 - \delta)\underline{p}[t_q + h] \\ &> \delta(\underline{P} + \varepsilon_h) + (1 - \delta)(\underline{P} - \varepsilon) \\ &= \underline{P} + \varepsilon_{h+1}, \end{aligned}$$

which means that

$$\mathcal{Y}_1(t_q + h, \varepsilon_h) \cap \mathcal{Z}_2(t_q + h + 1, \varepsilon_{h+1}) = \emptyset.$$

Thus,

$$\mathcal{Y}_1(t_q + h, \varepsilon_h) \subset \mathcal{Y}_5(t_q + h + 1, \varepsilon_{h+1}).$$

For any agent $i \in \mathcal{Y}_2(t_q + h, \varepsilon_h),$ *the neighbors with* $p_j[t_q + h] \leq \overline{P} - \varepsilon_h, j \in \mathcal{N}_i[t_q + h]$ *are removed, and one has*

$$p_i[t_q + h + 1] > \overline{P} - \varepsilon_h > \overline{P} + \varepsilon_{h+1}.$$

Thus,

$$\mathcal{Y}_2(t_q + h, \varepsilon_h) \cap \mathcal{Z}_2(t_q + h + 1, \varepsilon_{h+1}) = \emptyset.$$

Similar to the above analysis, one can also obtain

$$\mathcal{Y}_3(t_q + h, \varepsilon_h) \subset \mathcal{Y}_5(t_q + h + 1, \varepsilon_{h+1})$$

and

$$\mathcal{Y}_4(t_q + h, \varepsilon_h) \cap \mathcal{Z}_1(t_q + h + 1, \varepsilon_{h+1}) = \emptyset.$$

For any agent $i \in \mathcal{Y}_5(t_q + h, \varepsilon_h),$ *with* $p_i[t_q + h] \geq \underline{p}[t_q + h],$ *one has*

$$p_i[t_q + h + 1] \leq \delta(\overline{P} - \varepsilon_h) + (1 - \delta)\overline{p}[t_q + h] < \underline{P} - \varepsilon_{h+1}$$

and

$$p_i[t_q + h + 1] \geq \delta(\underline{P} + \varepsilon_h) + (1 - \delta)\underline{p}[t_q + h] > \underline{P} + \varepsilon_{h+1}.$$

Thus,

$$\mathcal{Y}_5(t_q + h, \varepsilon_h) \subset \mathcal{Y}_5(t_q + h + 1, \varepsilon_{h+1}).$$

It is derived from the above analysis that the normal agents in $\mathcal{Z}_1(t_q + h, \varepsilon_h)$ would be in $\mathcal{V} \setminus \mathcal{Z}_2(t_q + h + 1, \varepsilon_{h+1})$, the normal agents in $\mathcal{Z}_2(t_q + h, \varepsilon_h)$ would lie in $\mathcal{V} \setminus \mathcal{Z}_1(t_q + h + 1, \varepsilon_{h+1})$, and the normal agents in

$$\mathcal{V} \setminus \{\mathcal{Z}_1(t_q + h, \varepsilon_h) \cup \mathcal{Z}_2(t_q + h, \varepsilon_h)\}$$

would still stay in

$$\mathcal{V} \setminus \{\mathcal{Z}_1(t_q + h + 1, \varepsilon_{h+1}) \cup \mathcal{Z}_2(t_q + h + 1, \varepsilon_{h+1})\}.$$

Therefore, it is concluded that

$$\{\mathcal{Z}_1(t_q + h, \varepsilon_h) \cap \mathcal{N}\} \supset \{\mathcal{Z}_1(t_q + h + 1, \varepsilon_{h+1}) \cap \mathcal{N}\}$$
$$\{\mathcal{Z}_2(t_q + h, \varepsilon_h) \cap \mathcal{N}\} \supset \{\mathcal{Z}_2(t_q + h + 1, \varepsilon_{h+1}) \cap \mathcal{N}\}$$

and thus

$$\{(\mathcal{Z}_1(t_l, \varepsilon_{t_l - t_q}) \cup \mathcal{Z}_2(t_l, \varepsilon_{t_l - t_q})) \cap \mathcal{N}\}$$
$$\supset \{(\mathcal{Z}_1(t_{l+1}, \varepsilon_{t_{l+1} - t_q}) \cup \mathcal{Z}_2(t_{l+1}, \varepsilon_{t_{l+1} - t_q})) \cap \mathcal{N}\}$$

which yields

$$|(\mathcal{Z}_1(t_l, \varepsilon_{t_l - t_q}) \cup \mathcal{Z}_2(t_l, \varepsilon_{t_l - t_q})) \cap \mathcal{N}|$$
$$\geq |(\mathcal{Z}_1(t_{l+1}, \varepsilon_{t_{l+1} - t_q}) \cup \mathcal{Z}_2(t_{l+1}, \varepsilon_{t_{l+1} - t_q})) \cap \mathcal{N}|.$$

It remains to prove that the equality does not hold in the above inequality. Consider the pair of nonempty disjoint subsets $\mathcal{Z}_1(t_l, \varepsilon_{t_l - t_q})$ and $\mathcal{Z}_2(t_l, \varepsilon_{t_l - t_q})$. There exists a normal agent $i_j \in \mathcal{Z}_1(t_l, \varepsilon_{t_l - t_q}) \cup \mathcal{Z}_2(t_l, \varepsilon_{t_l - t_q})$ and a time instant $T_l \in [t_l, t_{l+1})$ such that the agent i_j has at least $F + 1$ neighbors outside the set which it belongs to at the time instant T_l. Without loss of generality, assume that $i_j \in \mathcal{Z}_1(t_l, \varepsilon_{t_l - t_q})$. If $i_j \notin \mathcal{Z}_1(T_l, \varepsilon_{T_l - t_q})$, then, one has

$$|(\mathcal{Z}_1(t_l, \varepsilon_{t_l - t_q}) \cup \mathcal{Z}_2(t_l, \varepsilon_{t_l - t_q})) \cap \mathcal{N}|$$
$$> |(\mathcal{Z}_1(T_l, \varepsilon_{T_l - t_q}) \cup \mathcal{Z}_2(T_l, \varepsilon_{T_l - t_q})) \cap \mathcal{N}|$$
$$\geq |(\mathcal{Z}_1(t_{l+1}, \varepsilon_{t_{l+1} - t_q}) \cup \mathcal{Z}_2(t_{l+1}, \varepsilon_{t_{l+1} - t_q})) \cap \mathcal{N}|.$$

For the case $i_j \in \mathcal{Z}_1(T_l, \varepsilon_{T_l - t_q})$, *one has that* $i_j \notin \mathcal{Z}_1(T_l + 1, \varepsilon_{T_l + 1 - t_q})$, *and thus*

$$|(\mathcal{Z}_1(t_l, \varepsilon_{t_l - t_q}) \cup \mathcal{Z}_2(t_l, \varepsilon_{t_l - t_q})) \cap \mathcal{N}|$$
$$\geq |(\mathcal{Z}_1(T_l, \varepsilon_{T_l - t_q}) \cup \mathcal{Z}_2(T_l, \varepsilon_{T_l - t_q})) \cap \mathcal{N}|$$
$$> |(\mathcal{Z}_1(t_{l+1}, \varepsilon_{t_{l+1} - t_q}) \cup \mathcal{Z}_2(t_{l+1}, \varepsilon_{t_{l+1} - t_q})) \cap \mathcal{N}|.$$

Up till now, (7.5) has been obtained, which immediately leads to the contradiction since there always holds that

$$|(\mathcal{Z}_1(t_l, \varepsilon_{t_l - t_q}) \cup \mathcal{Z}_2(t_l, \varepsilon_{t_l - t_q})) \cap \mathcal{N}| > 0,$$
$$|(\mathcal{Z}_1(t_q, \varepsilon_0) \cup \mathcal{Z}_2(t_q, \varepsilon_0)) \cap \mathcal{N}| \leq N_0.$$

Therefore, one has $\overline{P} = \underline{P}$, *i.e., the resilient consensus is realized.*

Remark 7.10 *It follows from Theorem 7.1 that the resilient consensus of the first-order NAS under F-total malicious model can be realized if and only if the time-varying communication graph $\mathcal{G}[t]$ is jointly $(F + 1, F + 1)$-robust. For the case of fixed graph, Theorem 7.1 degenerates into the resilient consensus of the F-total malicious model presented in [49]. By Lemma 7.9, the result presented in Theorem 7.1 is consistent with the case studied in [153] without malicious nodes. Compared with the result in [49], the proposed joint $(F + 1, F + 1)$-robustness captures the property of general switching topologies, where the graph $\mathcal{G}[t]$ at any time instant t may be not $(F + 1, F + 1)$-robust. Note that the condition of the joint $(F + 1, F + 1)$-robustness reduces both the connectivity requirement and the communication load, thereby expanding the robust graph theory to more real applications.*

Remark 7.11 *Compared with the results using the SW-MSR algorithm in [55, 57], where the global step interval T with the union of the graphs satisfying certain robust condition is required and all the information of neighbors within T steps needs to be stored, Theorem 7.1 does not need any global graph information, and it also removes the constraint of the storage of the neighboring information. But the price paid is that the requirement of the joint $(F+1, F+1)$-robustness may be a little bit tighter than the $(T, 2F+1)$-robustness proposed in [55, 57]. Different from the results for the random graphs in [162, 163] requiring $(2F+1)$-robust or $(2F+1)$-excess robust graph, Theorem 7.1 presents a lower conservative graph condition. Furthermore, it should be clarified that this is the first to present the necessary and sufficient condition for the resilient consensus of time-varying graph. Note that [162, 163] considered the high-order multi-agent systems. The proposed joint $(F+1, F+1)$-robustness will be introduced to investigate the resilient consensus of general linear multi-agent systems with time-varying topologies in the future.*

7.3.3   Resilient consensus for F-local malicious model

The previous subsection shows the resilient consensus of the NAS with time-varying communication graph $\mathcal{G}[t]$ for the F-total malicious model. To study the case with a large number of malicious nodes, the F-local malicious model is further considered. A sufficient condition and a necessary condition are provided separately for the resilient consensus of the F-local malicious model in the next theorem.

Theorem 7.2 *Consider the first-order F-local malicious NAS with time-varying communication graph $\mathcal{G}[t]$.*

1) *A necessary condition for the resilient consensus of the first-order F-local malicious NAS with $\mathcal{G}[t]$ under the W-MSR algorithm is that $\mathcal{G}[t]$ is jointly $(F+1)$-robust.*

2) *If the time-varying communication graph $\mathcal{G}[t]$ is jointly $(2F+1)$-robust, then the resilient consensus of the first-order NAS with $\mathcal{G}[t]$ under the W-MSR algorithm is realized with the safety interval $\Upsilon = [\underline{p}[0], \overline{p}[0]]$.*

Proof 7.8 *1) The necessary condition is first shown. Suppose that the network is not jointly $(F+1)$-robust, then there exist two disjoint subsets of $\mathcal{V}$ and a time instant $\hat{t}$, such that for $t \geq \hat{t}$, there are at most F neighbors outside its set for each node in these two sets. Let the values of the nodes in these two sets at time instant $\hat{t}$ be the maximum and minimum values, respectively. Then each node of these two sets removes all values different from itself, and keeps its value unchanged, i.e., the resilient consensus cannot be realized.*

2) Then, the sufficient condition is demonstrated. The proof is similar to that of Theorem 7.1. Consider the nonempty disjoint subsets $\mathcal{Z}_1(t_l, \varepsilon_{t_l - t_q}) \cap \mathcal{N}$ and $\mathcal{Z}_2(t_l, \varepsilon_{t_l - t_q}) \cap \mathcal{N}$ defined in Theorem 7.1. Since the graph $\mathcal{G}[t]$ is jointly $(2F+1)$-robust, there is a time instant T_l in each time interval $[t_l, t_{l+1})$ such that at least one normal node in these two sets has at least $2F+1$ neighbors outside its set, including at least $F+1$ normal neighbors as the total number of the malicious nodes is no more *than F. Therefore, at least one normal node in these two sets will use at least one of its normal neighbors' values outside at one time instant of each time interval, which results in*

$$|(\mathcal{Z}_1(t_l, \varepsilon_{t_l - t_q}) \cup \mathcal{Z}_2(t_l, \varepsilon_{t_l - t_q})) \cap \mathcal{N}|$$
$$> |(\mathcal{Z}_1(t_{l+1}, \varepsilon_{t_{l+1} - t_q}) \cup \mathcal{Z}_2(t_{l+1}, \varepsilon_{t_{l+1} - t_q})) \cap \mathcal{N}|.$$

Therefore, the resilient consensus can be realized.

Since the F-total malicious model is included in the case of the F-local malicious model, the following property is straightforward for the relationship between the joint r-robustness and the joint (r, s)-robustness.

Lemma 7.10 *The time-varying network $\mathcal{G}[t]$ is jointly $(F + 1, F + 1)$-robust, if it is jointly $(2F + 1)$-robust.*

The next result declares that the sufficient condition in Theorem 7.2 is sharp.

Lemma 7.11 *There exists an NAS with jointly $2F$-robust time-varying communication graph which fails to realize the resilient consensus under the F-local malicious model by using the W-MSR algorithm.*

Note that the graph $\mathcal{G}[t]$ is jointly $2F$-robust if it is $2F$-robust. Thus, the counterexample constructed in Proposition 2 of [49] for fixed graphs is also a counterexample to demonstrate the above lemma. So the detailed proof of Lemma 7.11 is omitted here for brevity.

Define the normal graph as $\mathcal{G}_{\mathcal{N}}[t] = (\mathcal{N}, \mathcal{E}_{\mathcal{N}}[t])$, which is induced by the normal node set and the edge set with edges among normal nodes. One has the following sufficient condition for the resilient consensus of the F-local malicious model.

Theorem 7.3 *For the F-local malicious model, the resilient consensus of the first-order NAS with time-varying communication graph $\mathcal{G}[t]$ under the W-MSR algorithm is realized with the safety interval $\Upsilon = [\underline{p}[0], \overline{p}[0]]$ if the normal graph $\mathcal{G}_{\mathcal{N}}[t]$ is jointly $(F + 1)$-robust.*

Proof 7.9 *Since the normal network $\mathcal{G}_{\mathcal{N}}[t]$ is jointly $(F+1)$-robust, it is straightforward to get that there is a time instant T_l in each time interval $[t_l, t_{l+1})$, such that at least one normal node in the sets $\mathcal{Z}_1(t_l, \varepsilon_{t_l - t_q}) \cap \mathcal{N}$ and $\mathcal{Z}_2(t_l, \varepsilon_{t_l - t_q}) \cap \mathcal{N}$ has at least $F + 1$ normal neighbors outside $\mathcal{Z}_1(t_l, \varepsilon_{t_l - t_q})$ or $\mathcal{Z}_2(t_l, \varepsilon_{t_l - t_q})$. Therefore, the resilient consensus can be realized by following the similar proof of Theorem 7.2.*

The sufficient condition presented in Theorem 7.3 is more straightforward than the condition in Theorem 7.2. Moreover, it can be verified by the following lemma that the condition in Theorem 7.3 is relaxed when compared with that in Theorem 7.2.

Lemma 7.12 *The normal network $\mathcal{G}_{\mathcal{N}}[t]$ is jointly $(F+1)$-robust, if the time-varying network $\mathcal{G}[t]$ is jointly $(2F + 1)$-robust under the F-local or F-total attack model.*

Proof 7.10 *The converse-negative proposition is proved. If the normal network $\mathcal{G}_\mathcal{N}[t]$ is not jointly $(F+1)$-robust, then there exist two disjoint subsets of the normal nodes and a time instant $\hat{t}$ such that for $t \geq \hat{t}$, each node in these two sets will have at most F normal neighbors outside its set. Since each node has at most F compromised neighbors, each node in these two sets has at most $2F$ neighbors outside its set after the time instant $\hat{t}$. Therefore, the time-varying network $\mathcal{G}[t]$ is not jointly $(2F+1)$-robust.*

7.4 RESILIENT CONSENSUS OF SECOND-ORDER NASS WITH DIRECTED SWITCHING TOPOLOGIES

7.4.1 Problem formulation

The second-order NAS is described by

$$
\begin{aligned}
p_i[t+1] &= p_i[t] + Tv_i[t], \\
v_i[t+1] &= v_i[t] + Tu_i[t], \ i = 1, \cdots, N,
\end{aligned}
\tag{7.6}
$$

in which $p_i[t]$, $v_i[t]$ and $u_i[t]$ are respectively the position, the velocity and the control input of the i-th agent at time instant t, and T is the sampling period.

The following DP-MSR algorithm is modified from [51]:

1) At every time step t, each normal agent i receives the position values of its neighbors $p_j[t]$, $j \in \mathcal{N}_i$, and sorts the relative position values $p_j[t] - p_i[t]$ in an increasing order.

2) Each normal agent i removes the smallest F negative values $p_j[t] - p_i[t]$ and the largest F positive values $p_j[t] - p_i[t]$. Otherwise, it removes all the negative (positive) values if there are less than F negative (positive) ones. Denote by $\mathcal{S}_i[t]$ the removed neighboring set of agent i.

3) Each normal agent i employs the following control protocol

$$
u_i[t] = \sum_{j=1}^{N} a_{ij}[t](p_j[t] - p_i[t]) - \beta v_i[t]
\tag{7.7}
$$

in which $a_{ij}[t] > 0$ if $j \in \mathcal{N}_i[t] \backslash \mathcal{S}_i[t]$; otherwise, $a_{ij}[t] = 0$. Particularly, $\sum_{j=1}^{N} a_{ij}[t] < \frac{1}{T^2}$ with β being a positive constant satisfying

$$
\frac{1}{T} + T \sum_{j=1}^{N} a_{ij}[t] \leq \beta \leq \frac{2}{T}.
$$

The resilient consensus problem discussed in this section is formulated in Definition 7.6.

7.4.2 Resilient consensus against malicious model

Let

$$\underline{z}[t] = \min(p_i[t], p_i[t-1])$$
$$\overline{z}[t] = \max(p_i[t], p_i[t-1]), \ \forall i \in \mathcal{N}.$$

The following result is provided for the resilient consensus of second-order time-varying NASs with F-total malicious model under the DP-MSR algorithm.

Theorem 7.4 *For the F-total malicious model, the resilient consensus of the second-order NAS with time-varying communication graph $\mathcal{G}[t]$ under the DP-MSR algorithm is realized if and only if $\mathcal{G}[t]$ is jointly $(F+1, F+1)$-robust. Moreover, the safety interval is given by $\overline{\Upsilon} = [\underline{z}[1], \overline{z}[1]]$ with*

$$\underline{z}[1] = \min(p_i[0], p_i[0] + Tv_i[0])$$
$$\overline{z}[1] = \max(p_i[0], p_i[0] + Tv_i[0]).$$

Proof 7.11 *Necessity. If $\mathcal{G}[t]$ is not jointly $(F+1, F+1)$-robust, then there are nonempty disjoint subsets $\mathcal{D}_1, \mathcal{D}_2 \subseteq \mathcal{V}$ such that after some finite time instant $\hat{t}$, the following three conditions hold:*

1) $|\Phi_{\mathcal{D}_1}^{F+1}[\hat{t}, \infty)| < |\mathcal{D}_1|$;

2) $|\Phi_{\mathcal{D}_2}^{F+1}[\hat{t}, \infty)| < |\mathcal{D}_2|$;

3) $|\Phi_{\mathcal{D}_1}^{F+1}[\hat{t}, \infty)| + |\Phi_{\mathcal{D}_2}^{F+1}[\hat{t}, \infty)| \leq F$.

Suppose that

$$p_i[\hat{t}] = a, \ \forall i \in \mathcal{D}_1$$
$$p_j[\hat{t}] = b, \ \forall j \in \mathcal{D}_2$$
$$v_i[\hat{t}] = 0, \ \forall i \in \mathcal{D}_1 \cup \mathcal{D}_2$$

in which $a < b$. By Condition 3), assume that all the agents in $\Phi_{\mathcal{D}_1}^{F+1}[\hat{t}, \infty) \cup \Phi_{\mathcal{D}_2}^{F+1}[\hat{t}, \infty)$ are malicious, which implement zero control inputs. Besides, assume that the agents in $\mathcal{D}_1 \backslash \Phi_{\mathcal{D}_1}^{F+1}[\hat{t}, \infty)$ and $\mathcal{D}_2 \backslash \Phi_{\mathcal{D}_2}^{F+1}[\hat{t}, \infty)$ are normal. By Conditions 1) and 2), there exists at least one normal node in both subsets $\mathcal{D}_1$ and $\mathcal{D}_2$. There are at most F neighbors with positions different from a for each node $i \in \mathcal{D}_1 \cap \mathcal{N}$, which implies that

$$p_j[t] = a, \ \forall j \in \{i\} \cup \mathcal{N}_i[t] \backslash \mathcal{S}_i[t].$$

Thus, one has that

$$p_i[t] = a, \ v_i[t] = 0, \ \forall i \in \mathcal{D}_1 \cap \mathcal{N}, \ \forall t.$$

Similarly, one has that

$$p_j[t] = a, \ v_j[t] = 0, \ \forall j \in \mathcal{D}_2 \cap \mathcal{N}, \ \forall t.$$

This means that the resilient consensus cannot be realized.

Sufficiency. *Substituting 7.7 into 7.6 yields*

$$p_i[t+1] = p_i[t] + Tv_i[t],$$

$$v_i[t+1] = (1 - T\beta)v_i[t] + T\sum_{j=1}^{N} a_{ij}[t](p_j[t] - p_i[t]). \tag{7.8}$$

Note that

$$p_i[t+1] - (1 - T\beta)p_i[t]$$
$$= p_i[t] - (1 - T\beta)p_i[t-1] + T(v_i[t] - (1 - T\beta)v_i[t-1]).$$

Then, one has that

$$p_i[t+1] = (2 - T\beta)p_i[t] + T^2 \sum_{j=1}^{N} a_{ij}[t-1]p_j[t-1]$$

$$+ \left(T\beta - 1 - T^2 \sum_{j=1}^{N} a_{ij}[t-1] \right) p_i[t-1].$$

Since

$$1 + T^2 \sum_{j=1}^{N} a_{ij}[t-1] \leq \beta T \leq 2$$

and $a_{ij} \geq 0$ with $\sum_{j=1}^{N} a_{ij}[t-1] < \frac{1}{T^2}$, one can derive that for each normal agent i, $p_i[t+1]$ lies in the convex combination of the positions of itself at time instants t and $t-1$, and its neighbors in the set $\mathcal{N}_i[t-1] \backslash \mathcal{S}_i[t-1]$ at time instant $t-1$. Since $p_i[t] \in [\underline{z}[t], \overline{z}[t]]$ and $p_j[t-1] \in [\underline{z}[t], \overline{z}[t]]$ if $j \in \{i\} \cup \mathcal{N}_i[t-1] \backslash \mathcal{S}_i[t-1]$ for all normal agent i, one has $p_i[t+1] \in [\underline{z}[t], \overline{z}[t]]$ for all normal agent i, implying that

$$[\underline{z}[1], \overline{z}[1]] \supset \cdots \supset [\underline{z}[t], \overline{z}[t]] \supset \cdots .$$

Thus,

$$p_i[t] \subseteq [\underline{z}[1], \overline{z}[1]], \ \forall i \in \mathcal{N}.$$

The proof of the resilient consensus achievement is omitted for brevity.

Remark 7.12 *As shown in Theorem 7.4, the resilient consensus of second-order NASs under the F-total malicious model can be realized if and only if the time-varying communication graph $\mathcal{G}[t]$ is jointly $(F+1, F+1)$-robust. For the case that the communication graph is fixed, Theorem 7.4 degenerates into the result presented in [51].*

For the case of the F-local malicious model, the next result is provided for the resilient consensus of second-order NASs.

Theorem 7.5 *Consider the second-order F-local malicious NAS with time-varying communication graph $\mathcal{G}[t]$.*

1) A necessary condition for the resilient consensus of the second-order NAS with $\mathcal{G}[t]$ under the DP-MSR algorithm is that $\mathcal{G}[t]$ is jointly $(F+1)$-robust.

2) If $\mathcal{G}[t]$ is jointly $(2F+1)$-robust, then the resilient consensus of the second-order NAS with $\mathcal{G}[t]$ under the DP-MSR algorithm is realized with the safety interval $\bar{\Upsilon} = [\underline{z}[1], \overline{z}[1]]$.

Proof 7.12 *The sufficiency is similar to the proof in Theorem 7.4. So the necessity is only shown. If the network is not jointly $(F+1)$-robust, then there exist two disjoint subsets of $\mathcal{V}$ and a time instant $\hat{t}$ such that for $t \geq \hat{t}$, each node in these two sets would have at most F neighbors outside its set. Let the positions of the nodes in these two sets at the time instant $\hat{t}$ be respectively the maximum and minimum values, and the velocities of these nodes be zero. Then each normal node in these two sets removes all position values different from itself, and keeps its position unchanged, which suggests that the consensus cannot be realized.*

Remark 7.13 *The resilient consensus of second-order NASs under the F-local malicious model was studied in [50] for the fixed communication graphs. Theorem 7.5 extends the result into the general time-varying network case, which is coincident with the result in [50].*

7.4.3 Resilient consensus against Byzantine attack model

For the cases of the F-total or F-local Byzantine model, the following result can be derived, and the proof is omitted here for brevity.

Theorem 7.6 *For the F-local or F-total Byzantine model, the resilient consensus of the second-order NAS with time-varying communication graph $\mathcal{G}[t]$ under the DP-MSR algorithm is realized with the safety interval $\bar{\Upsilon} = [\underline{z}[1], \overline{z}[1]]$ if and only if the normal graph $\mathcal{G}_{\mathcal{N}}[t]$ is jointly $(F+1)$-robust.*

And the following corollary follows from Theorem 7.6 by noticing Lemma 7.12.

Corollary 7.1 *For the F-local or F-total Byzantine model, the resilient consensus of the second-order NAS with time-varying communication graph $\mathcal{G}[t]$ under the DP-MSR algorithm is realized with the safety interval $\bar{\Upsilon} = [\underline{z}[1], \overline{z}[1]]$, if $\mathcal{G}[t]$ is jointly $(2F+1)$-robust.*

Remark 7.14 *It can be seen that the graph conditions for the resilient consensus of second-order time-varying NASs under the DP-MSR algorithm are the same as those for the resilient consensus of first-order time-varying NASs under the W-MSR*

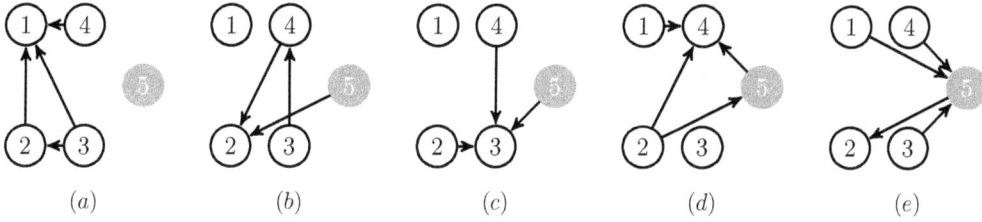

Figure 7.1   The jointly $(2,2)$-robust graph.

algorithm. The main reason is that only the static resilient consensus is realized under the DP-MSR algorithm based on the information of absolute velocity and relative positions. Based on the structure of the DP-MSR algorithm, it is not difficult to design certain MSR algorithms for high-order integrators, and the notion of joint robustness proposed in this chapter can be used to develop appropriate conditions ensuring the resilient consensus of high-order time-varying graph.

7.5 NUMERICAL SIMULATIONS

This section provides numerical examples to illustrate the theoretical results of this chapter. The time-varying communication graph of the NAS consisting of five agents is described in Fig. 7.1, which is jointly $(2,2)$-robust by checking every nonempty disjoint pair of subsets of $\mathcal{V}$. For example, in the two subsets $\{1,\ 3\}$ and $\{2,\ 5\}$, node 1 has neighbors indexed 2 and 4 in Fig. 7.1(a), and nodes $2, 4, 5$ are the neighbors of node 3 in Fig. 7.1(c), indicating that these two subsets meet the condition of Definition 7.5. Obviously, it is not a jointly $(3,3)$-robust graph.

Assume that agent 5 is malicious. First, consider the resilient consensus of the first-order NAS with time-varying communication graph. The input of agent 5 is randomly chosen in $[-1,1]$. The normal agents update by the W-MSR algorithm with

$$a_{ij}[t] = \frac{1}{1 + |\mathcal{N}_i[t] \backslash \mathcal{S}_i[t]|}.$$

The initial values of the agents are randomly set from -5 to 5. The trajectory of each agent is demonstrated in Fig. 7.2, where the trajectories of the malicious node 5 is represented by the line with dots, and the normal nodes $1-4$ are depicted by the other lines, respectively. Clearly, the resilient consensus is achieved.

Then, the edge $(1,4)$ in Fig. 7.1(d) is removed, and the time-varying topology switches among the graphs in Fig. 7.3, which is no longer jointly $(2,2)$-robust. It is shown in Fig. 7.4 that the resilient consensus can no longer be realized. Note that the value of node 4 remains unchanged since it removes all the values of its neighbors at any time instant. On the other hand, node 4 can never affect the values of other nodes, since all the other nodes remove its value at every time instant as well. As a result, the values of the normal nodes cannot realize the resilient consensus.

Finally, the effectiveness of the DP-MSR algorithm is illustrated for the second-order NASs with time-varying communication graphs. Choose the sampling period

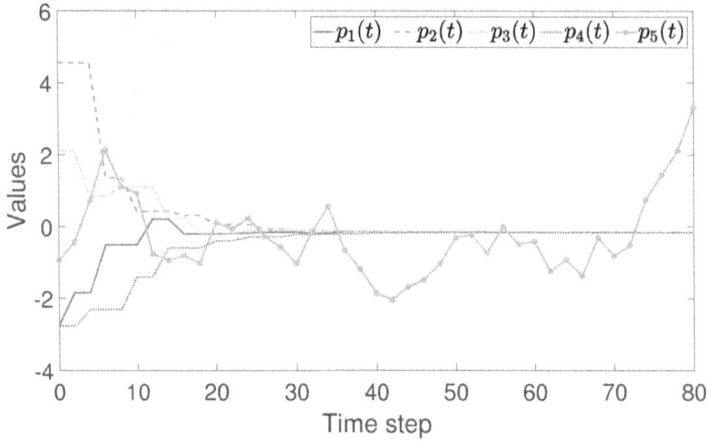

Figure 7.2   The resilient consensus of normal agents is realized under the W-MSR algorithm.

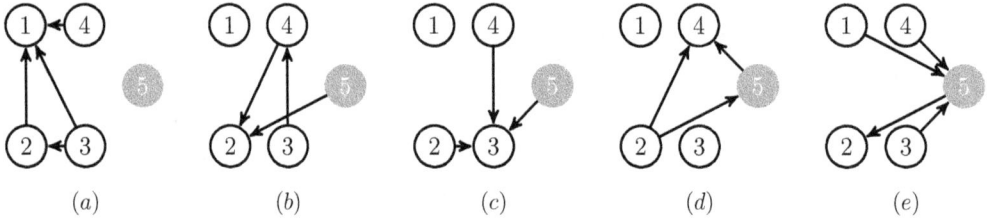

Figure 7.3   The topology is not jointly $(2,2)$-robust since the pair of subsets $\{4\}$ and $\{1,2,3,5\}$ do not satisfy the condition.

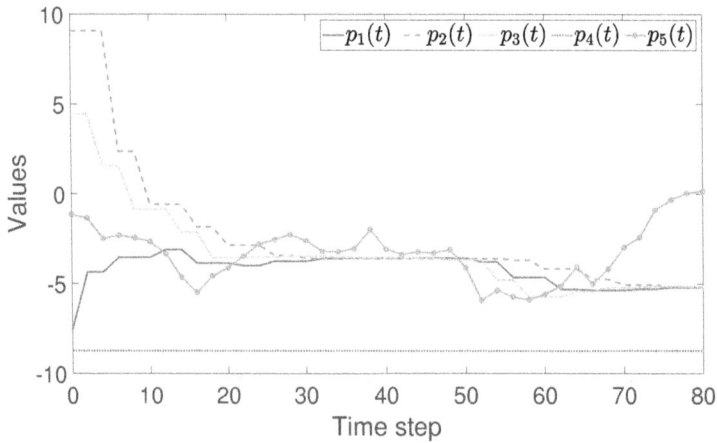

Figure 7.4   The resilient consensus of normal agents cannot be realized under the W-MSR algorithm, since the value of agent 4 maintains unchanged.

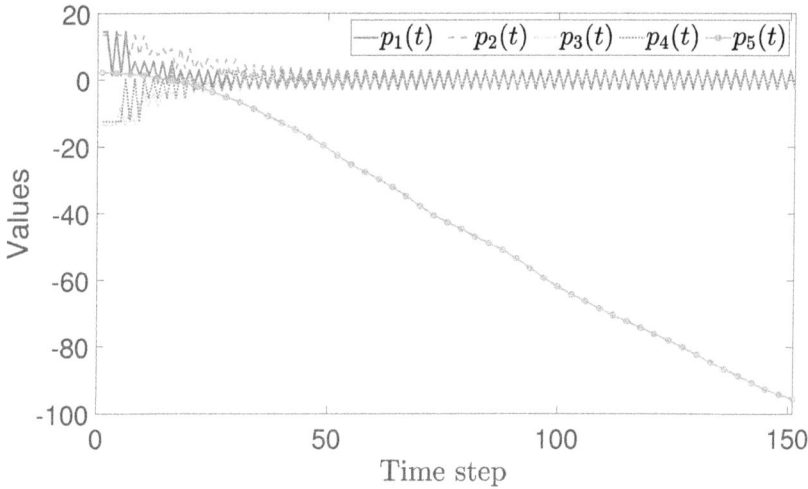

Figure 7.5   Resilient consensus of normal second-order agents is realized under the DP-MSR algorithm.

$T = 1$. The input of the malicious node 5 is randomly chosen in $[-0.1, 0.1]$. The control parameters are chosen as

$$w_{ij}[t] = \frac{1}{1 + \mathcal{N}_i[t] \backslash \mathcal{S}_i[t]}$$

and $\beta = 2$. The initial positions and velocities of the agents are randomly set in the intervals $[-15, 15]$ and $[-0.1, 0.1]$, respectively. The trajectories of the five agents are depicted in Fig. 7.5 to show the achievement of the resilient consensus, where the position trajectory of the malicious node 5 is represented by the line with dots, and the normal nodes $1 - -4$ are depicted by the other lines, respectively. For the time-varying topology in Fig. 7.3, the position trajectories of the second-order agents are presented in Fig. 7.6, where the resilient consensus cannot be reached under certain initial values.

7.6   EXPERIMENTS

In this section, two practical experiments are provided to validate the applicability of the theoretical findings based on the self-developed experiment platform. The platform consists of an Ultra Wide Band (UWB) positioning system, a ground control station, and quadcopters. The motion of each quadcopter is generated based on its own position information obtained from the UWB positioning system and the position commands received from the ground control station. Our goal is to ensure that the quadcopters achieve the preset arrow-shaped formation and maintain this formation while continuously flying along the x-axis.

Consider the quadcopter platform with communication topology shown in Fig. 7.7, which is a joint $(2, 2)$-robust graph and can withstand 1-total malicious attacks. For the i-th quadcopter, the position commands for x-, y-, and z-axis at

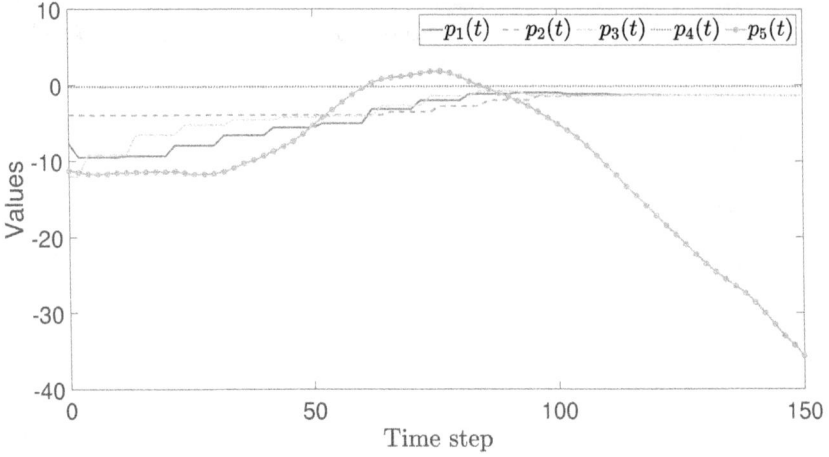

Figure 7.6 Resilient consensus of normal second-order agents cannot be realized under the DP-MSR algorithm, since the value of agent 4 maintains unchanged.

time step t is denoted by $x_i^c[t]$, $y_i^c[t]$, and $z_i^c[t]$, respectively. Each $x_i^c[t]$ is the sum of three components, i.e., $x_i^c[t] = x_i^{c_1}[t] + x_i^{c_2}[t] + x_i^{c_3}[t]$, where $x_i^{c_1}[t]$ is a displacement that define the shape of the formation, $x^{c_2}[t]$ and $x_i^{c_3}[t]$, together, drive the quadcopters to achieve the formation. While $x^{c_2}[t] = \alpha t$, with α being a positive constant, remains the same for all quadcopters at every time step, $x_i^{c_3}[t]$ may differ among quadcopters at the initial time step and is the target of attacks. To mitigate the influence caused by the attacks, each normal quadcopter receives the $x_j^{c_3}[t]$ from its neighbors and updates $x_i^{c_3}[t]$ using the W-MSR algorithm at each time step. Additionally, to prevent potential collisions caused by resilient consensus, $y_i^c[t]$ is set as a fixed distinct value for each quadcopter for all t. And $z_i^c[t]$ is also time-invariant due to safety reason.

Specifically, the initial positions of the quadcopters are $X[0] = [2, 2, 4.5, 1]^T$, $Y[0] = [3.5, 5, 6.5, 8]^T$ and $Z[0] = [0.5, 0.5, 0.5, 0.5]^T$. The position commands for x-axis are given by $X^c[t] = [0 + 0.2t + x_1^{c_3}[t], 2 + 0.2t + x_2^{c_3}[t], 2 + 0.2t + x_3^{c_3}[t], 0 + 0.2t + x_4^{c_3}[t]]^T$ with the initial values of the third components assigned as $X^{c_3}[0] = [2, 0, 2.5, 1]^T$. The position commands for y- and z-axis are defined as $Y^c[t] = [3.5, 5, 6.5, 8]^T$ and $Z^c = [0.5, 0.5, 0.5, 0.5]^T$, respectively. Suppose quadcopter 2 is subject to malicious attack since $t = 8$, with $x_2^{c_3}[t] = x_2[8] - \alpha t, \forall t \geq 8$. Δt is chosen as $0.3s$. Before the experiments commence, each the quadcopter hovers at $X_{init} = [0, 0, 0, 0]^T$, $Y_{init} = [3.5, 5, 6.5, 8]^T$, $Z_{init} = [0.5, 0.5, 0.5, 0.5]^T$. After receiving the start signal from the ground control station, they move to their respective initial positions and then proceed to conduct formation flight in accordance with real-time position commands.

Within the aforementioned setup, one can conduct two experiments, where each normal quadcopter i updates $x_i^{c_3}[t]$ using the W-MSR algorithm with parameter $F = 0$ and $F = 1$, respectively. When the W-MSR algorithm with parameter $F = 0$ is applied, each normal quadcopter utilizes all the neighboring values. As shown in Fig. 7.8, the normal quadcopters are influenced by the attack and fail to realize the prescribed formation. However, as shown in Fig. 7.9, when the W-MSR algorithm

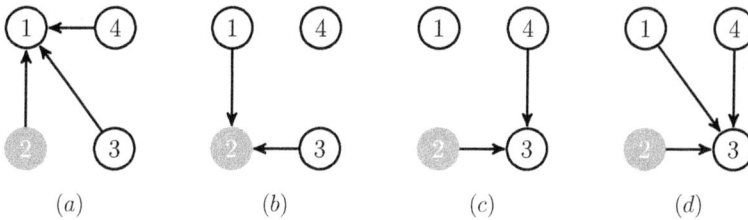

Figure 7.7   A $(2, 2)$-robust graph that can withstand 1-total malicious attacks.

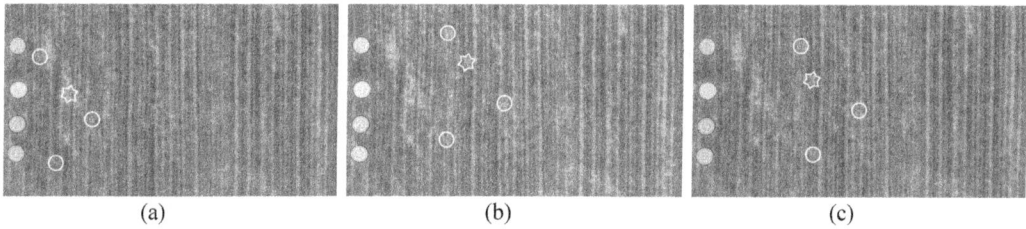

(a)   (b)   (c)

Figure 7.8   When the W-MSR algorithm with parameter $F = 0$ is used, normal quadcopters fail to realize the formation: (a) $t = 2$; (b) $t = 8$; (c) $t = 50$.

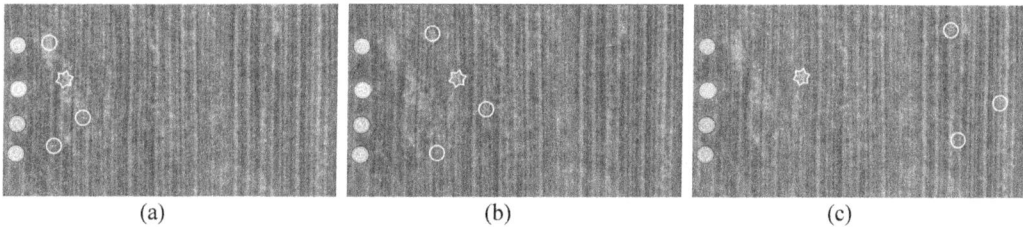

(a)   (b)   (c)

Figure 7.9   When the W-MSR algorithm with parameter $F = 1$ is used, normal quadcopters mitigate the influence of the attack and continue flying along the x-axis in the predefined formation: (a) $t = 2$; (b) $t = 8$; (c) $t = 50$.

with parameter $F = 1$ is applied, the normal quadcopters are able to mitigate the influence of the attack and continue flying along x-axis according to the predefined formation.

7.7   CONCLUSIONS

This chapter has studied the resilient consensus problem of NASs with misbehaving agents and time-varying communication graphs under certain MSR algorithms. To characterize the graph-theoretic properties of the considered NASs for realizing the resilient consensus in the presence of the influence of misbehaving agents, a new notion, called joint robustness, has been introduced herein. Within the framework of joint robustness, some necessary and sufficient conditions have been respectively derived for achieving the resilient consensus in NASs with first-order and second-order inherent agent dynamics. It has been proven that the resilient consensus under the F-total malicious model is realized if and only if the time-varying communication graph is jointly $(F+1, F+1)$-robust; while the resilient consensus under the F-total or F-local Byzantine model is guaranteed if and only if the time-varying subgraph consisting of normal agents is jointly $(F+1)$-robust. A remaining yet interesting issue deserving future study is how to systematically construct jointly robust time-varying graphs.

Furthermore, one has considered the resilient consensus problem of NASs with first-order and second-order inherent agent dynamics subject to time-varying communication graphs, for which some necessary and sufficient criteria have been derived. Extending these findings to NASs with nonlinear dynamics may require connecting our current approaches with complementary methods that consider the evolution of nonlinear systems.

CHAPTER 8

Distributed resilient cooperative control of high-order NASs based on attack isolation

This chapter studies the distributed resilient cooperative control problem of high-order NASs under both false data injection (FDI) attacks and covert attacks. Section 8.1 provides a review of relevant prior studies and outlines the motivations underlying this work. Section 8.2 introduces a distributed inclusive neighbor-based isolation algorithm for FDI attacks. To characterize a feasible communication topology capable of isolating compromised agents, the graph-theoretic concept of r-isolability is introduced. Based on this concept, sufficient conditions are established to ensure the successful isolation of compromised agents. Section 8.3 presents a distributed neighbor-based isolation algorithm for covert attacks, and introduces a new graph-theoretic notion of r^*-isolability to establish sufficient conditions for the success of attack isolation under this algorithm. Section 8.4 develops distributed resilient cooperative control protocols based on attack isolation, where each normal agent discards information from neighbors identified as compromised, thereby ensuring resilient cooperation among the normal agents. Section 8.5 provides some numerical examples to illustrate the effectiveness of the theoretical results. Section 8.6 concludes this chapter.

8.1 INTRODUCTION

Note that the MSR-based algorithms presented in the previous chapter primarily address NASs with lower-order nodal dynamics. These algorithms rely on sorting the values of neighboring nodes as an exclusion criterion, thereby restricting their applicability to NASs with integrator-type nodal dynamics. In [61, 62], safe-point-based resilient vector consensus algorithms were investigated. The core idea of these algorithms is to compute a safe point within the convex hull of the vector states

DOI: 10.1201/9781003669913-8

of normal neighbors, after which each normal agent updates its states by moving toward this safe point while treating the vector states as an integrated whole. Subsequently, [63, 64] utilized the idea of Tverberg partition to compute the safe point, leading to the development of the approximate distributed robust convergence algorithm for achieving resilient vector consensus. Furthermore, [65, 66] incorporated the notion of centerpoint to compute the safe point, demonstrating that centerpoint-based resilient algorithms can significantly enhance the resilience of NASs against adversarial attacks.

Given the coupled states in general high-order NASs, the algorithms proposed in [61–66] become inapplicable, necessitating the development of resilient cooperative control algorithms that exclude isolated compromised agents. This chapter addresses the problem of distributed resilient cooperative control in general high-order NASs under both FDI attacks and covert attacks. FDI attacks and covert attacks are two representative types of malicious attacks. FDI attacks compromise agent behavior by injecting deceptive data into actuator and/or sensor channels, whereas covert attacks manipulate the agent's state while concealing their impact on the output, thereby rendering the compromised output indistinguishable from that of a normal system. To address FDI attacks, a set of distributed fixed-time observers from [166] is employed to aggregate compromised agents within a subsystem comprising specific agents and their neighbors. Subsequently, a distributed inclusive neighbor-based attack isolation (DINAIs) algorithm is designed to isolate compromised agents. To prevent the mistaken isolation of normal agents, a novel graph-theoretic property, termed r-isolability, is introduced to facilitate the achievement of attack isolation under the DINAIs algorithm. To resist covert attacks, a set of two-stage fixed-time observers are constructed, allowing each agent to generate a residual capable of detecting attacks on its neighbors. By analyzing the number of nonzero patterns in the residuals of the neighbors for each agent, the distributed neighbor-set-based attack isolation (DNAIs) algorithm is developed to isolate covert attacks. Furthermore, a graph-theoretic notion of r^*-isolability is introduced to provide sufficient conditions for attack isolation under the DNAIs algorithm. Finally, two control algorithms are derived by deleting the isolated compromised agents to ensure the resilient cooperation of the normal agents.

8.2 DISTRIBUTED ISOLATION OF FDI ATTACKS

8.2.1 Problem formulation

Consider a high-order NAS with the dynamics of the ith agent being governed by

$$\begin{cases} \dot{x}_i\left(t\right) = Ax_i\left(t\right) + Bu_i\left(t\right), \\ y_i\left(t\right) = Cx_i\left(t\right) + Mf_i\left(x, t, T_i\right), \ i = 1, 2, \ldots, N, \end{cases} \tag{8.1}$$

where $x_i \in \mathbb{R}^n$, $u_i \in \mathbb{R}^m$, $y_i \in \mathbb{R}^p$ are the state, control input, and measurement output vectors, respectively. $x = (x_1, x_2, \ldots, x_N)^T$, $f_i\left(x, t, T_i\right) \in \mathbb{R}^q$ denotes the false data injected to the sensor channel, and T_i is the time at which the FDI attack occurs. A, B, C and M are constant system matrices with appropriate dimensions. For simplicity, the time index t will be omitted if there is no ambiguity.

Assumption 8.1 $rank(CB) = rank(B) = m.$

Assumption 8.2 $rank\begin{pmatrix} sI_n - A & B \\ C & 0 \end{pmatrix} = n + m, \ \forall s \in \mathbb{C}.$

Remark 8.1 *Assumption 8.1 indicates that the dimension of the output is no less than that of the input, while Assumption 8.2 asserts that the system has no transmission zeros. Assumptions 8.1 and 8.2 together provide the existence conditions of the distributed fixed-time observer developed in [166], which will be used in this section for attack detection.*

The following definitions are provided to characterize the number of compromised agents in this section.

Definition 8.1 (1-weak-local FDI attack model) *The NAS is said to be 1-weak-local compromised by FDI attacks if the compromised node set $\mathcal{V}_A$ contains at most one node from the neighborhood and two-hop neighborhood of any normal node, i.e., $|\mathcal{J}_i^* \cap \mathcal{V}_A| \le 1, \forall i \in \mathcal{V}_R$, with $\mathcal{J}_i^*$ is defined in Section 2.3.*

Definition 8.2 (F-total FDI attack model) *The NAS is said to be F-total compromised by FDI attacks if the compromised node set $\mathcal{V}_A$ contains at most F nodes, i.e., $|\mathcal{V}_A| \le F$.*

Remark 8.2 *The condition that there exists at most one compromised agent in each subsystem $\mathcal{G}_i(\mathcal{J}_i^*, \mathcal{E}_i)$ implies that the NAS is 1-weak-local compromised. Thus, these two terminologies '1-weak-local FDI attack model' and 'at most one agent is compromised by a FDI attack in each subsystem $\mathcal{G}_i(\mathcal{J}_i^*, \mathcal{E}_i)$' are used interchangeably in the following.*

Since the actual number and positions of the compromised agents are unknown to the normal agents, it is imperative to design a distributed attack isolation algorithm to accurately identify each compromised agent. The problem investigated in this section is formulated as follows.

Problem 8.1. Consider a high order NAS (8.1), with a communication topology represented by the graph $\mathcal{G}$. Suppose that Assumptions 8.1 and 8.2 hold. Under the 1-weak-local and F-total FDI attack models, design a distributed attack isolation algorithm such that i) all compromised agents are correctly isolated; ii) no normal agent is mistakenly isolated as compromised.

The following two-step design structure is proposed to solve the above attack isolation problem.

i) *Attack detection: Designing a distributed fixed-time observer to detect whether there are compromised agents in each subsystem $\mathcal{G}_i(\mathcal{J}_i, \mathcal{E}_i)$;*

ii) *Attack isolation: Developing a distributed attack isolation algorithm to correctly isolate each compromised agent, along with a graph condition to ensure that no normal agent is mistakenly isolated.*

8.2.2   Distributed detection of FDI attacks

In this subsection, a distributed attack detection strategy is developed to determine the presence of compromised agents in each subsystem $\mathcal{G}_i(\mathcal{J}_i, \mathcal{E}_i)$. This is achieved by incorporating a distributed fixed-time observer, adapted from [166].

Let $\varepsilon_i(t) = \sum_{j \in \mathcal{N}_i} (x_i(t) - x_j(t))$, $i = 1, 2, \ldots, N$, be the consensus error for each agent. A distributed observer with relative output information is introduced as follows:

$$\dot{\hat{z}}_i(t) = A_c \hat{z}_i(t) + B_c \sum_{j \in \mathcal{N}_i} (y_i(t) - y_j(t)), \tag{8.2}$$

where $z_i(t) \in \mathbb{R}^{2n}$, $A_c = \begin{pmatrix} GA - H_1C & 0 \\ 0 & GA - H_2C \end{pmatrix}$, $B_c = \begin{pmatrix} B_{c1} \\ B_{c2} \end{pmatrix}$, H_k is the observer gain which makes $GA - H_kC$ stable, $B_{ck} = H_k(I_p + CE) - GAE$, $k = 1, 2$, $G = I_n + EC$, $E = -B \left((CB)^T CB \right)^{-1} (CB)^T$, and $C_c = \begin{pmatrix} I_n & I_n \end{pmatrix}^T$.

Define $\tilde{z}_i(t) = \hat{z}_i(t) - C_c G \varepsilon_i(t)$, it can be derived from (8.1) and (8.2) that

$$\begin{aligned} \dot{\tilde{z}}_i(t) &= H_c \tilde{z}_i(t) + B_c C \varepsilon_i(t) + B_c F \sum_{j \in \mathcal{N}_i} a_{ij} (f_i(t) - f_j(t)) \\ &\quad - C_c G \left(A \varepsilon_i(t) + B \sum_{j \in \mathcal{N}_i} (u_i(t) - u_j(t)) \right) \\ &= H_c \tilde{z}_i(t) + B_c F \sum_{j \in \mathcal{N}_i} (f_i(t) - f_j(t)), \end{aligned} \tag{8.3}$$

where $GB = 0$.

Let $D_c = \begin{pmatrix} I_n & 0 \end{pmatrix} \begin{pmatrix} C_c & \exp(A_c \tau) C_c \end{pmatrix}^{-1}$, by introducing a time-delay τ on the observer state $\hat{z}_i(t)$, the fixed-time observer is developed as follows:

$$\begin{cases} \hat{\chi}_i(t) = D_c \left[\hat{z}_i(t) - \exp(A_c \tau) \hat{z}_i(t - \tau) \right], \\ \hat{\varepsilon}_i(t) = \hat{\chi}_i(t) - E \sum_{j \in \mathcal{N}_i} (y_i(t) - y_j(t)), \end{cases} \tag{8.4}$$

where $\hat{z}_i(t) = 0$ for $-\tau \leq t \leq 0$, $\hat{\varepsilon}_i(t)$ is the estimate of consensus error $\varepsilon_i(t)$.

The complete fixed-time observer based on relative output information can then be reformulated as follows:

$$\begin{cases} \dot{\hat{z}}_i(t) = A_c \hat{z}_i(t) + B_c \sum_{j \in \mathcal{N}_i} (y_i(t) - y_j(t)), \\ \hat{\chi}_i(t) = D_c \left[\hat{z}_i(t) - \exp(A_c \tau) \hat{z}_i(t - \tau) \right], \\ \hat{\varepsilon}_i(t) = \hat{\chi}_i(t) - E \sum_{j \in \mathcal{N}_i} (y_i(t) - y_j(t)), \end{cases} \tag{8.5}$$

Let $\tilde{\varepsilon}_i(t) = \varepsilon_i(t) - \hat{\varepsilon}_i(t)$ be the attack detection residual, one has

$$\begin{aligned} \tilde{\varepsilon}_i(t) &= -D_c \int_{t-\tau}^{t} \exp(A_c(t-s)) B_c M \sum_{j \in \mathcal{N}_i} (f_i(s) - f_j(s)) \, ds \\ &\quad + EM \sum_{j \in \mathcal{N}_i} (f_i(t) - f_j(t)), \ t \geq \tau, \end{aligned} \tag{8.6}$$

where $f_i(t)$ stands for $f_i(x, t, \mathcal{T}_i)$.

Then the attack detection result for the subsystem $\mathcal{G}_i(\mathcal{J}_i, \mathcal{E}_i)$ can be readily derived in accordance with (8.6), as presented below.

Algorithm 8.1 DINAIs algorithm

Input: $\hat{z}_i(0) = \mathbf{0}$, $\mathcal{I}_i(0) = 0$, $C_i(0) = 0$, $S_i(0) = 0$, $i \in \mathcal{V}$;
Output: $S_i(t)$, $i \in \mathcal{V}$;

1: **for** $t \geq \tau$ **do**
2:     **if** $\|\tilde{\varepsilon}_i(t)\| \neq 0$ **then**
3:         $\mathcal{I}_i(t) = 1$;
4:         $C_i(t) = 1$;
5:     **end if**
6:     i sends $\mathcal{I}_i(t)$ to all its neighbors within $\mathcal{N}_i$;
7:     **for** $j = i^1 : i^{|\mathcal{N}_i|}$ **do**
8:         **if** $\mathcal{I}_j(t) = 1$ **then**
9:             $C_i(t) = C_i(t) + 1$;
10:        **end if**
11:     **end for**
12:     **if** $C_i(t) = |\mathcal{J}_i|$ **then**
13:         $S_i(t) = i$;
14:         i sends $S_i(t)$ to all its neighbors within $\mathcal{N}_i$.
15:     **end if**
16: **end for**

Lemma 8.1 *For the subsystem $\mathcal{G}_i(\mathcal{J}_i, \mathcal{E}_i)$ described by (8.1), suppose that Assumptions 8.1 and 8.2 hold. There exist compromised agents if there exists a time $t \geq \tau$ such that $\|\tilde{\varepsilon}_i(t)\| \neq 0$ holds under the fixed-time observer (8.5).*

Remark 8.3 *Notably, if there is only a single compromised agent within $\mathcal{G}_i(\mathcal{J}_i, \mathcal{E}_i)$, one has that $\sum_{j \in \mathcal{N}_i} (f_i - f_j) \neq \mathbf{0}$. Consequently, the necessary and sufficient condition for the presence of a single compromised agent in $\mathcal{G}_i(\mathcal{J}_i, \mathcal{E}_i)$ is the existence of a time instant $t \geq \tau$ such that $\|\tilde{\varepsilon}_i(t)\| \neq 0$. However, if two or more agents within $\mathcal{G}_i(\mathcal{J}_i, \mathcal{E}_i)$ are compromised, they may collude to make $\sum_{j \in \mathcal{N}_i} (f_i - f_j) = \mathbf{0}$, thereby evading detection.*

8.2.3   Distributed isolation of FDI attacks

Motivated by the above analysis, a DINAIs algorithm is developed to tackle the attack isolation problem. To begin with, some notations used are introduced. The state index $\mathcal{I}_i(t) = 1$ means that there are compromised agents in the subsystem $\mathcal{G}_i(\mathcal{J}_i, \mathcal{E}_i)$, while $\mathcal{I}_i(t) = 0$ indicates that the subsystem $\mathcal{G}_i(\mathcal{J}_i, \mathcal{E}_i)$ is free of compromised agents. The counter $C_i(t)$ records the number of neighbors $j \in \mathcal{J}_i$ for which $\mathcal{I}_j(t) = 1$. The safety index $S_i(t) = i$ implies that agent i is identified as compromised, and $S_i(t) = 0$ indicates that agent i is considered normal.

Remark 8.4 *It can be seen from the DINAIs algorithm that the state index $\mathcal{I}_i$ only depends on $\tilde{\varepsilon}_i$, which relates to its inclusive neighbors $\mathcal{J}_i$. However, the derivation of the safety index S_i involves $\mathcal{I}_j$, $j \in \mathcal{N}_i$, which relates to the inclusive neighbors of j, e.g., the two-hop neighbors of i. Consequently, the DINAIs algorithm involves two-hop* *information of each agent. Such two-hop information can be obtained by exchanging $\mathcal{I}_j$ among neighboring agents.*

Remark 8.5 *The DINAIs algorithm contains three steps. Firstly, based on neighboring output information y_j, each agent i constructs the fixed-time observer (8.5), and obtains the estimation error $\bar{\varepsilon}_i$ as well as the state index $\mathcal{I}_i$. Secondly, each agent i receives the state index $\mathcal{I}_j$ from its neighbors and then calculates the safety index S_i. Finally, each compromised agent i broadcasts its safety index S_i to its neighbors. As such, the DINAIs algorithm is limited to undirected graphs.*

Remark 8.6 *For each compromised agent i, if there is no collusion among compromised agents, one has that there exists a time instant $t \geq \tau$ such that $\|\tilde{\varepsilon}_k\| \neq 0$ for all $k \in \mathcal{J}_i$. From the DINAIs algorithm, agent i is isolated as compromised if there exists a time instant $t \geq \tau$ such that $\|\tilde{\varepsilon}_k\| \neq 0$ for all $k \in \mathcal{J}_i$. As a result, all compromised agents can be isolated by the DINAIs algorithm. However, the normal agents may be mistakenly isolated as compromised. For example, if $\mathcal{J}_m \subseteq \mathcal{J}_i$ with the normal agent m and compromised agent i, the normal agent m will be mistakenly isolated as compromised since there exists a time instant $t \geq \tau$ such that $\|\tilde{\varepsilon}_k\| \neq 0$ for all $k \in \mathcal{J}_m$. Therefore, it needs to develop some conditions on graph structure to ensure that only the compromised agents are isolated.*

Intuitively, the communication topology structure plays a key role in attack isolation. More specifically, an appropriate communication topology structure ensures that no normal agent is mistakenly isolated as compromised. To find a sufficient condition for zero false alarms that captures this essence, some graph-theoretic properties are first introduced.

Definition 8.3 (node isolability) *Given a nontrivial graph $\mathcal{G} = (\mathcal{V}, \mathcal{E})$, a node $i \in \mathcal{V}$, and its neighbors i^j with $j = 1, 2, \ldots, |\mathcal{N}_i|$, node i is isolable if $\mathcal{J}_i \cap \mathcal{J}_{i^1} \cap \mathcal{J}_{i^2} \cap \ldots \cap \mathcal{J}_{i^{|\mathcal{N}_i|}} = \{i\}$.*

Remark 8.7 *If two nodes i and m satisfy $\mathcal{J}_i \cap \mathcal{J}_{i^1} \cap \mathcal{J}_{i^2} \cap \ldots \cap \mathcal{J}_{i^{|\mathcal{N}_i|}} \supseteq \{i, m\}$, then it follows that $i^k \in \mathcal{J}_m$, $k = 1, \cdots, |\mathcal{N}_i|$, and $i \in \mathcal{J}_m$, which further implies $\mathcal{J}_i \subseteq \mathcal{J}_m$. On the other hand, it is obvious that $\mathcal{J}_i \cap \mathcal{J}_{i^1} \cap \cdots \cap \mathcal{J}_{i^{|\mathcal{N}_i|}} \supseteq \{i, m\}$ if $\mathcal{J}_i \subseteq \mathcal{J}_m$. Consequently, node i is isolable if and only if $\mathcal{J}_i \nsubseteq \mathcal{J}_m, \forall m \neq i$.*

Intuitively, Definition 8.3 captures the node isolability. The graph isolability is defined as follows.

Definition 8.4 (r-isolability) *A nontrivial graph $\mathcal{G} = (\mathcal{V}, \mathcal{E})$ with N nodes is r-isolable if, for every node i and any other r nodes, represented by $R_1, R_2, \ldots, R_r$, their inclusive neighbor sets satisfy $\mathcal{J}_i \nsubseteq \{\mathcal{J}_{R_1} \cup \mathcal{J}_{R_2} \cup \cdots \cup \mathcal{J}_{R_r}\}$.*

Remark 8.8 *From Remark 8.7 and Definition 8.4, it follows that the graph $\mathcal{G}$ is 1-isolable if and only if every node in $\mathcal{G}$ is isolable. Besides, it is easy to derive from Definition 8.4 that a cycle graph with N ($N \geq 4$) nodes is 1-isolable.*

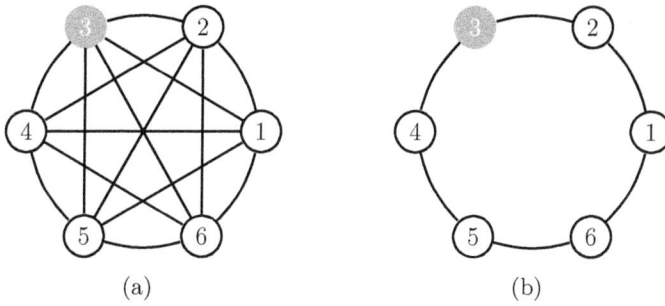

Figure 8.1 Two distinct graph topologies, each consisting of six nodes, with node 3 being compromised by an FDI attack. (a) The complete graph. (b) The cycle graph.

Remark 8.9 *Since $\mathcal{J}_i \nsubseteq \{\mathcal{J}_{i^1} \cup \mathcal{J}_{i^2} \cup \cdots \cup \mathcal{J}_{i|\mathcal{N}_i|}\}$, it follows from Definition 8.4 that $|\mathcal{N}_i| \geq r+1$ for all $i \in \mathcal{V}$ if graph $\mathcal{G}$ is r-isolable. Moreover, if a graph $\mathcal{G}$ is r-isolable, it is also r'-isolable for all $1 \leq r' \leq r$.*

Note that it requires an adequate number of neighbors outside the considered set in graph robustness [49] and vertex expanders [171] to achieve resilient consensus, while the inclusion relation among the inclusive neighbors of any $F+1$ nodes may not hold, even in graphs with high robustness or expansion ratio. Thus, the concept of r-isolability is introduced to characterize the topological structure from the perspective of attack isolation, rather than relying on robustness or connectivity redundancy.

The advantage of the notion of *r-isolability* and its application is illustrated through the example in Fig. 8.1. The cycle graph, with simple structure and low resource cost, is widely used in industrial systems such as the transmission network of telecom carries [172] and metro networks [173]. In general, attack isolation in a cycle graph is challenging, as an attack on a single node can potentially disrupt the functionality of the entire network. For the complete graph in Fig. 8.1(a), by examining the relationship between the inclusive neighbors of node 3 ($\mathcal{J}_3$) and those of its neighbors ($\mathcal{J}_1, \mathcal{J}_2, \mathcal{J}_4, \mathcal{J}_5, \mathcal{J}_6$), it can be derived that $\mathcal{J}_3 \cap \mathcal{J}_1 \cap \mathcal{J}_2 \cap \mathcal{J}_4 \cap \mathcal{J}_5 \cap \mathcal{J}_6 = \{1, 2, 3, 4, 5, 6\}$, which violates Definition 8.3, thus node 3 is not isolable. Therefore, all nodes will be isolated as compromised by Algorithm 8.1 if the residuals of all inclusive neighbors of node 3 are nonzero, i.e., $\|\tilde{\varepsilon}_i\| \neq 0$ for $i \in \{1, 2, 3, 4, 5, 6\}$, which indicates the failure of attack isolation. For the cycle graph in Fig. 8.1(b), by analyzing the relationship among the inclusive neighbors of node 3 ($\mathcal{J}_3$) and its neighbors ($\mathcal{J}_2, \mathcal{J}_4$), one has $\mathcal{J}_3 \cap \mathcal{J}_2 \cap \mathcal{J}_4 = \{3\}$, which satisfies Definition 8.3. Therefore, the compromised node 3 can be isolated by Algorithm 8.1 when $\|\tilde{\varepsilon}_3\| \neq 0$, $\|\tilde{\varepsilon}_2\| \neq 0$, and $\|\tilde{\varepsilon}_4\| \neq 0$. Furthermore, it is important to note that if only one node is compromised by an FDI attack in the cycle graph, it can be surely isolated, as the graph is 1-isolable. This means that the compromised node cannot be isolated even in a complete graph, while can be easily isolated in a cycle graph using the DINAIs algorithm. In the following, the above concepts will be applied to characterize the performance of the DINAIs algorithm.

Theorem 8.1 *Consider a high-order NAS described by (8.1), with a communication topology represented by the graph $\mathcal{G}$. Suppose that Assumptions 8.1 and 8.2 hold. In the 1-weak-local FDI attack model, attack isolation can be achieved under the DINAIs algorithm if and only if the followings hold simultaneously:*

i) $\exists t \geq \tau$ s.t. $\|\tilde{\varepsilon}_k(t)\| \neq 0$ for all $k \in \mathcal{J}_i, i \in \mathcal{V}_A$;

ii) Graph $\mathcal{G}$ is 1-isolable.

Proof 8.1 *For the 1-weak-local FDI attack model, it follows from (8.6) and Lemma 8.1 that there exists a single compromised agent in $\mathcal{G}_i(\mathcal{J}_i^*, \mathcal{E}_i)$ if and only if there exists a time instant $t \geq \tau$ such that $\|\tilde{\varepsilon}_i(t)\| \neq 0$.*

(Sufficiency). Consider two cases, Case 1): There exists at least one compromised agent not being isolated, and Case 2): All compromised agents are successfully isolated, however, at least one normal agent is mistakenly isolated as compromised.

Case 1): Suppose that the compromised agent i is not isolated. Then, one has that there exists a time instant $t \geq \tau$ such that $\|\varepsilon_k(t)\| = 0$, $k \in \mathcal{J}_i$ holds, which contradicts condition i).

Case 2): Suppose that both the compromised agent i and the normal agent m are isolated as compromised within the same subsystem $\mathcal{G}_i(\mathcal{J}_i^*, \mathcal{E}_i)$. Then, there exists a time instant $t \geq \tau$ such that $\|\tilde{\varepsilon}_k(t)\| \neq 0$ for all $k \in \mathcal{J}_i \cup \mathcal{J}_m$. According to the definition of the 1-weak-local FDI attack model, at most one agent can be compromised within the subsystem $\mathcal{G}_m(\mathcal{J}_m^*, \mathcal{E}_m)$, which implies that all neighbors of m must also be neighbors of i, i.e., $\mathcal{J}_m \subseteq \mathcal{J}_i$. This contradicts the definition of 1-isolability.

(Necessity). Case 1): Suppose that condition i) is not satisfied. Then, at least one agent $m \in \mathcal{J}_i \cap \mathcal{V}_A$ satisfies $\|\tilde{\varepsilon}_m(t)\| = 0$ for $t \geq \tau$, which means that the compromised agent i cannot be isolated by the DINAIs algorithm.

Case 2): Suppose that $\mathcal{J}_m \subseteq \mathcal{J}_i$, where agent i is compromised and agent m is normal under the 1-weak-local FDI attack model. It follows that there exists a time instant $t \geq \tau$ such that $\|\tilde{\varepsilon}_k(t)\| \neq 0$ for all $k \in \mathcal{J}_m$, which implies that the normal agent m is mistakenly isolated by the DINAIs algorithm.

Note that Theorem 8.1 gives a necessary and sufficient condition for attack isolation under the 1-weak-local FDI attack model. It is more general and practical to consider the case that the positions of compromised agents are random. And the challenge lies in isolating any F compromised agents under the F-total FDI attack model. With Definition 8.4, the following main result on attack isolation under the F-total FDI attack model can be established.

Theorem 8.2 *Consider a high-order NAS described by (8.1), with a communication topology represented by the graph $\mathcal{G}$. Suppose that Assumptions 8.1 and 8.2 hold. In the F-total FDI attack model with $F \geq 2$, attack isolation can be achieved under the DINAIs algorithm if the followings hold simultaneously:*

i) $\exists t \geq \tau$ s.t. $\|\tilde{\varepsilon}_k(t)\| \neq 0$ for all $k \in \mathcal{J}_i, i \in \mathcal{V}_A$;

ii) Graph $\mathcal{G}$ is F-isolable.

Proof 8.2 *Under condition i), it is clear that all compromised agents can be isolated by the DINAIs algorithm. It remains to show that no mistaken isolation of normal agent happens if the graph $\mathcal{G}$ is F-isolable, which is straightforward by noticing that the normal agent i is mistakenly isolated only if $\mathcal{J}_i \subseteq \{\mathcal{J}_{R_1} \cup \mathcal{J}_{R_2} \cup \ldots \cup \mathcal{J}_{R_F}\}$ with $R_k \in \mathcal{V}_A, \ k = 1, \cdots, F$.*

Remark 8.10 *Notably, both [174] and [175] resort to information set with historical data to detect the compromised agents. In contrast, the DINAIs algorithm enables attack isolation without requiring access to historical data, which significantly reduces computational burden and associated economic costs.*

Remark 8.11 *A typical application of the DINAIs algorithm is the cycle topology, which is commonly used in economic dispatch of smart grids [170]. Any single compromised generator can be isolated under the DINAIs algorithm. Moreover, the proposed graph property provides guidance for constructing a robust communication topology, thereby enhancing the resilience of NASs against attacks.*

The following result reveals that, for an F-isolable graph (but not $(F+1)$-isolable), the upper bound of the allowable compromised agents is F.

Proposition 8.1 *There exists an F-isolable graph for which the DINAIs algorithm fails to achieve attack isolation under $(F + 1)$-total FDI attacks.*

Proof 8.3 *The result is proved by contradiction. Consider an F-isolable graph where there exists a node i and other $F + 1$ nodes $R_1, R_2, \cdots, R_{F+1}$ satisfying $\mathcal{J}_i \subseteq \{\mathcal{J}_{R_1} \cup \mathcal{J}_{R_2} \cup \cdots \cup \mathcal{J}_{R_{F+1}}\}$. Suppose that i is normal and $R_1, R_2, \cdots, R_{F+1}$ are compromised. Then, there exists a time instant $t \geq \tau$ such that $\|\bar{\varepsilon}_k(t)\| \neq 0$ for all $k \in \mathcal{J}_i$ holds, and the normal node i will be mistakenly isolated as compromised by the DINAIs algorithm, indicating the failure of attack isolation.*

8.3 DISTRIBUTED ISOLATION OF COVERT ATTACKS

8.3.1 Problem formulation

Consider a physically coupled NAS with heterogeneous dynamics, in which the dynamics of each agent is governed by

$$\begin{cases} \dot{x}_i(t) = A_i x_i(t) + B_i u_i(t) + \sum_{j=1}^{N} a_{ij} F_{ij} x_j(t), \\ y_i(t) = C_i x_i(t), \ i = 1, 2, \ldots, N, \end{cases} \tag{8.7}$$

where $x_i \in \mathbb{R}^n$, $u_i \in \mathbb{R}^m$, $y_i \in \mathbb{R}^p$ respectively denote the state, the control input, and the output of the ith agent. A_i, B_i, C_i are known system matrices, and $\sum_{j=1}^{N} a_{ij} F_{ij} x_j(t)$ describes the physical interconnection with neighboring agents. Denote $F_i = [a_{i1}F_{i1}, \cdots, a_{iN}F_{iN}]$, and the assumptions on system (8.7) are made.

Assumption 8.3 $rank\ (B_i, F_i) = rank\ (B_i) = m.$

Assumption 8.4 $rank\ (C_i \cdot [B_i \ F_i]) = rank\ (B_i) = m.$

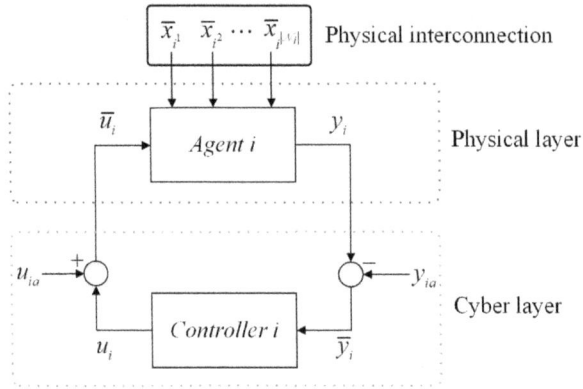

Figure 8.2   An agent under a covert attack.

Assumption 8.5 *rank* $\begin{pmatrix} sI_n - A_i & B_i & F_i \\ C_i & \mathbf{0} & \mathbf{0} \end{pmatrix} = n + m, \ \forall s \in \mathbb{C}.$

Remark 8.12 *Assumptions 8.3–8.5 together ensure the existence of the unknown input observer to be introduced in the next subsection, and Assumption 8.5 indicates that the pair (A_i, C_i) is detectable.*

In a physically coupled NAS, the wide distribution of agents requires their information to be transmitted through a communication network, rendering the system vulnerable to various types of cyber-attacks. If the cyber-attack occurs on agent i at time instant T_{ia}, then the corrupted control input $\bar{u}_i$ and measured output $\bar{y}_i$ can be expressed as follows:

$$\begin{aligned} \bar{u}_i(t) &= u_i(t) + u_{ia}(t), \\ \bar{y}_i(t) &= y_i(t) - y_{ia}(t), \ t \geq T_{ia}. \end{aligned} \tag{8.8}$$

In this section, assume that the covert attacks occur on the agents. An illustrative example of an agent subject to a covert attack is shown in Fig. 8.2. Consider the worst case that the attacker is omniscient and the covert attack model in [155, 167] is adopted, in which the dynamics of the covert attack on the ith agent are governed by

$$\begin{cases} \dot{x}_{ia}(t) = A_i x_{ia}(t) + B_i u_{ia}(t), \\ y_{ia}(t) = C_i x_{ia}(t), \ t \geq T_{ia}, \ x_{ia}(T_{ia}) = \mathbf{0}, \end{cases} \tag{8.9}$$

where $x_{ia} \in \mathbb{R}^n$, $u_{ia} \in \mathbb{R}^m$, $y_{ia} \in \mathbb{R}^p$ are the state, the control input, and the output of the covert attack, respectively.

Remark 8.13 *It derives from Proposition 1 of [155] that the attack (8.8) and (8.9) is covert since the measured output $\bar{y}_i$ is identical to the nominal output y_i.*

Combing (8.7), (8.8), and (8.9), the dynamics of agent i under a covert attack turn into

$$\begin{cases} \dot{\bar{x}}_i(t) = A_i\bar{x}_i(t) + B_i\bar{u}_i(t) + \sum_{i=1}^{N} a_{ij}F_{ij}\bar{x}_j(t), \\ \bar{u}_i(t) = u_i(t) + u_{ia}(t), \\ \bar{y}_i(t) = C_i\bar{x}_i(t) - y_{ia}(t), \\ \dot{x}_{ia}(t) = A_i x_{ia}(t) + B_i u_{ia}(t), \\ y_{ia}(t) = C_i x_{ia}(t), \ t \geq T_{ia}, \ x_{ia}(T_{ia}) = \mathbf{0}. \end{cases} \quad (8.10)$$

Similar to Definition 8.2, the following definition is given to characterize the number of compromised agents subject to cover attacks in this section.

Definition 8.5 (F-total covert attack model) *The NAS is said to be F-total compromised by covert attacks if the compromised node set $\mathcal{V}_A$ contains at most F nodes, i.e., $|\mathcal{V}_A| \leq F$.*

The detection of covert attacks proves challenging under single-observer schemes, rendering Algorithm 8.1, which is designed to counter FDI attacks, ineffective against such threats. This section aims at designing a novel algorithm to detect and isolate covert attacks. The problem investigated in this section can be formulated as follows.

Problem 8.2. Consider the physically coupled NAS under covert attacks described by (8.10), with a communication topology represented by the graph $\mathcal{G}$. Suppose that Assumptions 8.3–8.5 hold. Under the F-total covert attack model, the objective is to design an attack detection and isolation algorithm capable of identifying all compromised agents, while ensuring that no normal agents are mistakenly isolated.

8.3.2 Distributed detection of covert attacks

In this subsection, a distributed detection mechanism for covert attacks is developed based on a two-stage fixed-time observer, as illustrated in Fig. 8.3. The proposed observer structure, adapted from [155, 167], consists of two fixed-time observers: a decentralized unknown input observer and a distributed Luenberger observer. In the first stage, the unknown input observer treats the couplings from neighboring agents as unknown inputs, enabling the estimated state $\hat{x}_i^{\mathbf{u}}$ to be computed independently of any information from neighboring agents. In the second stage, the Luenberger observer incorporates the estimated states $\hat{x}_j^{\mathbf{u}}$ from neighboring agents $j \in \mathcal{N}_i$ to generate the final estimate output $\hat{y}_i$. The residual between the actual output and the estimate output of the two-stage observer is then used as the basis for attack detection. The detailed design of the two-stage fixed-time observer is presented in the following.

As the first stage of the two-stage observer, the structure of the decentralized fixed-time unknown input observer, stemming from [166], is described as follows:

$$\begin{cases} \dot{\omega}_i^{\mathbf{u}}(t) = A_{ci}^{\mathbf{u}}\omega_i^{\mathbf{u}}(t) + B_{ci}^{\mathbf{u}}\bar{y}_i(t), \\ \chi_i(t) = D_{ci}^{\mathbf{u}}\left[\omega_i^{\mathbf{u}}(t) - \exp\left(A_{ci}^{\mathbf{u}}\tau_{1i}\right)\omega_i^{\mathbf{u}}(t - \tau_{1i})\right], \\ \hat{x}_i^{\mathbf{u}}(t) = \chi_i(t) - E_i\bar{y}_i(t), \\ \hat{y}_i^{\mathbf{u}}(t) = C_i\hat{x}_i^{\mathbf{u}}(t), \end{cases} \quad (8.11)$$

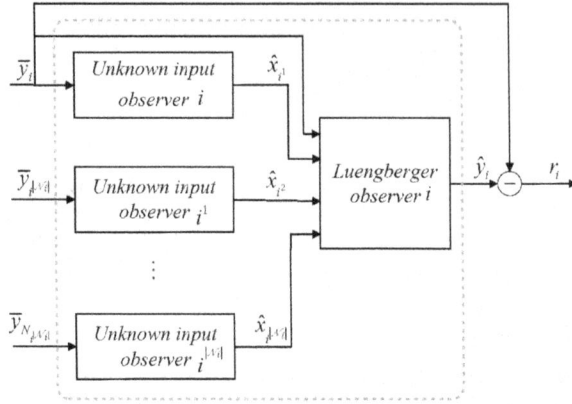

Figure 8.3   The attack detection mechanism based on the two-stage fixed-time observer.

where $\omega_i^{\mathbf{u}}(t) \in \mathbb{R}^{2n}$ and $\chi_i(t) \in \mathbb{R}^n$ are auxiliary variables with $\omega_i^{\mathbf{u}}(t) = \mathbf{0}$ for $t \in [-\tau_{1i}, 0]$, $\tau_{1i} > 0$ is the predefined convergent time of the unknown input observer, $\hat{x}_i^{\mathbf{u}}(t)$ and $\hat{y}_i^{\mathbf{u}}(t)$ are the state and the output of the unknown input observer, respectively. $A_{ci}^{\mathbf{u}} = \begin{pmatrix} G_i A_i - H_i^1 C_i & \mathbf{0} \\ \mathbf{0} & G_i A_i - H_i^2 C_i \end{pmatrix}$, $B_{ci}^{\mathbf{u}} = \begin{pmatrix} B_{ci}^1 \\ B_{ci}^2 \end{pmatrix}$, H_i^k is the gain matrix such that $G_i A_i - H_i^k C_i$ is stable, $B_{ci}^k = H_i^k(I_p + C_i E_i) - G_i A_i E_i, k = 1, 2$, $E_i = -\bar{B}_i\left(C_i \bar{B}_i\right)^\dagger$, $G_i = I_n + E_i C_i$, $\bar{B}_i = (B_i \ F_i)$, $C_c = (I_n \ I_n)^T$, $D_{ci}^{\mathbf{u}} = \begin{pmatrix} I_n & \mathbf{0} \end{pmatrix}\begin{pmatrix} C_c \ \exp\left(A_{ci}^{\mathbf{u}} \tau_{1i}\right) C_c \end{pmatrix}^{-1}$. Then, one has $G_i \bar{B}_i = \mathbf{0}$, $D_{ci}^{\mathbf{u}} C_c = I_n$, and $D_{ci}^{\mathbf{u}} \exp(A_{ci}^{\mathbf{u}} \tau_{1i}) C_c = \mathbf{0}$.

The state estimation error and the output estimation error between agent i and the observer are defined as $\tilde{x}_i^{\mathbf{u}}(t) = x_i(t) - \hat{x}_i^{\mathbf{u}}(t)$ and $r_i^{\mathbf{u}}(t) = \bar{y}_i(t) - \hat{y}_i^{\mathbf{u}}(t)$, respectively. Then, the following result can be derived.

Proposition 8.2 *Consider the physically coupled NAS under covert attacks described by (8.10), with a communication topology represented by the graph $\mathcal{G}$. Suppose that Assumptions 8.3–8.5 hold. Under the fixed-time observer (8.11), the dynamics of the state estimation error and the output estimation error of each agent $i \in \mathcal{V}_A$ are described as*

$$\begin{cases} \dot{\tilde{x}}_i^{\mathbf{u}}(t) = A_i x_{ia}(t) + B_i u_{ia}(t), \\ r_i^{\mathbf{u}}(t) = \mathbf{0}, \ \forall t \geq \tau_{1i}. \end{cases} \tag{8.12}$$

Proof 8.4 *Define $\tilde{\omega}_i^{\mathbf{u}}(t) = \omega_i^{\mathbf{u}}(t) - C_c G_i\left(x_i(t) - x_{ia}(t)\right)$, one has*

$$\dot{\tilde{\omega}}_i^{\mathbf{u}}(t) = A_{ci}^{\mathbf{u}} \omega_i^{\mathbf{u}}(t) + B_{ci}^{\mathbf{u}} C\left(x_i(t) - x_{ia}(t)\right)$$
$$- C_c G_i \left[A_i x_i(t) + B_i \bar{u}_i(t) + \sum_{i=1}^N a_{ij} F_{ij} x_j(t) - A_i x_{ia}(t) - B_i u_{ia}(t) \right]$$
$$= A_{ci}^{\mathbf{u}} \tilde{\omega}_i^{\mathbf{u}}(t),$$

where $G_i \bar{B}_i = \mathbf{0}$ and $B_{ci}^{\mathbf{u}} C_i - C_c G_i A_i = -A_{ci}^{\mathbf{u}} C_c G_i$ have been used to get the above *equality. Then, it derives that*

$$\tilde{\omega}_i^u(t) = \exp\left(A_{ci}^u \tau_{1i}\right) \tilde{\omega}_i^u(t - \tau_{1i}), \quad \forall t \geq \tau_{1i}. \tag{8.13}$$

Combining (8.11) and (8.13), one has

$$
\begin{aligned}
\chi_i(t) &= D_{ci}^u [C_c G_i \left(x_i(t) - x_{ia}(t)\right) - \exp\left(A_{ci}^u \tau_{1i}\right) C_c G_i \left(x_i(t - \tau_{1i}) - x_{ia}(t - \tau_{1i})\right)] \\
&= G_i \left(x_i(t) - x_{ia}(t)\right), \quad \forall t \geq \tau_{1i},
\end{aligned}
$$

where $D_{ci}^u C_c = I_n$ and $D_{ci}^u \exp(A_{ci}^u \tau_{1i}) C_c = 0$ have been used. Thus, direct calculation gives that

$$
\begin{aligned}
\dot{\tilde{x}}_i^u(t) &= \dot{x}_{ia}(t) = A_i x_{ia}(t) + B_i u_{ia}(t), \\
r_i^u(t) &= C_i \left(x_i(t) - x_{ia}(t) - \hat{x}_i^u(t)\right) = 0, \quad \forall t \geq \tau_{1i}.
\end{aligned}
$$

It shows in Proposition 8.2 that the decentralized unknown input observer cannot detect the cover attack since the output error $r_i^u(t) = 0$ when $t \geq \tau_{1i}$, while the state estimation error $\tilde{x}_i^u(t)$ reflects some information about the attack. Observing this, the distributed Luenberger observer in the second stage uses the state of the above unknown input observer $\hat{x}_j^u(t)$, $j \in \mathcal{N}_i$ to expose the attack on its neighboring agents.

The second stage of the two-stage observer is activated when $t \geq \tau_{1i}$, and the distributed fixed-time Luenberger observer, originating from [165], is introduced in the following.

$$
\begin{cases}
\dot{\omega}_i^l(t) = A_{ci}^l \omega_i^l(t) + B_{ci}^l u_i(t) + L_{ci} \bar{y}_i(t) + C_c \sum_{i=1}^N a_{ij} F_{ij} \hat{x}_j^u(t), \\
\hat{x}_i^l(t) = D_{ci}^l \left[\omega_i^l(t) - \exp\left(A_{ci}^l \tau_{2i}\right) \omega_i^l(t - \tau_{2i})\right], \\
\hat{y}_i(t) = C_i \hat{x}_i^l(t),
\end{cases}
\tag{8.14}
$$

where $\omega_i^l(t) \in \mathbb{R}^{2n}$ is the auxiliary variable with $\omega_i^l(t) = 0$ for $t \in [\tau_{1i} - \tau_{2i}, \tau_{1i}]$, $\tau_{2i} > 0$ is the predefined convergent time of the Luenberger observer, $\hat{x}_i^l(t)$ and $\hat{y}_i(t)$ are the state of the Luenberger observer and the output of the two-stage observer, respectively. $A_{ci}^l = \begin{pmatrix} A_i - L_{1i} C_i & 0 \\ 0 & A_i - L_{2i} C_i \end{pmatrix}$, L_{ki} is the gain matrix such that $A_i - L_{ki} C_i$ is stable, $k = 1, 2$. $B_{ci}^l = \left(B_i^T \ B_i^T\right)^T$, $L_{ci} = \left(L_{1i}^T \ L_{2i}^T\right)^T$, $D_{ci}^l = \left(\begin{array}{cc} I_n & 0 \end{array}\right)\left(C_c \ \exp\left(A_{ci}^l \tau_{2i}\right) C_c \ \right)^{-1}$. Then, one has $D_{ci}^l C_c = I_n$ and $D_{ci}^l \exp(A_{ci}^l \tau_{2i}) C_c = 0$.

The output estimation residual between the agent i and its two-stage observer is defined as $r_i(t) = \bar{y}_i(t) - \hat{y}_i(t)$, and the attack detection result is given in the following.

Proposition 8.3 *Consider the physically coupled NAS under covert attacks described by (8.10), with a communication topology represented by the graph $\mathcal{G}$. Suppose that Assumptions 8.3–8.5 hold. Under the two-stage fixed-time observer (8.11) and (8.14), there exist compromised agents in $\mathcal{N}_i$ if there exists a time instant $t \geq \tau_{1i} + \tau_{2i}$ such that $\|r_i(t)\| \neq 0$. Moreover, for the case that there is at most one compromised agent in $\mathcal{N}_i$, the attack can be detected if and only if there exists a time instant $t \geq \tau_{1i} + \tau_{2i}$ such that $\|r_i(t)\| \neq 0$.*

Proof 8.5 *Define $\tilde{\omega}_i^l(t) = \omega_i^l(t) - C_c\left(x_i(t) - x_{ia}(t)\right)$, one has*

$$
\begin{aligned}
\dot{\tilde{\omega}}_i^l(t) &= A_{ci}^l \omega_i^l(t) + B_{ci}^l u_i(t) + L_{ci} C_i \left(x_i(t) - x_{ia}(t)\right) + C_c \sum_{i=1}^N a_{ij} F_{ij} \hat{x}_j^u(t) - \\
&\quad C_c \left(A_i x_i(t) + B_i \bar{u}_i(t) + \sum_{i=1}^N a_{ij} F_{ij} x_j(t) - A_i x_{ia}(t) - B_i u_{ia}(t) \right) \\
&= A_{ci}^l \tilde{\omega}_i^l(t) - C_c \sum_{i=1}^N a_{ij} F_{ij} \tilde{x}_j^u(t).
\end{aligned}
\tag{8.15}
$$

If there is no attack in $\mathcal{N}_i$, one has $\tilde{x}_j^u(t) = \mathbf{0}$ for all $j \in \mathcal{N}_i$ and $t \geq \tau_{1i}$. Then, it derives that

$$
\tilde{\omega}_i^l(t) = \exp\left(A_{ci}^l \tau_{2i}\right) \tilde{\omega}_i^l(t - \tau_{2i}), \ \forall t \geq \tau_{1i} + \tau_{2i}.
\tag{8.16}
$$

Besides, direct calculation gives that

$$
\begin{aligned}
\hat{x}_i^l(t) &= D_{ci}^l[C_c\left(x_i(t) - x_{ia}(t)\right) - \exp\left(A_{ci}^l \tau_{2i}\right) C_c \left(x_i(t - \tau_{2i}) - x_{ia}(t - \tau_{2i})\right)] \\
&= \left(x_i(t) - x_{ia}(t)\right), \ \forall t \geq \tau_i,
\end{aligned}
\tag{8.17}
$$

where $\tau_i = \tau_{1i} + \tau_{2i}$, $D_{ci}^l C_c = I_n$ and $D_{ci}^l \exp(A_{ci}^l \tau_{2i}) C_c = \mathbf{0}$ have been used to obtain the above equality. Thus, one has $r_i(t) = \mathbf{0}$, $\forall t \geq \tau_i$.

If only one agent in $\mathcal{N}_i$ is subject to a covert attack, one has $\|r_i(t)\| \neq 0$ since $\tilde{x}_j^l(t) \neq \mathbf{0}$ for $j \in \mathcal{N}_i \cap \mathcal{V}_A$ while $\tilde{x}_j^u(t) = \mathbf{0}$ for $j \in \mathcal{N}_i \backslash \mathcal{V}_A$, and vice versa. However, if there exists more than one compromised agent within $\mathcal{N}_i$, then $\|r_i(t)\| \neq 0$ implies that there exist compromised agents within $\mathcal{N}_i$. Nevertheless, $\|r_i(t)\| = 0$ does not necessarily imply the absence of attacks, as multiple covert attacks may collude to make $\sum_{i=1}^N a_{ij} F_{ij} \tilde{x}_j^u(t) = \mathbf{0}$, thereby evading detection.

Remark 8.14 *Compared with the distributed fixed-time observer based attack detection method in Section 8.2 which fails to detect matched actuator attacks, the above two-stage fixed-time observer based attack detection strategy presents a more universal characteristic, which is effective for FDI attacks [168], matched actuator attacks [169], and covert attacks.*

Remark 8.15 *As is known, the detection threshold and other issues such as noise and disturbance have influence on attack detection. In this section, the fixed-time observers are proposed to exclude the effect of detection threshold. Besides, the system model is free of noise and disturbance, which supplies an ideal case to investigate the basic graph conditions for successful attack isolation.*

8.3.3 Distributed isolation of covert attacks

The following DNAIs algorithm is developed in Algorithm 8.2 to identify agents subject to covert attacks. In the case where $\mathcal{N}_j \subseteq \mathcal{N}_i$, with agent i compromised and agent j normal, the normal agent j may be mistakenly isolated as compromised by Algorithm 8.2. This means that additional graph-theoretic conditions are required to ensure zero false alarms under Algorithm 8.2.

To prevent the mistaken isolation of normal agents, a novel graph property in terms of r^*-*isolability* is introduced first.

Algorithm 8.2 DNAIs algorithm

Input: $S_i(0) = 0, i \in \mathcal{V}, r_j(t), j \in \mathcal{N}_i$;
Output: $S_i(t), i \in \mathcal{V}$;

1: **for** $t \geq \max_{i \in \mathcal{V}} \{T_{ia}, \tau_i\}$ **do**

2:      $\Upsilon_i(t) = \left(\|r_{i^1}(t)\|, \|r_{i^2}(t)\|, ..., \|r_{i^{|\mathcal{N}_i|}}(t)\| \right)^T \in \mathbb{R}^{|\mathcal{N}_i|}, \forall i \in \mathcal{V}$;

3:      **if** $\|\Upsilon_i(t)\|_{l_0} = |\mathcal{N}_i|$ **then**

4:          $S_i(t) = i$.

5:      **end if**

6: **end for**

Definition 8.6 (r^*-isolable node) *Given a nontrivial graph $\mathcal{G} = (\mathcal{V}, \mathcal{E})$ and a node $i \in \mathcal{V}$, the node i is r^*-isolable if, for any other r nodes, denoted as $R_1, R_2, \ldots, R_r$, one has $\mathcal{N}_i \not\subseteq \{\mathcal{N}_{R_1} \cup \mathcal{N}_{R_2} \cup \cdots \cup \mathcal{N}_{R_r}\}$.*

The reason that the neighbor sets of any other r nodes are considered in Definition 8.6 is that the normal agent i will be mistakenly isolated if the r nodes are compromised and their neighbor sets satisfy $\mathcal{N}_i \subseteq \{\mathcal{N}_{R_1} \cup \mathcal{N}_{R_2} \cup \cdots \cup \mathcal{N}_{R_r}\}$. Based on this observation, the following definition is introduced to characterize the notion of r^*-isolability.

Definition 8.7 (r^*-isolable graph) *A nontrivial graph $\mathcal{G} = (\mathcal{V}, \mathcal{E})$ is r^*-isolable if, for every node $i \in \mathcal{V}$ and any other r nodes, $R_1, R_2, \ldots, R_r$, one has $\mathcal{N}_i \not\subseteq \{\mathcal{N}_{R_1} \cup \mathcal{N}_{R_2} \cup \cdots \cup \mathcal{N}_{R_r}\}$.*

Remark 8.16 *It derives from Definition 8.7 that $|\mathcal{N}_i| \geq r + 1, \forall i \in \mathcal{V}$ if $\mathcal{G}$ is r^*-isolable. Besides, $\mathcal{G}$ is r'^*-isolable if it is r^*-isolable with $1 \leq r' \leq r$.*

Remark 8.17 *Compared with the definition of r-isolability in Section 8.2, the notion of r^*-isolability only focuses on the neighbor sets of the $r + 1$ nodes. Some simple properties can be easily derived: 1) the cycle graph with $N = 3$ is 1^*-isolable but not 1-isolable; 2) the cycle graph with $N = 4$ is 1-isolable but not 1^*-isolable; 3) any cycle graph with $N \geq 5$ is both 1^*-isolable and 1-isolable.*

Subsequently, the necessary and sufficient conditions for attack isolation are established. In particular, if there exists at most one compromised agent in each subsystem $\mathcal{G}_i(\mathcal{J}_i^*, \mathcal{E}_i)$, the following result on attack isolation can be derived.

Theorem 8.3 *Consider a physically coupled NAS under covert attacks described by (8.10), with a communication topology represented by the graph $\mathcal{G}$. Suppose that Assumptions 8.3–8.5 hold. Assume that the path length of any two compromised agents satisfies $|l_{ij}| \geq 3, \forall i, j \in \mathcal{V}_A$, and there exists a time instant $t \geq \max_{i \in \mathcal{V}} \{T_{ia}, \tau_i\}$ such that $\|r_k(t)\| \neq 0$ for all $i \in \mathcal{V}_A, k \in \mathcal{N}_i$, under the two-stage fixed-time observer (8.11) and (8.14). Attack isolation can be achieved under Algorithm 8.2 if and only if the communication topology $\mathcal{G}$ is 1^*-isolable.*

Proof 8.6 *For any two compromised agents i and j, $|l_{ij}| \geq 3$ implies that there exists at most one compromised agent in each subsystem $\mathcal{G}_i(\mathcal{J}_i^*, \mathcal{E}_i)$. It follows from Algorithm 8.2 that all the compromised agents can be isolated if $\|r_k(t)\| \neq 0$, $\forall i \in \mathcal{V}_A, k \in \mathcal{N}_i$. On the other hand, Algorithm 8.2 shows that the normal agent k is mistakenly isolated as compromised if and only if $\mathcal{N}_k \subseteq \mathcal{N}_i$.*

The following attack isolation result is provided for general F-total covert attack model.

Theorem 8.4 *Consider the physically coupled NAS under covert attacks described by (8.10), with a communication topology represented by the graph $\mathcal{G}$. Suppose that Assumptions 8.3–8.5 hold. Assume that there exists a time instant $t \geq \max_{i \in \mathcal{V}} \{T_{ia}, \tau_i\}$ such that $\|r_k(t)\| \neq 0$, $\forall i \in \mathcal{V}_A, k \in \mathcal{N}_i$, under the two-stage fixed-time observer (8.11) and (8.14). For any F-total covert attack model, attack isolation can be accomplished by Algorithm 8.2 if the communication graph $\mathcal{G}$ is F-isolable.*

The proof is similar to that of Theorem 8.3, and is hence omitted here.

8.4 RESILIENT COOPERATIVE CONTROL BASED ON ATTACK ISOLATION

8.4.1 Resilient consensus of high-order NASs

For the NAS (8.1), a fully distributed adaptive control protocol borrowed from [176] is given as follows:

$$\begin{cases} u_i = \rho_i K \hat{\varepsilon}_i, \\ \dot{\rho}_i = \hat{\varepsilon}_i^T Q B B^T Q \hat{\varepsilon}_i, \end{cases} \tag{8.18}$$

where $K = -B^T Q$, Q is a positive matrix satisfying $Q^{-1} A^T + A Q^{-1} - 2 B B^T < 0$.

Lemma 8.2 *[176] Consider a high-order NAS (8.1) without any compromised agent, suppose that Assumptions 8.1 and 8.2 hold. Consensus can be achieved by protocol (8.18) if and only if its communication topology $\mathcal{G}$ is connected.*

The control protocol in [176] fails to reach consensus in the presence of compromised agents. To realize resilient consensus with compromised agents in the NAS, an extra resilient consensus algorithm based on DINAIs (DINAIs-RC) algorithm is designed as follows.

Remark 8.18 *Once all compromised agents are removed by setting $a_{ij} = 0$, resilient consensus can be reached by the above DINAIs-RC algorithm if and only if the induced graph consisting of normal agents is connected.*

Remark 8.19 *In [49], [51] and [160], MSR algorithms were used to remove at most $2F$ extreme state values from the neighbor set $\mathcal{N}_i$ so as to avoid the influence of compromised agents, which however cannot be applied to general linear NASs. In this section, the DINAIs-RC algorithm only removes the compromised agents isolated by the DINAIs algorithm, thus relaxing the condition of graph robustness. Furthermore, it can be applied to general high-order linear NASs.*

Algorithm 8.3 DINAIs-RC algorithm

1: For each normal agent i with $S_i = 0$ and its neighbors $j \in \mathcal{N}_i$ **do**
2: **for** $j = N_{i^1} : N_{i|\mathcal{N}_i|}$ **do**
3:     **if** $S_j = j$ **then**
4:         $a_{ij} = 0$;
5:         $\mathcal{N}_i = \mathcal{N}_i \setminus j$;
6:     **end if**
7: **end for**
8: Design control protocol u_i as in (8.18).

Leveraging on Theorem 8.1, the main result about resilient consensus can be established as follows.

Corollary 8.1 *Consider a high-order NAS described by (8.1), with a communication topology represented by the graph $\mathcal{G}$. Suppose that Assumptions 8.1 and 8.2 holds. In the 1-weak-local FDI attack model, resilient consensus can be achieved under the DINAIs-RC algorithm if and only if the followings hold simultaneously:*

i) $\exists t \geq \tau$ s.t. $\|\tilde{\varepsilon}_k(t)\| \neq 0$ for all $k \in \mathcal{J}_i, i \in \mathcal{V}_A$;

ii) Graph $\mathcal{G}$ is 1-isolable;

iii) The subgraph $\mathcal{G}'$ induced by removing all the compromised agents is connected.

Proof 8.7 *The proof straightforwardly follows from Theorem 8.1 for attack isolation and Lemma 8.2 for resilient consensus. Conditions i) and ii) ensure that attack isolation can be achieved. Condition iii) ensures that resilient consensus of the normal agents can be reached after removing the compromised agents.*

Corollary 8.2 *Consider a high-order NAS described by (8.1), with a communication topology represented by the graph $\mathcal{G}$. Suppose that Assumptions 8.1 and 8.2 holds. In the F-total FDI attack model with $F \geq 2$, resilient consensus can be achieved under the DINAIs-RC algorithm if the followings hold simultaneously:*

i) $\exists t \geq \tau$ s.t. $\|\tilde{\varepsilon}_k(t)\| \neq 0$ for all $i \in \mathcal{V}_A, k \in \mathcal{J}_i$;

ii) Graph $\mathcal{G}$ is F-isolable;

iii) The subgraph $\mathcal{G}'$ induced by removing the F compromised agents is connected.

The proof is similar to that of Corollary 8.1, and hence is omitted here.

8.4.2   Resilient stability of physically coupled NASs

Borrowed from [190], the control protocol for the physically coupled NAS (8.11) is given as follows:

$$u_i = K\hat{x}_i^l + \sum_{j=1}^{N} \bar{a}_{ij} K_{ij} \hat{x}_j^l, \tag{8.19}$$

Algorithm 8.4 NAIs-RS algorithm

Input: $S_i, i = 1, 2, \ldots, N$;
 1: **for** $i = 1 : N$ **do**
 2: **if** $S_i = i$ **then**
 3:    **for** $j = 1 : N$ **do**
 4:      $\bar{a}_{ij} = 0$;
 5:      Unplug agent i from the physically coupled NAS;
 6:    **end for**
 7: **else**
 8:    **for** $j = 1 : N$ **do**
 9:      $\bar{a}_{ij} = a_{ij}$;
10:    **end for**
11: **end if**
12: Design control protocol u_i as in (8.19).
13: **end for**

---

where K and K_{ij} can optimally designed as [177] to ensure the stability of the whole system, $\bar{a}_{ij}$ is the updated weight coefficient which will be given in Algorithm 8.4.

Before going on, the following assumption is made.

Assumption 8.6 *The physical unplugging of agents is allowed in the physically coupled NAS (8.11).*

If some agents are compromised, it is imperative to unplug them to reduce the propagation of their influence in the whole system. Based on this, the resilient stability algorithm based on DNAIs (NAIs-RS) algorithm is described in Algorithm 8.4.

Based on attack isolation, the result on resilient stability are directly derived in the following.

Corollary 8.3 *Consider the physically coupled NAS under covert attacks described by (8.10), with a communication topology represented by the graph $\mathcal{G}$. Suppose that Assumptions 8.3–8.6 hold. Assume that there exists a time instant $t \geq \max_{i \in \mathcal{V}} \{T_{ia}, \tau_i\}$ such that $\|r_k(t)\| \neq 0$ for all $i \in \mathcal{V}_A, k \in \mathcal{N}_i$ under the two-stage fixed-time observer (8.11) and (8.14). For the F-total covert attack model, the resilient stability of all normal agents can be achieved by Algorithms 8.2 and 8.4 if the followings hold simultaneously:*

 (i) Graph $\mathcal{G}$ is F^*-isolable;

 (ii) The subgraph $\mathcal{G}'$ induced by removing the isolated compromised agents is connected.

8.5 NUMERICAL SIMULATIONS

In this chapter, three numerical examples are provided to verify the above theoretical results in Sections 8.2–8.4.

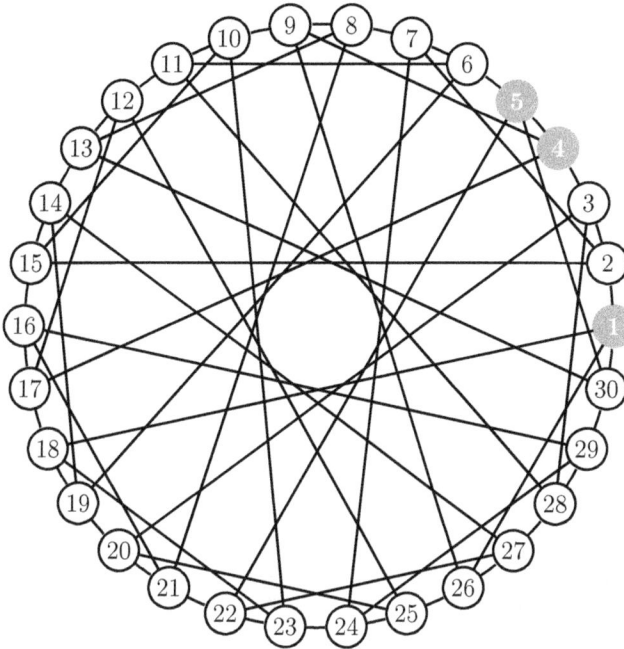

Figure 8.4   The communication topology of a NAS consisting of thirty agents.

Example 1: This example demonstrates the graph-theoretic conditions required to achieve resilient consensus in general high-order NASs under FDI attacks. Consider the 3-isolable graph as shown in Fig. 8.4, where agents 1, 4 and 5 are compromised by FDI attacks. The dynamics of each agent are described by (8.1) and the system matrices are given as follows:

$$A = \begin{bmatrix} 2.25 & 9 & 0 \\ 1 & -1 & 0 \\ 0 & -18 & 0 \end{bmatrix}, \ B = \begin{bmatrix} 1 \\ 0 \\ 0 \end{bmatrix}, \ C = \begin{bmatrix} 1 & 0 & 0 \\ 0 & 1 & 0 \end{bmatrix}, \ M = \begin{bmatrix} 1 \\ 0 \end{bmatrix}.$$

The time delay $\tau = 1$, choose $H_1 = \begin{bmatrix} 2.6426 & -9.1084 \\ 1.0810 & 0.0392 \\ -6.4073 & 1.4995 \end{bmatrix}$, $H_2 =$

$\begin{bmatrix} 3.3919 & -4.0151 \\ 4.7846 & 3.9039 \\ 4.7477 & 4.0377 \end{bmatrix}$, $Q = \begin{bmatrix} 0.4851 & 0.1186 & 0.3075 \\ 0.1186 & 1.7495 & 0.0211 \\ 0.3075 & 0.0211 & 0.5212 \end{bmatrix}$. Assume that the FDI at-

tacks occur at $T_1 = T_4 = T_5 = 1.7$, the influence of attacks on agents 1, 4, 5 are described as $f_1 = x_{1_2} \sin(x_{2_1}) + \sin(0.2(t-1))$, $f_4 = 3\cos(x_{2_3}) x_{1_1} - \cos(0.5(t-1))$, $f_5 = 2\cos(x_{7_2})x_{2_2} + \cos(0.1(t-1))$.

The consensus estimation error trajectory of each agent is shown in Fig. 8.5, where $\tilde{\varepsilon}_i = \mathbf{0}$ for all $i \in \mathcal{V}$ when $1 \leq t \leq 1.7$, indicating the accurate estimation of consensus error. When $t > 1.7$, the attacks occur, and $\tilde{\varepsilon}_i \neq \mathbf{0}$ for $i \in \{1, 2, 3, 4, 5, 6, 9, 17, 18, 22, 26, 30\}$, one has $C_i = |\mathcal{J}_i|, i \in \{1, 4, 5\}$ and $C_i < |\mathcal{J}_i|, i \in \mathcal{V}\backslash\{1, 4, 5\}$. The safety indices of each agent are illustrated in Fig. 8.6, where the safety indices corresponding to agents 1, 4, and 5 clearly indicate that they have

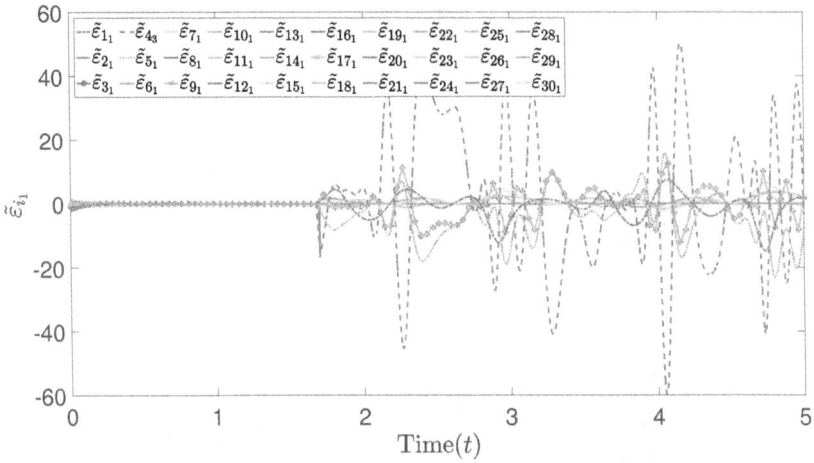

(a)

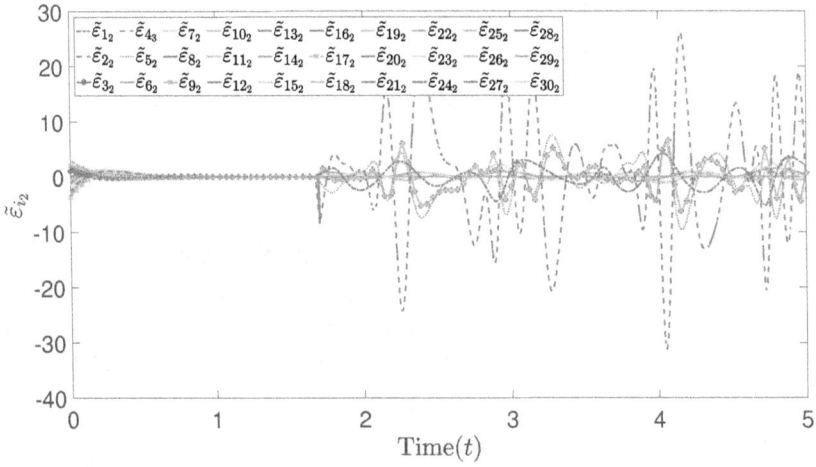

(b)

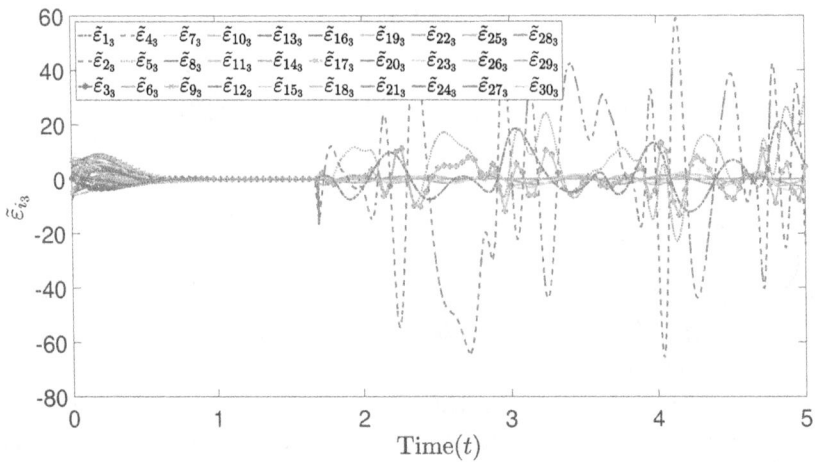

(c)

Figure 8.5　The consensus estimation error trajectory $\tilde{\varepsilon}_i$ of each agent.

been isolated as compromised. This observation verifies the correctness of Theorem 8.2. Finally, Fig. 8.7 presents the consensus error trajectories of all agents, which demonstrate that the normal agents successfully achieve resilient consensus under Algorithm 8.3, thereby confirming the effectiveness of Corollary 8.1.

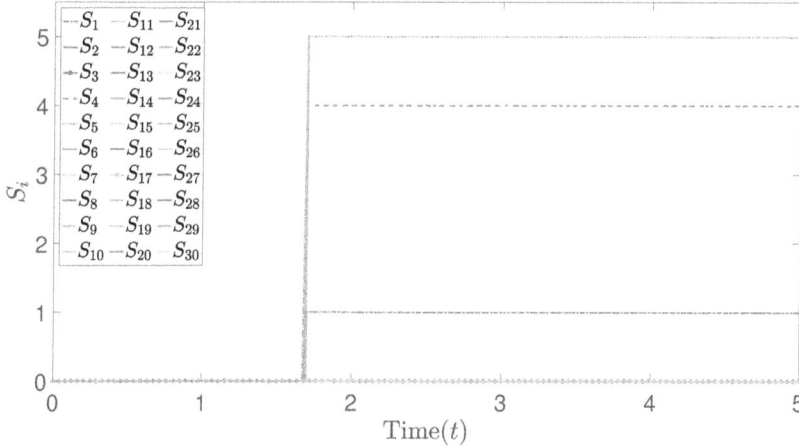

Figure 8.6   The safety index S_i of each agent.

Example 2: This example compares the DINAIs-RC algorithm with the DP-MSR algorithm in [51]. Consider the NAS consisting of six agents, with its communication topology is shown in Fig. 8.1(b). With Definitions 2.5 and 8.4, it is easy to verify that the graph given in Fig. 8.1(b) is 1-isolable but not $(2,2)$-robust. Assume that the normal agents follow double-integrator dynamics and agent 3 is compromised at $T_3 = 0$, and the attack signal $Mf_3 = (2 + 2\cos(3t), 3 + 3\sin(4t))^T$. The initial value of each agent is given as in [51].

The trajectory of each agent under the DP-MSR algorithm in [51] and Algorithm 8.3 are illustrated in Fig. 8.8 and Fig. 8.9, respectively. Note that resilient consensus in Fig. 8.8 cannot be reached since the graph is not $(2,2)$-robust. However, Fig. 8.9 illustrates that all normal agents reach resilient consensus since conditions in Corollary 8.1 are satisfied. This shows that the notion of r-isolability can help reach resilient consensus when (r,s)-robustness fails in a NAS.

Now, the methods of DINAIs-based resilient consensus involving F-isolability and MSR-based resilient consensus with $(F+1, F+1)$-robustness are compared and summarized as follows. The notion of F-isolability emphasizes graph structure but not high connectivity, and thus needs less communication cost, while the construction of fixed-time observers and the execution of DINAIs algorithm increase the computation burden. On the other hand, the notion of $(F + 1, F + 1)$-robustness depends on higher connectivity and requires higher communication cost, while the pure local MSR-algorithm takes up less computation burden. Therefore, for integrator-typed networks, the selection of an appropriate algorithm depends on 1) the trade-off between communication cost and computation burden; 2) the given network properties.

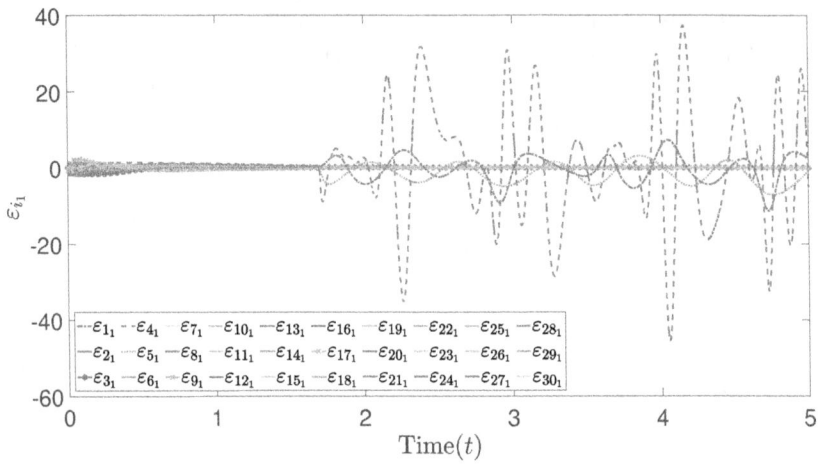

(a)

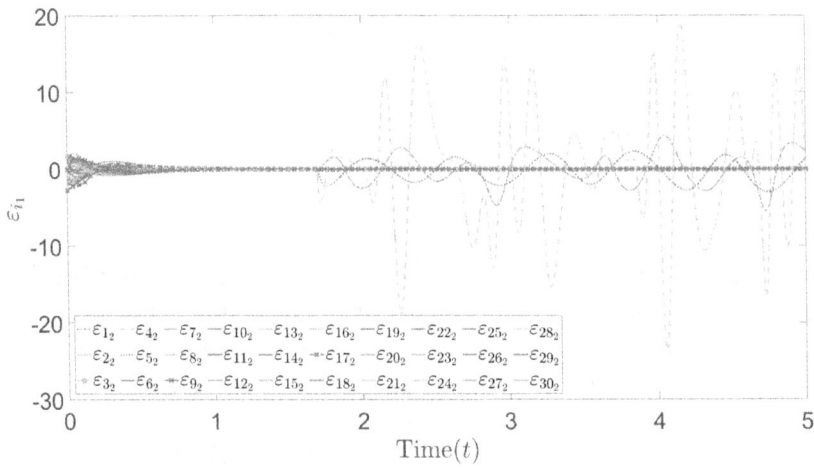

(b)

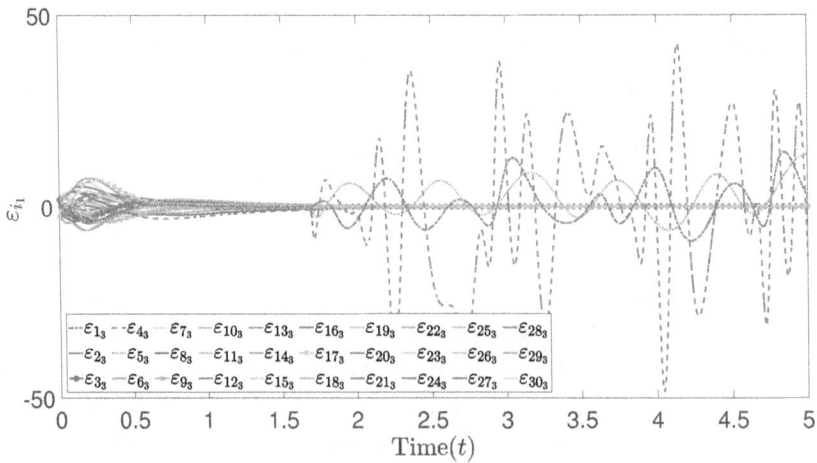

(c)

Figure 8.7   The consensus error trajectory ε_i of each agent.

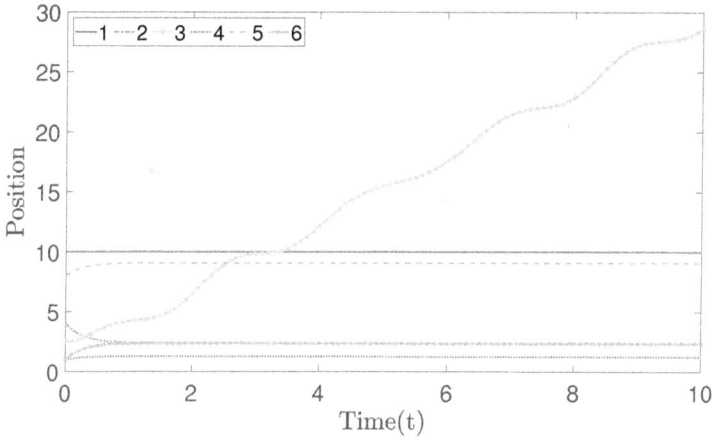

(a) Position.

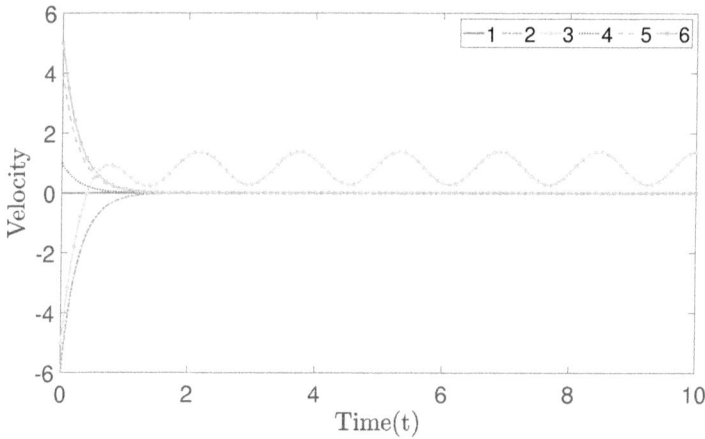

(b) Velocity.

Figure 8.8   The trajectory of each agent under the DP-MSR algorithm in [51].

Besides, the DINAIs-based resilient consensus algorithm can be applied to general high-order networks.

Example 3: This example demonstrates the graph-theoretic conditions required to achieve resilient stability in physically coupled NASs under covert attacks. Consider a physically coupled NAS consisting of ten agents with heterogeneous dynamics described by (8.7), the system matrices are given as follows:

$$A_i = \begin{bmatrix} -0.45 & 1.8 & 0 \\ 0.2 & -0.2 & 0.2 \\ 0 & -3.6 & 0 \end{bmatrix}, B_i = \begin{bmatrix} 1 \\ 0 \\ 0 \end{bmatrix}, C_i = \begin{bmatrix} 1 & 0 & 0 \\ 1 & 0 & 1 \end{bmatrix}, \ i = 1, 2, 3, 6 - 10,$$

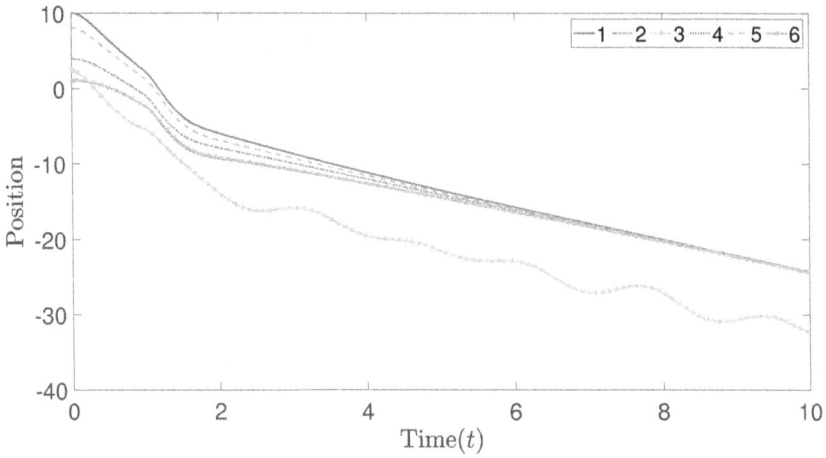

(a) Position.

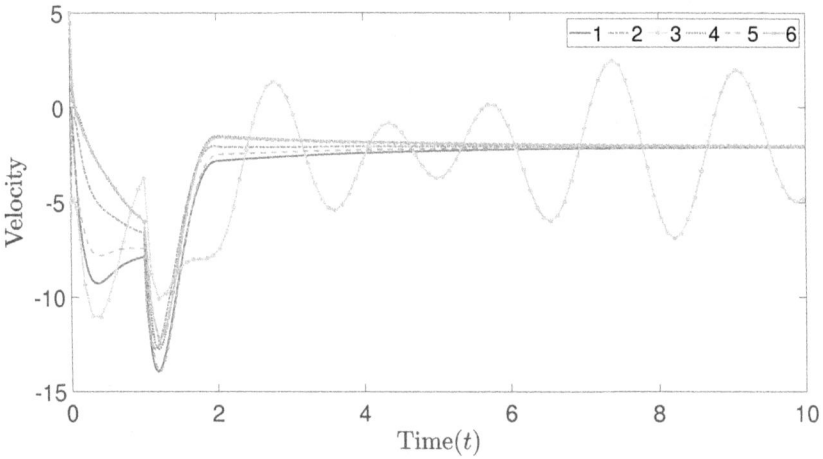

(b) Velocity.

Figure 8.9   The trajectory of each agent under Algorithm 8.3.

$$A_j = \begin{bmatrix} -0.225 & 0.9 & 0 \\ 0.1 & -0.1 & 0.1 \\ 0 & -1.8 & 0 \end{bmatrix}, B_j = \begin{bmatrix} 2 \\ 0 \\ 0 \end{bmatrix}, C_j = \begin{bmatrix} 2 & 0 & 0 \\ 1 & 0 & 1 \end{bmatrix}, \ j = 4, 5.$$

The time delays of the two-stage fixed-time observer are chosen as $\tau_{i1} = 1.5$, $\tau_{i2} = 0.5$, $i \in \mathcal{V}$, and relative parameters of the two-stage fixed-time observer are calculated as

$$H_i^1 = \begin{bmatrix} 3.8762 & 14.3249 \\ 5.4710 & 10.3434 \\ -10.5239 & 9.3397 \end{bmatrix}, H_i^2 = \begin{bmatrix} 11.2850 & 15.2306 \\ 17.7751 & 2.8519 \\ -17.3938 & 9.7091 \end{bmatrix},$$

$$L_{1i} = \begin{bmatrix} 1.2878 & -4.0240 \\ 0.8598 & -3.7252 \\ 3.2781 & 14.3944 \end{bmatrix}, L_{2i} = \begin{bmatrix} 9.0643 & 0.7217 \\ 5.8654 & -6.9583 \\ 1.6025 & 10.6746 \end{bmatrix}, i = 1, 2, 3, 6 - 10,$$

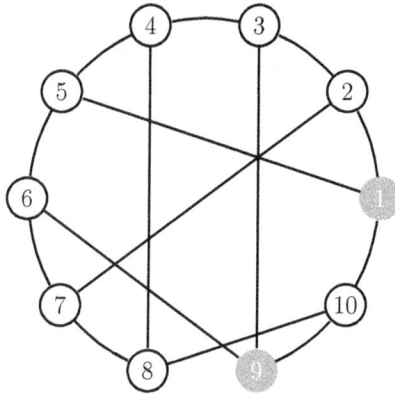

Figure 8.10   The communication topology of a physically coupled NAS consisting of ten agents.

$$H_j^1 = \begin{bmatrix} 7.4497 & -3.3363 \\ -3.0124 & -13.4267 \\ 1.7021 & 9.1620 \end{bmatrix}, H_j^2 = \begin{bmatrix} 6.8333 & 5.9756 \\ 9.7784 & -13.1848 \\ -11.7704 & 0.0695 \end{bmatrix},$$

$$L_{1j} = \begin{bmatrix} 3.7498 & -0.9108 \\ 12.1637 & -5.2812 \\ -6.2269 & 5.2889 \end{bmatrix}, L_{2j} = \begin{bmatrix} 2.1538 & 9.5640 \\ 11.7780 & 10.5162 \\ -4.6353 & -0.2625 \end{bmatrix}, j = 4, 5.$$

The communication topology of the physically coupled NAS with ten agents is shown in Fig. 8.10, which is 2*-isolable by Definition 8.6. Assume that agents 1 and 9 subject to covert attacks at $T_{1a}=T_{9a}=2.5$, and the control inputs of the attackers satisfy $u_{1a}=(100\ 150\ 200)^T$ and $u_{9a}=(150\ 100\ 200)^T$.

The output residuals for all agents are depicted in Fig. 8.11, demonstrating that the output of the observer $\hat{y}_i$ converges to the output of the agent y_i within $\tau_i = 2$. Once the covert attacks occur at $t = 2.5$, it shows that $\|r_i(t)\| \neq 0$ for $i \in \{2, 3, 5, 6, 10\}$, and one has $\|\Upsilon_1\|_{l_0} = |\mathcal{N}_1|$ and $\|\Upsilon_9\|_{l_0} = |\mathcal{N}_9|$, while other agents satisfy $\|\Upsilon_i\|_{l_0} < |\mathcal{N}_i|, i \in \mathcal{V}\backslash\{1,9\}$. The safety indices S_i in Fig. 8.12 indicate that the agents 1 and 9 are isolated as compromised. The state trajectories of all agents are shown in Fig. 8.13, demonstrating the effectiveness of Corollary 8.3.

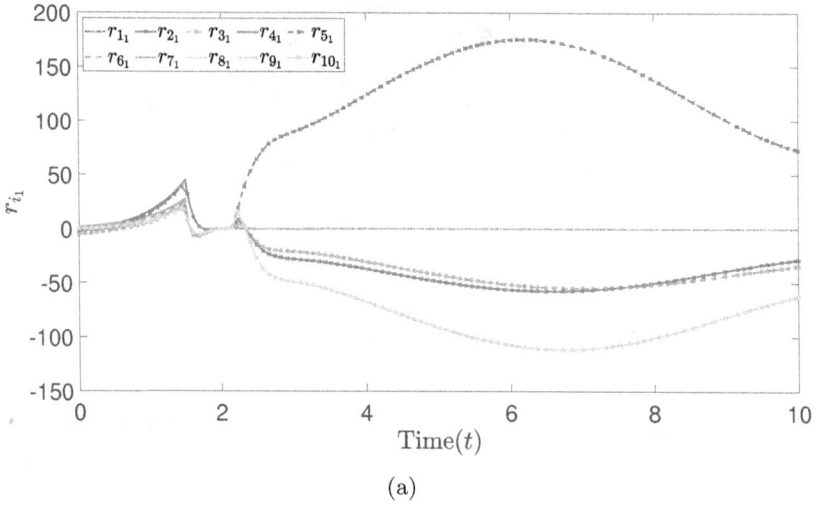

(a)

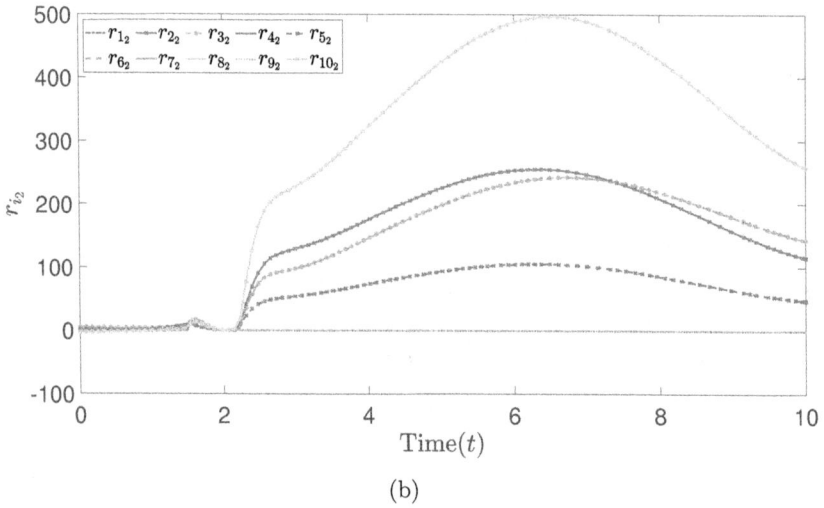

(b)

Figure 8.11   The residual trajectory of each agent r_i.

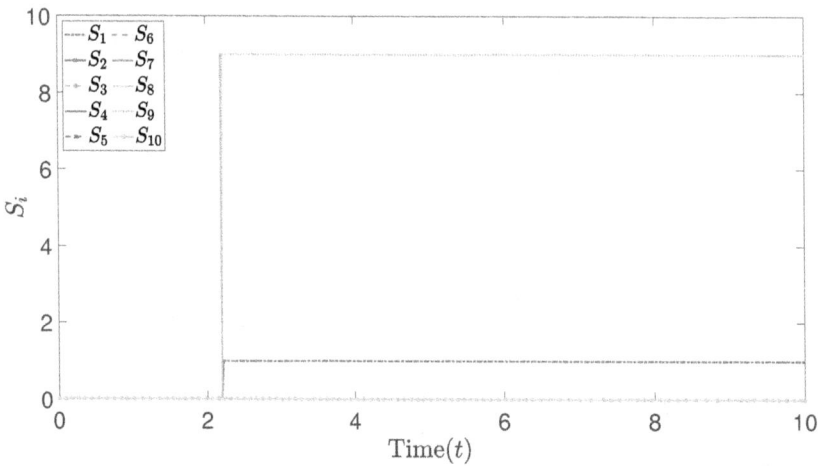

Figure 8.12   The safety index of each agent S_i.

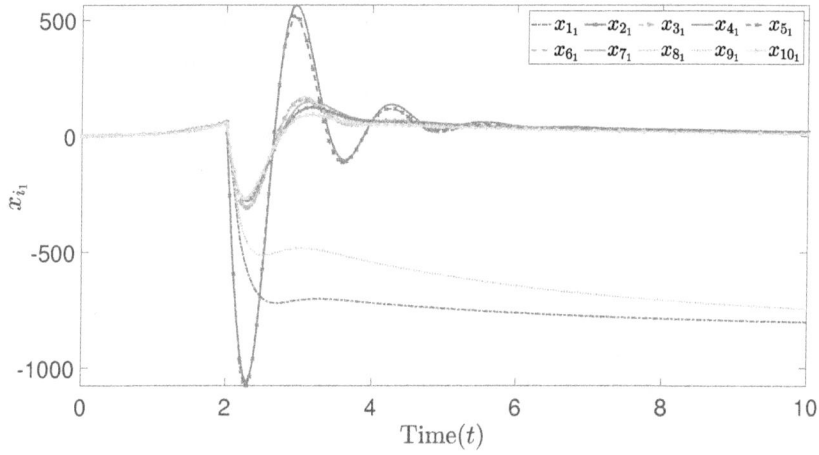

(a)

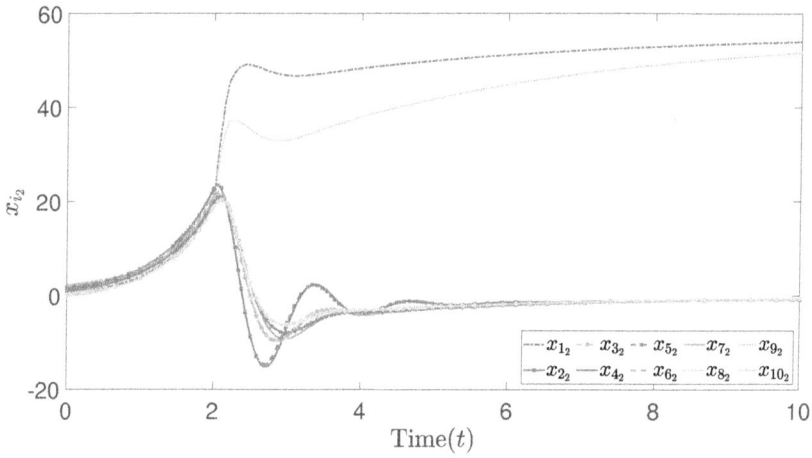

(b)

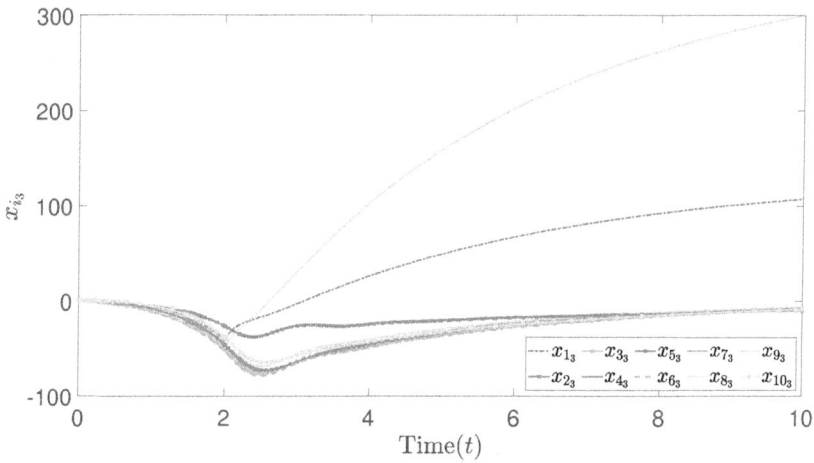

(c)

Figure 8.13   The state trajectory of each agent x_i.

8.6 CONCLUSIONS

This chapter investigates the resilient cooperative control problem in general high-order NASs under both FDI attacks and covert attacks. To mitigate the impact of FDI attacks, a distributed inclusive neighbor-based attack isolation algorithm based on a fixed-time observer is proposed to identify compromised agents. Furthermore, the concept of r-isolability is introduced to prevent the mistaken isolation of normal agents. To identify covert attacks, a distributed neighbor-based attack isolation algorithm is further developed, employing a two-stage fixed-time observer. A novel concept, referred to as r^*-isolability, is proposed to establish sufficient conditions for zero false alarms. Building upon these isolation mechanisms, two control algorithms are designed to guarantee resilient cooperation among the normal agents.

CHAPTER 9

Distributed resilient cooperative control of high-order NASs against collusive attacks

This chapter explores the distributed resilient cooperative control problem of high-order NASs, specifically addressing the challenges posed by collusive FDI attacks and covert attacks. Section 9.1 reviews relevant prior works and articulates the motivations that drive the current study. Section 9.2 introduces a distributed inclusive neighbor-based isolation (DINIs) algorithm against collusive attacks to identify collusive FDI attacks. The notion of (r, s)-isolability is introduced to provide sufficient conditions for attack isolation. Section 9.3 develops a distributed neighbor-based isolation (DNIs) algorithm against collusive attacks. A sufficient condition is derived to ensure successful isolation of collusive covert attacks. Section 9.4 provides distributed resilient cooperative control protocols such that all normal agents, including those mistakenly isolated, to successfully accomplish resilient control tasks. Section 9.5 provides some simulations to illustrate the effectiveness of the theoretical results. Section 9.6 concludes this chapter.

9.1 INTRODUCTION

The attacks on NASs can be described as data corruption of actuators or sensors [15]. With well-designed attacks, it is almost imperative to devise sophisticated attack detection and isolation algorithms to improve the reliability and security of NASs. The main idea of attack detection in model-based NASs is to construct observers to detect the attacks if the residuals between the actual systems and the observers exceed the tolerant thresholds. In [178–180], a bank of observers relying on state information was constructed. In engineering systems, full state information is often difficult to obtain. In [181], unknown input observers were designed based on the relative output information, and [166] extended them into finite-time observers by borrowing the structure in [165]. In [190], attack isolation was achieved by actively reconfiguring the controller. Note that only one compromised agent is allowed for the subsystems consisting of each agent and its neighbors in the above works. In reality, the normal agents cannot access the prior knowledge of the compromised agents and their attack signals [15, 182]. In [15, 183], a complete attack isolation algorithm was provided, which required substantial computational effort. In [67], a divide-and-conquer approach was proposed to reduce computational burden, at the expense of a more complicated communication structure. To relax the assumptions in [67, 190], a centralized framework was built in [15, 20] to isolate each compromised agent by enumerating and checking all possible compromised sets, which however was exactly an NP-hard problem. Considering the above mentioned conservative assumptions and high computational complexity, the attack isolation problem is nowhere near resolved.

Furthermore, with the high dependency of communication network and the rapid development of hacking, sophisticated attacks can be launched to manipulate the system while maintaining undetectable, such as replay attacks [184], zero-dynamics attacks [121], and covert attacks [185]. With the full knowledge of the system, covert attacks can tamper the system state while disguise their influence on the system output. One way to detect covert attacks is to prevent the attackers from obtaining the full information of the system. In [186], the system was extended to account for stochastic and time-varying parameters, while the auxiliary parameters increase the burden of attack detection. A time-varying modulation matrix was introduced in [187] to change the system input. Note that variations in the input can potentially deteriorate the control performance of the system. In [188], the watermarking signal [189] and the static auxiliary systems were combined to detect covert attacks without sacrificing the control performance. It's worth mentioning that the above approaches focus on the centralized system, in which, the detection of covert attacks is accomplished by actively changing the dynamics of the system. For the interconnected large-scale systems, two kinds of observers were designed in [155, 167] to achieve the detection of covert attacks in a distributed way. However, the above strategy assumes that the neighbors of each compromised agent are normal. Based on attack detection and isolation, fault-tolerant control strategies were designed to compensate the effect of the attacks [179, 180]. However, the above control strategies fail if the sophisticated attacks cannot be compensated.

In this chapter, we aim to deal with the resilient cooperative control problem of high-order NASs, in the presence of collusive attacks. With the coupled states in general high-order networks, the algorithms developed in [49, 51, 61–66, 160] are inapplicable, making it the only choice to design resilient cooperative control algorithms by excluding the isolated compromised agents. Furthermore, the sophisticated attacks may collude to disturb attack detection, resulting in the invalidation of the attack isolation strategy in Chapter 8. This forces us to seek novel attack isolation strategies to deal with collusive attacks. Here, we focus on F-total FDI attack model and F-total covert attack model, respectively. First, to against collusive FDI attacks, a distributed fixed-time Luenberger observer is introduced to derive the residual between each agent's consensus error and its observation, thereby achieving attack detection. Then, a distributed inclusive neighbor-based isolation (DINIs) algorithm against collusive attacks is developed. Specifically, each agent counts the number of non-zero residuals within its inclusive neighbors, isolates the agent with maximum number of non-zero residuals as potentially compromised, and subsequently resets the residuals of all its inclusive neighbors to non-zero values. This procedure is iteratively executed for F rounds, resulting in the isolation of F agents. To ensure that all compromised agents are included in the F isolated agents, the notion of (r, s)-isolability is introduced to provide sufficient conditions for successful isolation of collusive FDI attacks. Second, to address collusive covert attacks, the two-stage fixed-time observer enables the detection of covert attacks on a given agent by leveraging the residuals of its neighboring agents. A novel distributed neighbor-based isolation (DNIs) algorithm against collusive attacks is then presented, with the notion of (r, s)-isolability providing the sufficient condition for the achievement of attack isolation. Finally, two control algorithms are developed by removing the isolated compromised agents, ensuring the accomplishment of resilient control tasks even in the presence of mistakenly isolated normal agents.

9.2 DISTRIBUTED ISOLATION OF COLLUSIVE FDI ATTACKS

9.2.1 DINIs algorithm against collusive attacks

Consider a high-order NAS with the dynamics of the ith agent being governed by

$$\begin{cases} \dot{x}_i(t) = Ax_i(t) + Bu_i(t) + f_i(t, T_{f_i}), \\ y_i(t) = Cx_i(t) + \eta_i(t, T_{\eta_i}), \ i = 1, 2, \ldots, N, \end{cases} \tag{9.1}$$

where $x_i \in \mathbb{R}^n$, $u_i \in \mathbb{R}^g$, $y_i \in \mathbb{R}^p$ are the state, control input, and measurement output vectors, respectively. $f_i(t, T_{f_i}) \in \mathbb{R}^n$ and $\eta_i(t, T_{\eta_i}) \in \mathbb{R}^p$ denote the FDI attacks induced in actuator and sensor channels, respectively, T_{f_i} and T_{η_i} are respectively the time instants when actuator and sensor FDI attacks first occur, A, B and C are constant system matrices. The following assumption on system (9.1) is made.

Assumption 9.1 *The triple (A, B, C) is controllable and observable.*

In this chapter, we focus on node-based attacks, and thus, the assumption on communication network is provided to facilitate the analysis of attack isolation.

Assumption 9.2 *The communication among agents is secure.*

Let $\xi_i(t) = \sum_{j \in \mathcal{N}_i} (y_i(t) - y_j(t))$, $i \in \mathcal{V}$, be the output consensus error for each agent, and the distributed fixed-time observer is introduced as

$$\begin{cases} \dot{\hat{z}}_i(t) = A_c \hat{z}_i(t) + B_c \sum_{j \in \mathcal{N}_i} (u_i(t) - u_j(t)) + H_c \xi_i(t), \\ \hat{\varepsilon}_i(t) = D_c [\hat{z}_i(t) - \exp(A_c \tau) \hat{z}_i(t - \tau)], \end{cases} \tag{9.2}$$

where $\hat{z}_i(t) \in \mathbb{R}^{2n}$ is the auxiliary variable with $\hat{z}_i(t) = \mathbf{0}$ for $t \in [-\tau \ 0]$, $\tau > 0$ is the preset convergent time, $\hat{\varepsilon}_i(t)$ is the estimate of the state consensus error $\varepsilon_i(t)$ with $\varepsilon_i(t) = \sum_{j \in \mathcal{N}_i} (x_i(t) - x_j(t))$, $A_c = \begin{bmatrix} A - H_1 C & \mathbf{0} \\ \mathbf{0} & A - H_2 C \end{bmatrix}$, $B_c = \begin{bmatrix} B^T & B^T \end{bmatrix}^T$,

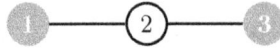

Figure 9.1   An example of attack collusion.

$H_c = \begin{bmatrix} H_1^T & H_2^T \end{bmatrix}^T$, H_k is the observer gain matrix which makes $A - H_k C$ stable, $k = 1, 2$, $C_c = \begin{bmatrix} I_n & I_n \end{bmatrix}^T$, $D_c = \begin{bmatrix} I_n & \mathbf{0} \end{bmatrix} \begin{bmatrix} C_c & \exp(A_c \tau) C_c \end{bmatrix}^{-1}$. It is clear that $D_c C_c = I_n$, and $D_c \exp(A_c \tau) C_c = \mathbf{0}$.

Define $\tilde{z}_i(t) = \hat{z}_i(t) - C_c \varepsilon_i(t)$, then one has

$$\tilde{z}_i(t) = \int_{t-\tau}^{t} \exp(A_c(t-s)) [H_c \epsilon_{\eta_i}(s) - C_c \epsilon_{f_i}(s)] ds + \exp(A_c \tau) \tilde{z}_i(t-\tau), \quad (9.3)$$

where $\epsilon_{\eta_i}(s) = \sum_{j \in \mathcal{N}_i} (\eta_i(s, T_{\eta_i}) - \eta_j(s, T_{\eta_j}))$, $\epsilon_{f_i}(s) = \sum_{j \in \mathcal{N}_i} (f_i(s, T_{f_i}) - f_j(s, T_{f_j}))$. It derives from (9.2) and (9.3) that

$$\hat{\varepsilon}_i(t) = D_c \int_{t-\tau}^{t} \exp(A_c(t-s)) [H_c \epsilon_{\eta_i}(s) - C_c \epsilon_{f_i}(s)] ds + \varepsilon_i(t), \quad t \geq \tau. \quad (9.4)$$

Clearly, one has $\hat{\varepsilon}_i(t) = \varepsilon_i(t), t \geq \tau$, if there is no attack. Let $\tilde{\xi}_i(t) = \xi_i(t) - C\hat{\varepsilon}_i(t)$ be the residual between the output consensus error and its observation for agent i. Combining (9.1), (9.2), and (9.4), one has

$$\tilde{\xi}_i(t) = C D_c \int_{t-\tau}^{t} \exp(A_c(t-s)) [C_c \epsilon_{f_i}(s) - H_c \epsilon_{\eta_i}(s)] ds + \epsilon_{\eta_i}(t), \quad t \geq \tau. \quad (9.5)$$

Lemma 9.1 *Consider the high-order NAS (9.1) with Assumptions 9.1–9.2. There exist compromised agents in $\mathcal{J}_i$ if there exists a time instant $t \geq \tau$ such that $\|\tilde{\xi}_i(t)\| \neq 0$ under the observer (9.2).*

The proof of above result is straightforward from (9.5). Meanwhile, it follows from (9.5) that multiple attacks may collude such that $\|\tilde{\xi}_i(t)\| = 0$ for $t \geq \tau$, which suggests that the attacks are undetectable. Take the graph in Fig. 9.1 for example. If the FDI attacks on agents 1 and 3 satisfy $f_1(t, T_{f_1}) = -f_3(t, T_{f_3})$ and $\eta_1(t, T_{\eta_1}) = -\eta_3(t, T_{\eta_3})$ with $T_{f_1} = T_{f_3}$ and $T_{\eta_1} = T_{\eta_3}$ such that $\|\epsilon_{f_2}(t)\| = 0$ and $\|\epsilon_{\eta_2}(t)\| = 0$, then one has $\|\tilde{\xi}_2(t)\| = 0$ for $t \geq \tau$. This suggests that the FDI attacks on agents 1 and 3 in subsystem $\mathcal{G}_2(\mathcal{J}_2, \mathcal{E}_2)$ can collude to make $\|\tilde{\xi}_2(t)\| = 0$, and thus, the residual $\|\tilde{\xi}_2(t)\|$ fails to detect whether or not there are attacks in $\mathcal{G}_2(\mathcal{J}_2, \mathcal{E}_2)$. The definition of collusive FDI attacks is given as follows.

Definition 9.1 (Collusive FDI attacks) *The FDI attacks $\bigcup_{j \in \mathcal{V}_A} (f_j(t), \eta_j(t))$ are collusive if there exists an agent i with $\mathcal{J}_i \cap \mathcal{V}_A \neq \varnothing$ satisfying $\|\epsilon_{f_i}(t)\| = 0$ and $\|\epsilon_{\eta_i}(t)\| = 0$.*

Remark 9.1 *Unlike the random faults or internal failures encountered by dynamic systems [37, 180], the external attacks are often elaborately designed to disrupt the systems while avoid detection [15]. For example, the line outages might not be detected under the collusive attacks on the bus load and line flow measurements [191].*

Based on attack detection, the DINIs algorithm against collusive attacks is developed. First, we provide some necessary notations.

Let $\Gamma_i = \left[\tilde{\xi}_i, \tilde{\xi}_{i^1}, \ldots, \tilde{\xi}_{i|\mathcal{N}_i|}\right] \in \mathbb{R}^{p \times |\mathcal{J}_i|}$ be the stack vector consisting of the residuals of agent i and its neighboring agents. The information set for each agent is defined as $\Phi_i = \left\{\|\Gamma_i\|_{l_0}, \|\tilde{\xi}_i\|, S_i\right\}$. Particularly, define $\Phi_i > \Phi_j$ if one of the followings holds: 1) $\|\Gamma_i\|_{l_0} > \|\Gamma_j\|_{l_0}$; 2) $\|\tilde{\xi}_i\| > \|\tilde{\xi}_j\|$ with $\|\Gamma_i\|_{l_0} = \|\Gamma_j\|_{l_0}$.

With collusive FDI attacks, there may exist $\|\tilde{\xi}_k(t)\| = 0$, $t \geq \tau$ for some $k \in \mathcal{J}_i$, $\mathcal{J}_i \cap \mathcal{V}_A \neq \varnothing$, which may result in $\|\Gamma_i\|_{l_0}, i \in \mathcal{V}_A$ being smaller than $\|\Gamma_j\|_{l_0}, j \in \mathcal{V}_N$ for some i, j. Consequently, not all compromised agents can be isolated in one round. Intuitively however, the largest $\Phi_i, i \in \mathcal{V}_A$ is expected to be greater than the largest $\Phi_j, j \in \mathcal{V}_N$. Thus, one can find the agent q with largest information set, i.e., $\Phi_q = \max_{j \in \mathcal{V}}\{\Phi_j\}$, and isolate agent q as the compromised one in each round. Moreover, the compromised agents isolated in earlier rounds will further help isolate other compromised agents in subsequent rounds. Motivated by the above analysis, the DINIs algorithm against collusive attacks is developed in Algorithm 9.1. The easy implementation of Algorithm 9.1 is illustrated as follows. In each round, the agent q with $\Phi_q = \max_{j \in \mathcal{V}}\{\Phi_j\}$ is isolated as compromised. To find the largest information set $\max_{j \in \mathcal{V}}\{\Phi_j\}$ in a distributed manner, the temporary set $M_i(k)$ is defined, which is initialized as $M_i(0) = \Phi_i, i \in \mathcal{V}$. And the temporary set $M_i(k)$ is updated by $M_i(k) = \max_{j \in \mathcal{J}_i}\{M_j(k-1)\}$ so that $M_i(N-1) = \max_{j \in \mathcal{V}}\{M_j(0)\} = \max_{j \in \mathcal{V}}\{\Phi_j\}$, $i \in \mathcal{V}$. Denote the agent q in $M_i(N-1), i \in \mathcal{V}$, and set $S_q = q$. With the isolation of agent q, set $\left\|\tilde{\xi}_m\right\| \neq 0$, $m \in \mathcal{J}_q$ to help uncover the collusion among attacks, and further update $\Phi_m, m \in \mathcal{V}\backslash q$ to obtain the isolated agent in the next round. Meanwhile, Φ_q is reset as $\Phi_q = \left\{0, \left\|\tilde{\xi}_q\right\|, q\right\}$ to avoid re-isolation of agent q. The above process is repeated F rounds such that there are F agents being isolated.

Remark 9.2 *Algorithm 9.1 identifies the F most suspicious agents as compromised, regardless of whether the actual number of compromised agents is less than F. In this sense, if the actual number of compromised agents is less than F, some normal agents will inevitably be mistakenly isolated; Conversely, if the number exceeds F, some compromised agents may remain unisolated.*

9.2.2 Analysis of DINIs algorithm against collusive attacks

In this subsection, we establish sufficient conditions for the proposed attack isolation strategy. Clearly, the graph structure plays a pivotal role in ensuring the successful isolation of all compromised nodes, as formalized below.

Definition 9.2 *((r, s)-isolable graph) A nontrivial graph $\mathcal{G} = (\mathcal{V}, \mathcal{E})$ is (r, s)-isolable $(r \geq 1, s \geq 3)$ if it holds that*

i) The graph $\mathcal{G}$ is connected;

ii) $|\mathcal{N}_i| \geq r+1, \forall i \in \mathcal{V}$;

iii) There is no cycle $\mathcal{C}_k$ with $k < s$ in $\mathcal{G}$.

Algorithm 9.1 DINIs algorithm against collusive attacks

Input: $S_i = 0$, Γ_i, $i \in \mathcal{V}$; $v = F$;
Output: $S_i, i \in \mathcal{V}$;
 1: **if** $t > \tau$ **then**
 2:     **while** $v > 0$ **do**
 3:         $M_i(0) = \Gamma_i, i \in \mathcal{V}$;
 4:         **for** $k = 1 : N - 1$ **do**
 5:             $M_i(k) = \max_{j \in \mathcal{J}_i} \{M_j(k - 1)\}, i \in \mathcal{V}$;
 6:         **end for**
 7:         Set $S_q = q$ if $\|\Gamma_q\|_{l_0} = M_1(N - 1)$;
 8:         **for** $\forall m \in \mathcal{J}_q$ **do**
 9:             $\|\tilde{\xi}_m\| \neq 0$;
10:         **end for**
11:         Update the relative parameters in Φ_m, $m \in \mathcal{V}\backslash q$;
12:         $\Phi_q = \{0, \|\tilde{\xi}_q\|, q\}$;
13:         $v = v - 1$.
14:     **end while**
15: **end if**

Remark 9.3 *Note that any cycle $\mathcal{C}_k$ satisfies $k \geq 3$. This implies that the connected graph $\mathcal{G} = (\mathcal{V}, \mathcal{E})$ is $(r, 3)$-isolable if and only if $|\mathcal{N}_i| \geq r + 1$, $\forall i \in \mathcal{V}$ hold. To certify the (r, s)-isolability property of a given graph, one only need to find the shortest cycle in the graph, the execution time of the latter is bounded by $O(N^2)$ under the MIN_CIRCUIT algorithm proposed in [192]. Therefore, the time complexity of certifying the (r, s)-isolability of a graph is $O(N^2)$.*

The following lemmas characterize the properties of (r, s)-isolability and establish its relationship with r-isolability and r^*-isolability, respectively.

Lemma 9.2 *It follows from Definition 9.2 that the graph $\mathcal{G}$ is (r', s')-isolable if it is (r, s)-isolable for $1 \leq r' \leq r$ and $3 \leq s' \leq s$.*

Lemma 9.3 *A nontrivial graph $\mathcal{G} = (\mathcal{V}, \mathcal{E})$ is 1-isolable if it is $(1, 4)$-isolable.*

Proof 9.1 *Suppose that $\mathcal{G}$ is not 1-isolable, i.e., there exists a node i and another node j, satisfying $\mathcal{J}_i \subseteq \mathcal{J}_j$. Since $\mathcal{G}$ is $(1, 4)$-isolable, one has $|\mathcal{N}_i| \geq 2$. Denote i^1 (distinct to j) as a neighbor of i, and one has $i^1 \in \mathcal{J}_j$, implying that nodes i, i^1, j form a cycle, which contradicts the definition of $(1, 4)$-isolability.*

Lemma 9.4 *A nontrivial graph $\mathcal{G} = (\mathcal{V}, \mathcal{E})$ is r-isolable $(r \geq 2)$ if it is $(r, 5)$-isolable.*

Proof 9.2 *Suppose that $\mathcal{G}$ is not r-isolable, i.e., there exists at least one node i and other r nodes, denoted by $R_1, R_2, \ldots, R_r$, satisfying $\mathcal{J}_i \subseteq \{\mathcal{J}_{R_1} \cup \mathcal{J}_{R_2} \cup \cdots \cup \mathcal{J}_{R_r}\}$. Since $|\mathcal{N}_i| \geq r+1$ and $\mathcal{N}_i \subseteq \mathcal{J}_i$, there must be two neighbors of node i, denoted by i^1, i^2, being included in the inclusive neighbor set of certain node R_k, i.e., $\mathcal{J}_{R_k} \supseteq \{i^1, i^2\}$.*

This contradicts the condition that all cycles in $\mathcal{G}$ contains no less than 5 nodes, as there exists a cycle formed by four nodes i, i^1, i^2, R_k (or three nodes i, i^1, i^2 if $R_k = i^1$ or $R_k = i^2$).

Lemma 9.5 *A nontrivial graph $\mathcal{G} = (\mathcal{V}, \mathcal{E})$ is r^*-isolable if it is $(r, 5)$-isolable.*

The proof is similar to that of Lemma 9.4, and is hence omitted here.

A connected graph $\mathcal{G}$ is (r, s)-isolable if every node has at least $r + 1$ neighbors, and all cycles $\mathcal{C}_k$ satisfy $k \geq s$. Intuitively, the (r, s)-isolability property is proposed based on the idea that all compromised agents are isolated before any normal one. It follows from Algorithm 9.1 that the agent with $\max_{i \in \mathcal{V}} \{\Phi_i\}$ will be isolated as compromised in each round, and the key to determining $\max_{i \in \mathcal{V}} \{\Phi_i\}$ is the value of $\|\Gamma_i\|_{l_0}$. However, the attacks may collude to reduce the value of $\|\Gamma_i\|_{l_0}, i \in \mathcal{V}_A$. In the following, we shall show that the $\max_{i \in \mathcal{V}} \{\Phi_i\}$ can be derived in a distributed manner under a connected graph, and the parameters r and s capture the graph structure to ensure a *sufficiently large* value of $\|\Gamma_i\|_{l_0}$ for the compromised agents.

Lemma 9.6 *[193] Given an NAS consisting of N agents, the dynamics of the ith agent is described by*

$$M_i(k) = \max_{j \in \mathcal{J}_i} \{M_j(k-1)\}, \quad i = 1, 2, \ldots, N,$$

and an initial vector of the states $M(0) = [M_1^T(0), M_2^T(0), \ldots, M_N^T(0)]^T$, there exists a time instant $l \leq N - 1$ such that the max-consensus is reached, i.e., $M_i(k) = M_j(k) = \max\{M_1(0), M_2(0), \ldots, M_N(0)\}, \forall k \geq l, \forall i, j \in \mathcal{V}$, if its communication topology $\mathcal{G}$ is connected.

Now, the graph conditions for isolation of FDI attacks are provided. In the simple case where each subsystem $\mathcal{G}_i(\mathcal{J}_i, \mathcal{E}_i)$ contains at most one compromised agent, the compromised agents are unable to collude according to (9.5). Then one can derive the following result on attack isolation.

Lemma 9.7 *Consider a high-order NAS described by (9.1), with a communication topology represented by the graph $\mathcal{G}$. Suppose that Assumptions 9.1–9.2 hold. Assume that the path length of any two compromised agents $|l_{ij}| \geq 3, \forall i, j \in \mathcal{V}_A$, and $\exists t \geq \tau$ s.t. $\|\tilde{\xi}_k(t)\| \neq 0, \forall i \in \mathcal{V}_A, k \in \mathcal{J}_i$ hold. Attack isolation can be achieved under Algorithm 9.1 if and only if the communication graph $\mathcal{G}$ is 1-isolable.*

Proof 9.3 *For any two compromised agents i and j, $|l_{ij}| \geq 3$ means that there is only one compromised agent in the subsystem $\mathcal{G}_i(\mathcal{J}_i, \mathcal{E}_i), \forall i \in \mathcal{V}$. The following proof is similar to that of Theorem 8.1, and is hence omitted here.*

Remark 9.4 *It follows from Remark 8.8 and Theorem 8.1 that under 1-isolable graph and Algorithm 8.1, attack isolation can be achieved if there is at most one compromised agent in each subsystem $\mathcal{G}_i(\mathcal{J}_i, \mathcal{E}_i)$, which means the path length of any two compromised agents satisfies $|l_{ij}| \geq 4$. While Lemma 9.7 shows that such condition can be relaxed into $|l_{ij}| \geq 3$. The main difference is that we do not require*

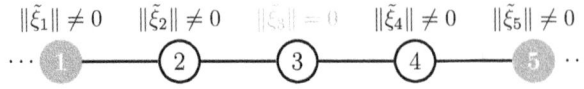

$$\|\tilde{\xi}_1\| \neq 0 \quad \|\tilde{\xi}_2\| \neq 0 \quad \|\tilde{\xi}_3\| = 0 \quad \|\tilde{\xi}_4\| \neq 0 \quad \|\tilde{\xi}_5\| \neq 0$$

Figure 9.2   An example of two compromised agents 1 and 5 satisfying $|l_{15}| = 3$.

(a)   (b)

Figure 9.3   Two strategies of collusion between two FDI attacks i and j, where '$= 0$' and '$\neq 0$' are short for '$\|\tilde{\xi}_i\| = 0$' and '$\|\tilde{\xi}_i\| \neq 0$,' respectively.

that $\mathcal{G}_i(\mathcal{J}_i, \mathcal{E}_i)$ contains at most one compromised agent for all i, but only for those $i \in \mathcal{V}_A$. We illustrate the above result by an example. It shows in Fig. 9.2 that $|l_{15}| = 3$ between two compromised agents 1 and 5 ensures $\|\tilde{\xi}_3\| = 0$, which makes that normal agents 2, 3, and 4 would not be mistakenly isolated by Algorithm 9.1.

For the case $F = 2$, it follows from (9.5) and Lemma 9.1 that the attacks in the same subsystem $\mathcal{G}_i(\mathcal{J}_i, \mathcal{E}_i)$ may collude to make $\|\tilde{\xi}_j\| = 0$ for some $j \in \mathcal{J}_i, \mathcal{J}_i \cap \mathcal{V}_A \neq \varnothing$. The following result on attack isolation under 2-total FDI attack model is provided.

Theorem 9.1 *Consider a high-order NAS described by (9.1), with a communication topology represented by the graph $\mathcal{G}$. Suppose that Assumptions 9.1–9.2 hold. In the 2-total FDI attack model, attack isolation can be achieved under Algorithm 9.1 if the communication graph $\mathcal{G}$ is $(2, 5)$-isolable.*

Proof 9.4 *Assume that there are F_0 ($F_0 \in \{1, 2\}$) compromised agents. For the case that $F_0 = 1$, there is no collusion among attacks, the proof is straightforward and is hence omitted.*

For the case that $F_0 = 2$, as is shown in Fig. 9.3, there are two ways that the attacks collude. It derives that $\|\Gamma_i\|_{l_0} \geq |\mathcal{J}_i| - 1 \geq F + 1 = 3, \forall i \in \mathcal{V}_A$. For Fig. 9.3(a), one has that $\|\Gamma_j\|_{l_0} \leq 2 < \|\Gamma_i\|_{l_0}, \forall j \in \mathcal{O}, i \in \mathcal{V}_A$. For Fig. 9.3(b), one has $\|\Gamma_j\|_{l_0} \leq 3, \forall j \in \mathcal{O}$. Denote by i_1 and i_2 the two compromised agents. Note that $\|\Gamma_j\|_{l_0} = \|\Gamma_{i_1}\|_{l_0} = \|\Gamma_{i_2}\|_{l_0} = 3$ only when $|\mathcal{N}_{i_1}| = |\mathcal{N}_{i_2}| = 3$, and j is the neighbor of one compromised agent and the two-hop neighbor of another. In this case, one has $\|\tilde{\xi}_{i_1}\| = \|\tilde{\xi}_{i_2}\| = 3\|\tilde{\xi}_j\|$. This implies that all compromised agents are isolated before any normal one.

Although the graph condition in Theorem 9.1 is not necessary, the following result shows that it is sharp.

Proposition 9.1 *Consider a high-order NAS described by (9.1), with a communication topology represented by the graph $\mathcal{G}$. Suppose that Assumptions 9.1–9.2 hold. For the 2-total FDI attack model, there exists a $(2, 4)$-isolable graph which fails to achieve attack isolation by Algorithm 9.1.*

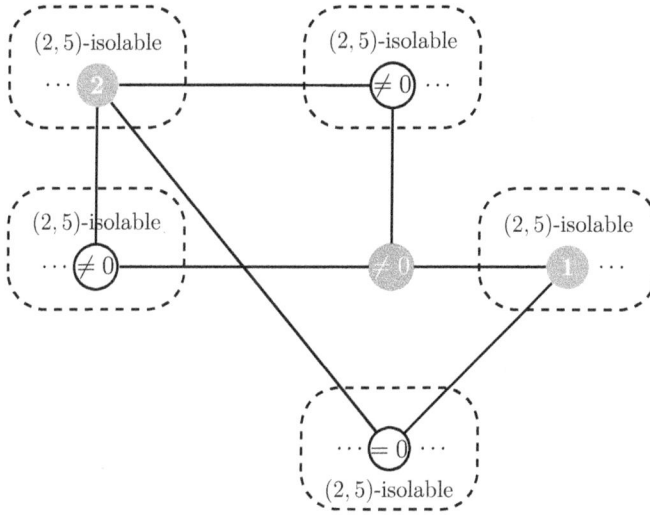

Figure 9.4   An illustration of $(2,4)$-isolable graph.

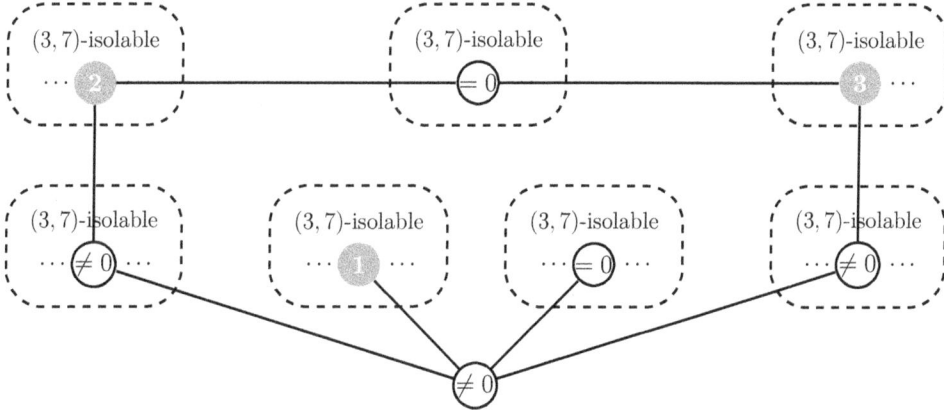

Figure 9.5   An illustration of $(3,6)$-isolable graph.

Proof 9.5 *It will be proved by constructing a counterexample, depicted in Fig. 9.4. There is only one cycle consisting of 4 nodes in the graph and the other cycles (not shown in the figure) satisfy $\mathcal{C}_k$ with $k \geq 5$, and the number of neighbors for each agent satisfies $|\mathcal{N}_i| = 3$, $\forall i \in \mathcal{V}$. The compromised agents 1 and 2 satisfy $\|\Gamma_1\|_{l_0} = \|\Gamma_2\|_{l_0} = 3$, while the normal agent (the filled blue circle) satisfies $\max_{j \in \mathcal{O}}\{\|\Gamma_j\|_{l_0}\} = 4$. Note that the normal agent j and the compromised agents 1 and 2 satisfy $\|\Gamma_j\|_{l_0} > \|\Gamma_1\|_{l_0} = \|\Gamma_2\|_{l_0}$. Consequently, the normal agent j is mistakenly isolated before the compromised agents 1 and 2, which implies the failure of attack isolation.*

For the case $F \geq 3$, the condition of $(F,5)$-isolability is far from sufficient for attack isolation. Fig. 9.5 gives a counterexample of $(3,6)$-isolable graph for 3-total FDI attack model. In Fig. 9.5, there is only one cycle consisting of 6 nodes in the graph and the other cycles (not shown in the figure) satisfy $\mathcal{C}_k$ with $k \geq 7$, and the number of neighbors for each agent satisfies $|\mathcal{N}_i| = 4$, $\forall i \in \mathcal{V}$. The residuals of

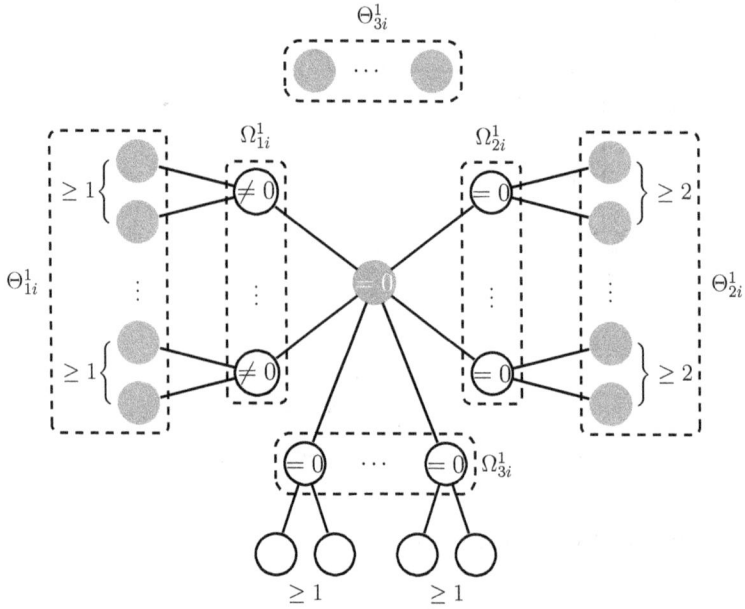

Figure 9.6   An illustration of Case 1 of Theorem 9.2.

the compromised agents 1, 2, 3 satisfy $\left\|\tilde{\xi}_1\right\| > 4\left\|\tilde{\xi}_2\right\|, \left\|\tilde{\xi}_2\right\| = \left\|\tilde{\xi}_3\right\|$. Note that there exists certain normal agent j (the filled blue one) which satisfies $\|\Gamma_j\|_{l_0} = 4$ with $\|\tilde{\xi}_j\| = \frac{1}{4}\|\tilde{\xi}_1\|$, while the compromised agents 2 and 3 satisfy $\|\Gamma_2\|_{l_0} = \|\Gamma_3\|_{l_0} = 4$ with $\|\tilde{\xi}_2\| = \|\tilde{\xi}_3\| < \|\tilde{\xi}_j\|$. Consequently, the normal agent j is mistakenly isolated before the compromised agents 2 and 3 by Algorithm 9.1, which means the failure of attack isolation. This forces us to investigate some more rigorous conditions to tackle with the attack isolation problem for general F-total FDI attack model with $F \geq 3$, which is provided as follows.

Theorem 9.2 *Consider a high-order NAS described by (9.1), with a communication topology represented by the graph $\mathcal{G}$. Suppose that Assumptions 9.1–9.2 hold. In the F-total ($F \geq 3$) FDI attack model, attack isolation can be achieved under Algorithm 9.1 if the communication graph $\mathcal{G}$ is $(F, 7)$-isolable.*

Proof 9.6 *The result will be proved by showing that all compromised agents are isolated before any normal one. It follows from Lemma 9.6 that $\max_{i \in \mathcal{V}}\{\Phi_i\}$ can be derived in a distributed manner by Algorithm 9.1 under $(F, 7)$-isolable graph. Define $\bar{M}_{\mathcal{O}} = \max_{i \in \mathcal{O}}\{\|\Gamma_i\|_{l_0}\}$. Without loss of generality, assume $\bar{M}_{\mathcal{O}} = |\mathcal{J}_i| - h$ with $h \in [0, |\mathcal{J}_i|]$ and there are F_0 compromised agents with $F_0 \leq F$. There are three cases for the normal agent i satisfying $\|\Gamma_i\|_{l_0} = \bar{M}_{\mathcal{O}}$, including: 1) $\|\tilde{\xi}_i\| = 0$, and all its neighbors are normal, i.e., $j \in \mathcal{O}, \forall j \in \mathcal{N}_i$; 2) $\|\tilde{\xi}_i\| = 0$, and some of its neighbors are compromised, i.e., $\mathcal{V}_A \cap \mathcal{N}_i \neq \varnothing$; 3) $\|\tilde{\xi}_i\| \neq 0$. In the following, these three cases will be discussed separately.*

Case 1): An illustration of this situation is shown in Fig. 9.6, where $\Omega_{1i}^1 = \{j | j \in \mathcal{O}, j \in \mathcal{N}_i, \|\tilde{\xi}_j\| \neq 0\}, |\Omega_{1i}^1| = |\mathcal{J}_i| - h; \Omega_{2i}^1 = \{j | j \in \mathcal{O}, j \in \mathcal{N}_i \cap \mathcal{N}_k, k \in$

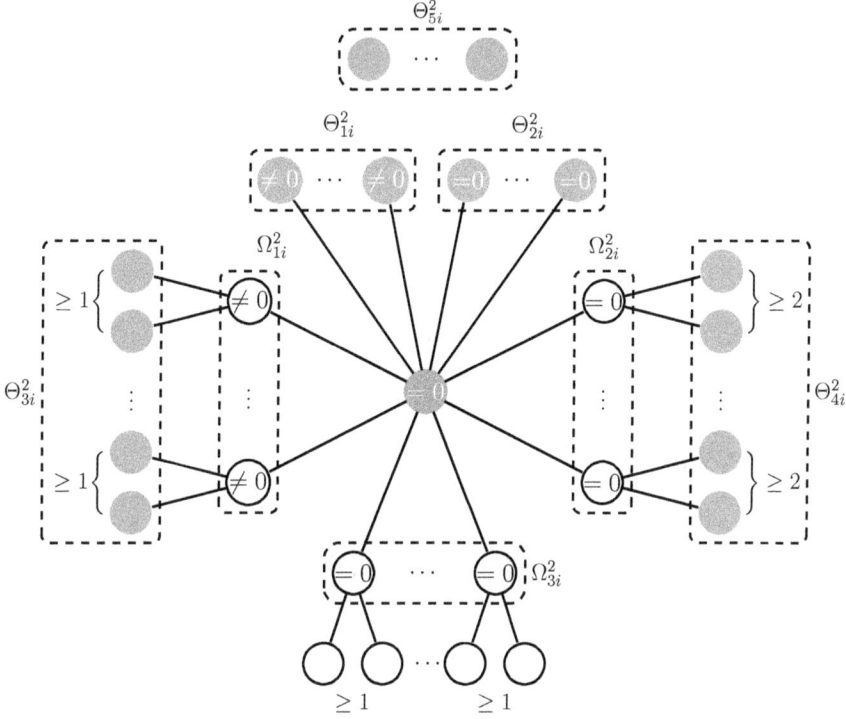

Figure 9.7   An illustration of Case 2 of Theorem 9.2.

$\mathcal{V}_A, \|\tilde{\xi}_j\| = 0\}, |\Omega_{2i}^1| = s; \ \Omega_{3i}^1 = \{j | j \in \mathcal{O}, j \in \mathcal{N}_i, j \notin \Omega_{1i}^1 \cup \Omega_{2i}^1, \|\tilde{\xi}_j\| = 0\}, |\Omega_{3i}^1| = |\mathcal{N}_i| - |\Omega_{1i}^1| - |\Omega_{2i}^1|; \ \Theta_{1i}^1 = \{m | m \in \mathcal{V}_A, m \in \mathcal{N}_j, j \in \Omega_{1i}^1\}, |\Theta_{1i}^1| \geq |\mathcal{J}_i| - h; \ \Theta_{2i}^1 = \{m | m \in \mathcal{V}_A, m \in \mathcal{N}_j, j \in \Omega_{2i}^1\}, |\Theta_{2i}^1| = s^* \geq 2s; \ \Theta_{3i}^1 = \{q | q \in \mathcal{V}_A, q \notin \Theta_{1i}^1 \cup \Theta_{2i}^1\}, |\Theta_{3i}^1| = F_0 - |\Theta_{1i}^1| - |\Theta_{2i}^1|.$ *Since there is no cycle* $\mathcal{C}_k$ *with* $k < 7$, *each compromised agent in* $\Theta_{1i}^1 \cup \Theta_{2i}^1$ *has only one neighbor in* $\Omega_{1i}^1 \cup \Omega_{2i}^1$, *and there is no common neighbor outside* $\Omega_{1i}^1 \cup \Omega_{2i}^1$ *for any two compromised agents in* $\Theta_{1i}^1 \cup \Theta_{2i}^1$. *For any compromised agent* $m \in \Theta_{1i}^1 \cup \Theta_{2i}^1$, *only the compromised agent* $q \in \Theta_{3i}^1$ *is able to decrease* $\|\Gamma_m\|_{l_0}$ *by colluding with* m. *Then, one has* $\|\Gamma_m\|_{l_0} \geq |\mathcal{J}_m| - |\Theta_{3i}^1| - 1 \geq |\mathcal{J}_m| - (F_0 - (|\mathcal{J}_i| - h) - s^*) - 1.$ *With* $|\mathcal{J}_m| \geq F + 2, F \geq F_0$, *it derives that* $\|\Gamma_m\|_{l_0} \geq |\mathcal{J}_i| - h + s^* + 1 > \bar{M}_{\mathcal{O}}$. *On the other hand, it follows from Algorithm 9.1 that the isolation of compromised agents* $m \in \Theta_{2i}^1$ *makes* $\|\tilde{\xi}_j\| \neq 0$ *for* $j \in \Omega_{2i}^1$, *resulting in the increase of* $\|\Gamma_i\|_{l_0}$, *while the isolation of compromised agents in* $\Theta_{1i}^1 \cup \Theta_{3i}^1$ *makes no difference to* $\|\Gamma_i\|_{l_0}$. *One then has* $\max\left\{\|\Gamma_i\|_{l_0}\right\} \leq |\mathcal{J}_i| - h + s < |\mathcal{J}_i| - h + s^* + 1 \leq \|\Gamma_m\|_{l_0}$ *for any compromised agent* $m \in \Theta_{1i}^1 \cup \Theta_{2i}^1$. *Therefore, all compromised agents in* $\Theta_{1i}^1 \cup \Theta_{2i}^1$ *are isolated before normal agent* i.

For any compromised agent $q \in \Theta_{3i}^1$, *the isolation of compromised agents in* $\Theta_{1i}^1 \cup \Theta_{2i}^1$ *makes* $\|\Gamma_q\|_{l_0} \geq |\mathcal{J}_q| - |\Theta_{3i}^1| \geq |\mathcal{J}_i| - h + s^* + 2 > \max\left\{\|\Gamma_i\|_{l_0}\right\}.$ *This implies that all compromised agents* $q \in \Theta_{3i}^1$ *are isolated before normal agent* i.

Case 2): *An illustration of this situation is shown in Fig. 9.7. Assume that there are* k *neighbors of agent* i *being compromised, in which* k_1 *neighbors satisfy* $\|\tilde{\xi}_m\| \neq$

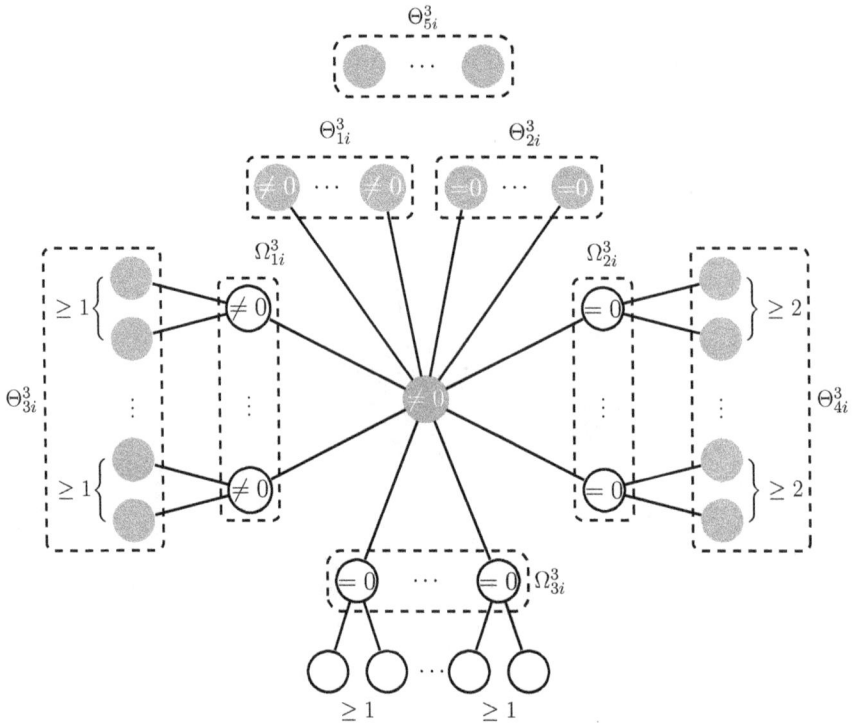

Figure 9.8   An illustration of Case 3 of Theorem 9.2.

0, $m \in \mathcal{N}_i \cap \mathcal{V}_A$. Then, one has $\Theta_{1i}^2 = \{m | m \in \mathcal{V}_A, m \in \mathcal{N}_i, \|\tilde{\xi}_m\| \neq 0\}$, $|\Theta_{1i}^2| = k_1$; $\Theta_{2i}^2 = \{m | m \in \mathcal{V}_A, m \in \mathcal{N}_i, \|\tilde{\xi}_m\| = 0\}$, $|\Theta_{2i}^2| = k - k_1$; $\Omega_{1i}^2 = \{j | j \in \mathcal{O}, j \in \mathcal{N}_i, \|\tilde{\xi}_j\| \neq 0\}$, $|\Omega_{1i}^2| = |\mathcal{J}_i| - h - k_1$; $\Omega_{2i}^2 = \{j | j \in \mathcal{O}, j \in \mathcal{N}_i \cap \mathcal{N}_k, k \in \mathcal{V}_A, \|\tilde{\xi}_j\| = 0\}$, $|\Omega_{2i}^2| = s$; $\Omega_{3i}^2 = \{j | j \in \mathcal{O}, j \in \mathcal{N}_i, j \notin \Omega_{1i}^2 \cup \Omega_{2i}^2, \|\tilde{\xi}_j\| = 0\}$, $|\Omega_{3i}^2| = |\mathcal{N}_i| - |\Theta_{1i}^2| - |\Theta_{2i}^2| - |\Omega_{1i}^2| - |\Omega_{2i}^2|$; $\Theta_{3i}^2 = \{m | m \in \mathcal{V}_A, m \in \mathcal{N}_j, j \in \Omega_{1i}^2\}$, $|\Theta_{3i}^2| \geq |\mathcal{J}_i| - h - k_1$; $\Theta_{4i}^2 = \{m | m \in \mathcal{V}_A, m \in \mathcal{N}_j, j \in \Omega_{2i}^2\}$, $|\Theta_{4i}^2| = s^* \geq 2s$; $\Theta_{5i}^2 = \{q | q \in \mathcal{V}_A, q \notin \Theta_{1i}^2 \cup \Theta_{2i}^2 \cup \Theta_{3i}^2 \cup \Theta_{4i}^2\}$, $|\Theta_{5i}^2| = F_0 - |\Theta_{1i}^2| - |\Theta_{2i}^2| - |\Theta_{3i}^2| - |\Theta_{4i}^2|$. *Similar to the analysis in Case 1), one has* $\|\Gamma_m\|_{l_0} \geq |\mathcal{J}_m| - |\Theta_{5i}^2| - 1 \geq |\mathcal{J}_i| - h + k - k_1 + s^* + 1$ *for any compromised agent* $m \in \Theta_{1i}^2 \cup \Theta_{2i}^2 \cup \Theta_{3i}^2 \cup \Theta_{4i}^2$, *and* $\max\{\|\Gamma_i\|_{l_0}\} \leq |\mathcal{J}_i| - h + k - k_1 + s + 1 \leq \|\Gamma_m\|_{l_0}$. *Note that* $\max\{\|\Gamma_i\|_{l_0}\} = |\mathcal{J}_i| - h + k - k_1 + s + 1$ *only when* $\|\tilde{\xi}_l\| \neq 0, \forall l \in \{i\} \cup \Theta_{2i}^2 \cup \Omega_{2i}^2$, *in which, one has* $\|\Gamma_m\|_{l_0} \geq |\mathcal{J}_m| - |\Theta_{5i}^2| \geq |\mathcal{J}_i| - h + k - k_1 + s^* + 2 > \max\{\|\Gamma_i\|_{l_0}\}$.
For any compromised agent $q \in \Theta_{5i}^2$, *the analysis is similar to that of Case 1).*

Case 3): *An illustration of this situation is shown in Fig. 9.8, where* $\Theta_{1i}^3 = \{m | m \in \mathcal{V}_A, m \in \mathcal{N}_i, \|\tilde{\xi}_m\| \neq 0\}$, $|\Theta_{1i}^3| = k_1$; $\Theta_{2i}^3 = \{m | m \in \mathcal{V}_A, m \in \mathcal{N}_i, \|\tilde{\xi}_m\| = 0\}$, $|\Theta_{2i}^3| = k - k_1$; $\Omega_{1i}^3 = \{j | j \in \mathcal{O}, j \in \mathcal{N}_i, \|\tilde{\xi}_j\| \neq 0\}$, $|\Omega_{1i}^3| = |\mathcal{J}_i| - h - k_1 - 1$; $\Omega_{2i}^3 = \{j | j \in \mathcal{O}, j \in \mathcal{N}_i \cap \mathcal{N}_k, k \in \mathcal{V}_A, \|\tilde{\xi}_j\| = 0\}$, $|\Omega_{2i}^3| = s$; $\Omega_{3i}^3 = \{j | j \in \mathcal{O}, j \in \mathcal{N}_i, j \notin \Omega_{1i}^3 \cup \Omega_{2i}^3, \|\tilde{\xi}_j\| = 0\}$, $|\Omega_{3i}^3| = |\mathcal{N}_i| - |\Theta_{1i}^3| - |\Theta_{2i}^3| - |\Omega_{1i}^3| - |\Omega_{2i}^3|$; $\Theta_{3i}^3 = \{m | m \in \mathcal{V}_A, m \in \mathcal{N}_j, j \in \Omega_{1i}^3\}$, $|\Theta_{3i}^3| \geq |\mathcal{J}_i| - h - k_1 - 1$; $\Theta_{4i}^3 = \{m | m \in \mathcal{V}_A, m \in \mathcal{N}_j, j \in \Omega_{2i}^3\}$, $|\Theta_{4i}^3| = s^* \geq 2s$; $\Theta_{5i}^3 = \{q | q \in \mathcal{V}_A, q \notin \Theta_{1i}^3 \cup \Theta_{2i}^3 \cup \Theta_{3i}^3 \cup \Theta_{4i}^3\}$, $|\Theta_{5i}^3| = F_0 - |\Theta_{1i}^3| - |\Theta_{2i}^3| - |\Theta_{3i}^3| - |\Theta_{4i}^3|$. The following analysis is similar to that of Case 2), and is hence omitted.

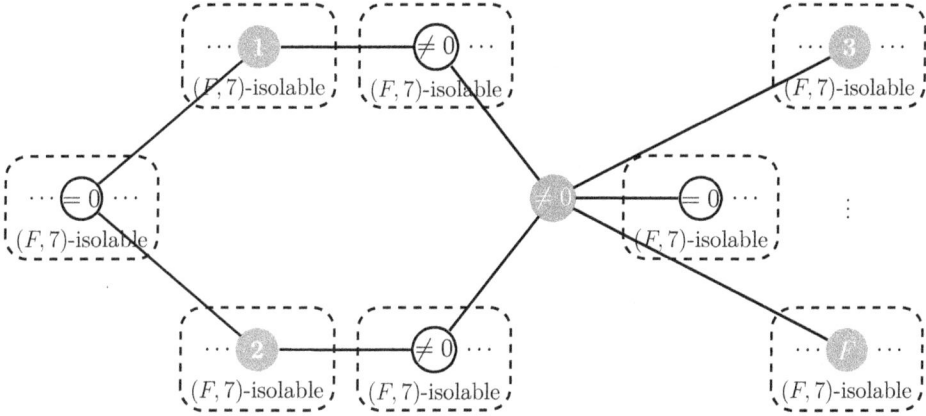

Figure 9.9   An illustration of $(F,6)$-isolable graph.

Similarly, the following result shows that the graph condition in Theorem 9.2 is sufficient yet sharp.

Proposition 9.2 *Consider a high-order NAS described by (9.1), with a communication topology represented by the graph $\mathcal{G}$. Suppose that Assumptions 9.1–9.2 hold. For the F-total FDI attack model with $F \geq 3$, there exists an $(F,6)$-isolable graph containing only one cycle $\mathcal{C}_k$ with $k = 6$ which fails to achieve attack isolation by Algorithm 9.1.*

Proof 9.7 *It will be proved by constructing a counterexample, depicted in Fig. 9.9. In Fig. 9.9, there is only one cycle consisting of six nodes, the other cycles (not shown in the figure) satisfy $\mathcal{C}_k$ with $k \geq 7$, and the neighbor number of each agent satisfies $|\mathcal{N}_i| = F + 1, \forall i \in \mathcal{V}$. The compromised agents satisfy $\|\Gamma_1\|_{l_0} = \|\Gamma_2\|_{l_0} = F + 1, \|\Gamma_i\|_{l_0} = F + 2, i \in \mathcal{V}_A \backslash \{1,2\}, \left\|\tilde{\xi}_1\right\| = \left\|\tilde{\xi}_2\right\|, \left\|\tilde{\xi}_3\right\| = \cdots = \left\|\tilde{\xi}_F\right\|$ with $\left\|\tilde{\xi}_3\right\| > \frac{F+1}{F-2}\left\|\tilde{\xi}_1\right\|$. Note that there exists a normal agent j satisfying $\|\Gamma_j\|_{l_0} = F + 1$ with $\left\|\tilde{\xi}_j\right\| = \frac{1}{F+1}\sum_{i=3}^{F}\left\|\tilde{\xi}_i\right\|$. Consequently, the normal agent j is mistakenly isolated before the compromised agents 1 and 2, which implies the failure of attack isolation.*

9.3   DISTRIBUTED ISOLATION OF COLLUSIVE COVERT ATTACKS

Consider the physically coupled NAS under covert attacks (8.10) and the two-stage fixed-time observers (8.11) and (8.14) in Section 8.3, if there are multiple attacks within $\mathcal{N}_i$, $\|r_i(t)\| = 0$ does not mean that there is no compromised agent since attacks can collude to make $\sum_{i=1}^{N} a_{ij} F_{ij} \tilde{x}_j^{\mathbf{u}}(t) = \mathbf{0}$. The definition of collusive covert attacks is given as follows.

Definition 9.3 (Collusive covert attacks) *The   covert   attacks $\bigcup_{k \in \mathcal{V}_A}(u_{ka}(t), y_{ka}(t))$ satisfying (8.8) and (8.9) are collusive if there exists an agent i with $\mathcal{N}_i \cap \mathcal{V}_A \neq \varnothing$ satisfying $\sum_{i=1}^{N} a_{ij} F_{ij} \tilde{x}_j^{\mathbf{u}}(t) = \mathbf{0}$.*

Algorithm 9.2 DNIs algorithm against collusive attacks

Input: $S_i = 0$, Υ_i, $i \in \mathcal{V}$; $v = F$;
Output: $S_i, i \in \mathcal{V}$;

  1: **if** $t \geq \max_{i \in \mathcal{V}}\{\tau_i\}$ **then**

  2:      **while** $v > 0$ **do**

  3:          $M_i(0) = \|\Upsilon_i\|_{l_0}, i \in \mathcal{V}$;

  4:          **for** $k = 1 : N - 1$ **do**

  5:              $M_i(k) = \max_{j \in \mathcal{J}_i}\{M_j(k - 1)\}, i \in \mathcal{V}$;

  6:          **end for**

  7:          Set $S_q = q$ if $\|\Upsilon_q\|_{l_0} = M_1(N - 1)$;

  8:          **for** $\forall m \in \mathcal{J}_q$ **do**

  9:              $\|r_m\| \neq 0$;

10:          **end for**

11:          Update $\|\Upsilon_i\|_0$, $i \in \mathcal{V} \backslash q$;

12:          $\|\Upsilon_q\|_{l_0} = 0$;

13:          $v = v - 1$.

14:      **end while**

15: **end if**

For the case with collusive covert attacks, there may exist a time instant $t \geq \tau_i$ such that $\|r_i(t)\| = 0$, $\mathcal{N}_i \cap \mathcal{V}_A \neq \varnothing$, which implies the failure of Algorithm 8.2. Thus, it is imperative to develop new algorithms to isolate the compromised agents subject to collusive covert attacks. Let $\Upsilon_i = \left(\|r_{i^1}\|, \|r_{i^2}\|, ..., \|r_{i|\mathcal{N}_i|}\|\right)^T \in \mathbb{R}^{|\mathcal{N}_i|}$ be the stack vector comprising the output residuals of all neighbors of agent i. Based on this, the following DNIs algorithm against collusive attacks is developed in Algorithm 9.2 to identify agents subject to collusive covert attacks. The main idea is summarized as follows: If there exist some compromised agents and all normal ones satisfying $\|\Upsilon_i\|_{l_0} \geq \|\Upsilon_j\|_{l_0}$, $\exists i \in \mathcal{V}_A$, $\forall j \in \mathcal{V}_R$ at the beginning of attack isolation, one can isolate the compromised agent by comparing the value of $\|\Upsilon_i\|_{l_0}$ for all $i \in \mathcal{V}$. Particularly, the agent i with largest $\|\Upsilon_i\|_{l_0}$ is isolated as being compromised by covert attacks in each round, and resets $\|r_j\| \neq 0, j \in \mathcal{J}_i$. Repeat the above operation F rounds until there are F compromised agents being isolated.

Remark 9.5 *It follows from Algorithm 9.2 that the isolation of agent i resets $\|r_j\| \neq 0$ for all $j \in \mathcal{J}_i$ since i may be the neighbor of another compromised agent k, i.e., $i \in \mathcal{N}_k \cap \mathcal{V}_A$, $k \in \mathcal{V}_A$, and resetting $\|r_i\| \neq 0$ facilitates the isolation of agent k. Besides, Algorithm 9.2 isolates F agents as compromised, even when there are only F_0 ($F_0 \leq F$) compromised agents without collusion, whereas Algorithm 8.2 can isolate these F_0 compromised agents without triggering false alarms.*

It derives from Proposition 8.3 that the compromised agents may collude to avoid being detected if there are multiple compromised agents in the subsystem $\mathcal{G}_i(\mathcal{N}_i, \mathcal{E}_i)$. As shown in Fig. 9.10, there are two strategies that any two compromised agents may collude. Consider the influence of compromised agents on the normal ones, one

Figure 9.10 Two collusive strategies between two covert attacks, where '= 0' and '≠ 0' are short for '$\|r_i\| = 0$' and '$\|r_i\| \neq 0$,' respectively.

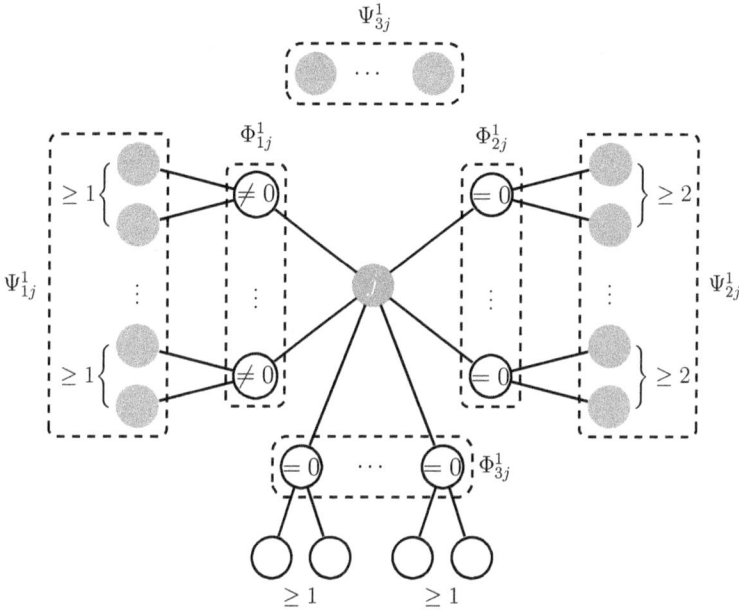

Figure 9.11 An illustration of Case 1 of Theorem 9.3.

concludes that the first strategy in Fig. 9.10(a) may increase $\|\Upsilon_i\|_{l_0}$, $i \in \mathcal{V}$ compared with the case without collusion, the second strategy in Fig. 9.10(b) decreases the value of $\|\Upsilon_i\|_{l_0}$ for $i \in \mathcal{V}_A$ while does not change $\|\Upsilon_j\|_{l_0}$ for $j \in \mathcal{V}_R$ compared with the case without collusion. Based on the above analysis, the isolation result against collusive attacks is derived as follows.

Theorem 9.3 *Consider the physically coupled NAS under covert attacks described by (8.10), with a communication topology represented by the graph $\mathcal{G}$. Suppose that Assumptions 8.3–8.5 hold. In the F-total covert attack model with $F \geq 2$, attack isolation can be achieved under Algorithm 9.2 if the communication graph $\mathcal{G}$ is $(F, 7)$-isolable.*

Proof 9.8 *Suppose that there are F_0 ($F_0 \leq F$) agents subject to covert attacks, and the normal agent j satisfies $\|\Upsilon_j\|_{l_0} = \max_{k \in \mathcal{V}_R}\{\|\Upsilon_k\|_{l_0}\}$ with $\|\Upsilon_j\|_{l_0} = |\mathcal{N}_j| - h$, $h \in [1, |\mathcal{N}_j|]$. We shall prove that all compromised agents are isolated before any normal agents. Two cases are needed to be considered for j: 1) All the neighbors of j are normal, i.e., $\mathcal{N}_j \cap \mathcal{V}_A = \varnothing$. 2): There exist some neighbors of j being compromised by covert attacks, i.e. $\mathcal{N}_j \cap \mathcal{V}_A \neq \varnothing$.*

Case 1): An illustration of Case 1) is shown in Fig. 9.11, where $\Phi_{1j}^1 = \{m|m \in$

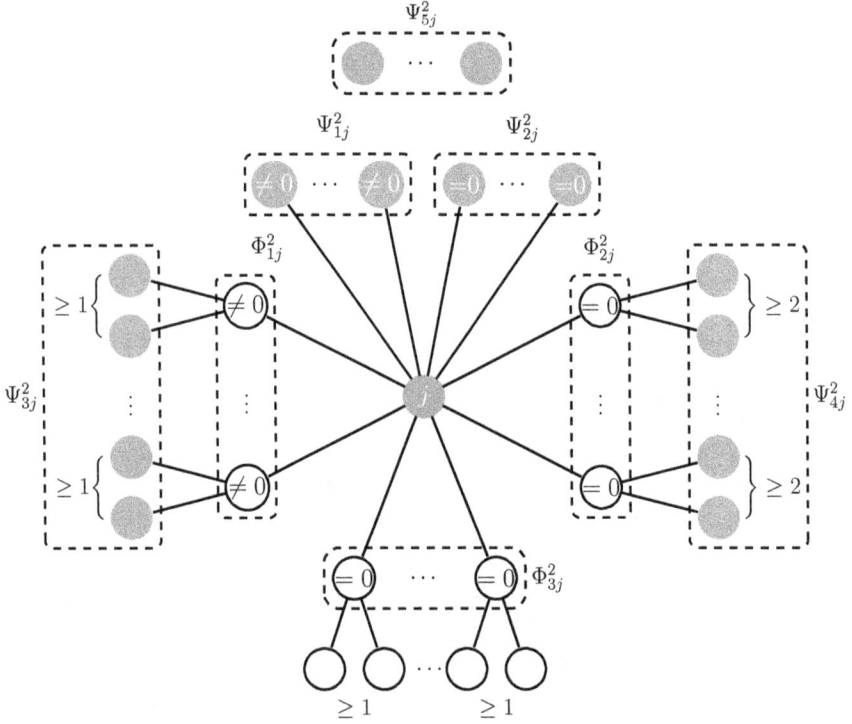

Figure 9.12 An illustration of Case 2 of Theorem 9.3.

$\mathcal{O}, m \in \mathcal{N}_j, \|r_m\| \neq 0\}, |\Phi^1_{1j}| = |\mathcal{N}_j| - h; \; \Phi^1_{2j} = \{m|m \in \mathcal{O}, m \in \mathcal{N}_j \cap \mathcal{N}_k, k \in \mathcal{V}_A, \|r_m\| = 0\}, |\Phi^1_{2j}| = s; \; \Phi^1_{3j} = \{m|m \in \mathcal{O}, m \in \mathcal{N}_j, m \notin \Phi^1_{1j} \cup \Phi^1_{2j}, \|r_m\| = 0\}, |\Phi^1_{3j}| = |\mathcal{N}_j| - |\Phi^1_{1j}| - |\Phi^1_{2j}|; \; \Psi^1_{1j} = \{i|i \in \mathcal{V}_A, i \in \mathcal{N}_m, m \in \Phi^1_{1j}\}, |\Psi^1_{1j}| \geq |\mathcal{N}_j| - h; \; \Psi^1_{2j} = \{i|i \in \mathcal{V}_A, i \in \mathcal{N}_m, m \in \Phi^1_{2j}\}, |\Psi^1_{2j}| = s^* \geq 2s; \; \Psi^1_{3j} = \{q|q \in \mathcal{V}_A, q \notin \Psi^1_{1j} \cup \Psi^1_{2j}\}, |\Psi^1_{3j}| = F_0 - |\Psi^1_{1j}| - |\Psi^1_{2j}|.$ As is shown in Fig. 9.11, the isolation of compromised agents in Ψ^1_{2j} makes the value of $\|\Upsilon_j\|_{l_0}$ increase, and $\max\{\|\Upsilon_j\|_{l_0}\} = |\mathcal{N}_j| - h + s.$ Besides, the isolation of other agents has no influence on $\|\Upsilon_j\|_{l_0}$. We shall prove that all compromised agents satisfy $\|\Upsilon_i\|_{l_0} > \max\{\|\Upsilon_j\|_{l_0}\}$ in the process of attack isolation.

Any two compromised agents $i_1, i_2 \in \Psi^1_{1j} \cup \Psi^1_{2j}$ have no common neighbor outside $\Phi^1_{1j} \cup \Phi^1_{2j}$ in the $(F, 7)$-isolable graph. Thus, for any compromised agent $i \in \Psi^1_{1j} \cup \Psi^1_{2j}$, the value of $\|\Upsilon_i\|_{l_0}$ decreases only when the compromised agent $q \in \Psi^1_{3j}$ colludes with i by the way in Fig. 9.10(b), meaning that $\|\Upsilon_i\|_{l_0} \geq |\mathcal{N}_i| - |\Psi^1_{3j}| \geq |\mathcal{N}_j| - h + s^* + 1 > \max\{\|\Upsilon_j\|_{l_0}\}$, where $F_0 \leq F$ and $|\mathcal{N}_i| \geq F + 1$ have been used to get the above inequality.

For any compromised agent $q \in \Psi^1_{3j}$, the isolation of compromised agents in $\Psi^1_{1j} \cup \Psi^1_{2j}$ makes $\|\Upsilon_q\|_{l_0} \geq |\mathcal{N}_q| - (|\Psi^1_{3j}| - 1) \geq |\mathcal{N}_j| - h + s^* + 2 > \max\{\|\Upsilon_j\|_{l_0}\}$.

The above analysis implies the achievement of attack isolation under Algorithm 9.2.

Case 2): An illustration of this case is depicted in Fig. 9.12, where $\Psi^2_{1j} = \{i|i \in \mathcal{V}_A, i \in \mathcal{N}_j, \|r_i\| \neq 0\}, |\Psi^2_{1j}| = k_1; \; \Psi^2_{2j} = \{i|i \in \mathcal{V}_A, i \in \mathcal{N}_j, \|r_i\| = 0\}, |\Psi^2_{2j}| = k - k_1;$

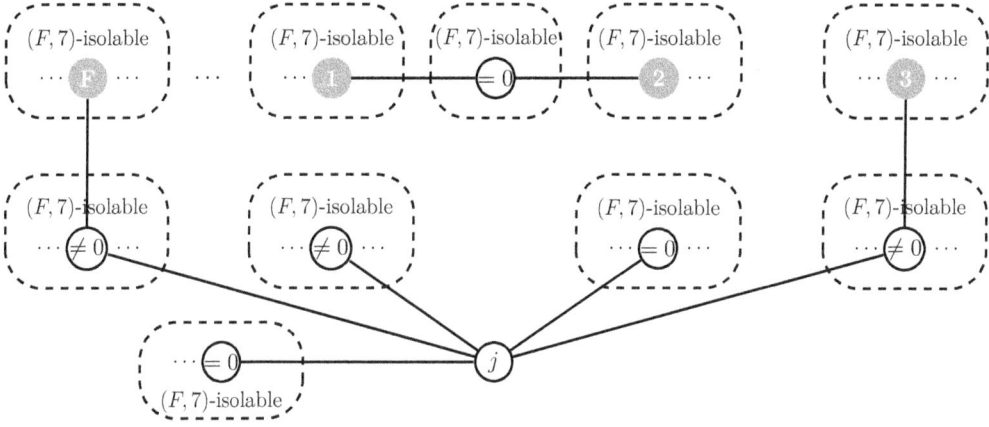

Figure 9.13   An illustration of $(F, 6)$-isolable graph.

$\Phi_{1j}^2 = \{m|m \in \mathcal{O}, m \in \mathcal{N}_j, \|r_m\| \neq 0\}, |\Phi_{1j}^2| = |\mathcal{N}_j| - h - k_1; \ \Phi_{2j}^2 = \{m|m \in \mathcal{N}_j \cap \mathcal{N}_k, k \in \mathcal{V}_A, \|r_m\| = 0\}, |\Phi_{2j}^2| = s; \ \Phi_{3j}^2 = \{m|m \in \mathcal{O}, m \in \mathcal{N}_j, m \notin \Phi_{1j}^2 \cup \Phi_{2j}^2, \|r_m\| = 0\}, |\Phi_{3j}^2| = |\mathcal{N}_j| - |\Phi_{1j}^2| - |\Psi_{2j}^2| - |\Phi_{1j}^2| - |\Phi_{2j}^2|; \ \Psi_{3j}^2 = \{i|i \in \mathcal{V}_A, i \in \mathcal{N}_m, m \in \Phi_{1j}^2\}, |\Psi_{3j}^2| \geq |\mathcal{N}_j| - h - k_1; \ \Psi_{4j}^2 = \{i|i \in \mathcal{V}_A, i \in \mathcal{N}_m, m \in \Phi_{2j}^2\}, |\Psi_{4j}^2| = s^* \geq 2s; \ \Psi_{5j}^2 = \{q|q \in \mathcal{V}_A, q \notin \Psi_{1j}^2 \cup \Psi_{2j}^2 \cup \Psi_{3j}^2 \cup \Psi_{4j}^2\}, |\Psi_{5j}^2| = F_0 - |\Psi_{1j}^2| - |\Psi_{2j}^2| - |\Psi_{3i}^2| - |\Psi_{4i}^2|.$
Similar to the analysis in Case 1)*, one has* $\max\{\|\Upsilon_j\|_{l_0}\} = |\mathcal{N}_j| - h + s + k - k_1$ *after the isolation of compromised agents in* $\Psi_{2j}^2 \cup \Psi_{4j}^2$. *For any compromised agent* $i \in \Psi_{1j}^2 \cup \Psi_{2j}^2 \cup \Psi_{3j}^2 \cup \Psi_{4j}^2$, $\|\Upsilon_i\|_{l_0}$ *decreases only when the compromised agent* $q \in \Psi_{5j}^2$ *colludes with* i *by the way in Fig. 9.10(b), meaning that* $\|\Upsilon_i\|_{l_0} \geq |\mathcal{N}_i| - |\Psi_{5j}^2| \geq |\mathcal{N}_j| - h + s^* + k - k_1 + 1 > \max\{\|\Upsilon_j\|_{l_0}\}$. *Furthermore, the isolation of compromised agents in* $\Psi_{1j}^2 \cup \Psi_{2j}^2 \cup \Psi_{3j}^2 \cup \Psi_{4j}^2$ *makes* $\|\Upsilon_q\|_{l_0} \geq |\mathcal{N}_q| - (|\Psi_{5j}^2| - 1) \geq |\mathcal{N}_j| - h + s^* + k - k_1 + 2 > \max\{\|\Upsilon_j\|_{l_0}\}, \forall q \in \Psi_{5j}^2$. *This implies the achievement of attack isolation by Algorithm 9.2.*

Remark 9.6 *Note that the graph condition of* $(F, 7)$*-isolability for attack isolation of F-total covert attack model is sufficient yet sharp, which is illustrated by a counterexample in Fig. 9.13. In Fig. 9.13, all agents satisfy* $|\mathcal{N}_i| = F + 1$, *and the compromised agents satisfy* $\|r_i\| = 0, \forall i \in \mathcal{V}_A$. *The normal agent* j *and the compromised agents 1, 2 satisfy* $\|\Upsilon_j\|_{l_0} = \|\Upsilon_1\|_{l_0} = \|\Upsilon_2\|_{l_0} = F$, *which implies the failure of attack isolation under Algorithm 9.2.*

Remark 9.7 *The demonstration of Theorem 9.3 is much simpler than that of Theorem 9.2 for FDI attacks. The main reason lies in the difference between Algorithms 9.1 and 9.2. Specifically, apart from the neighbors' residuals for each agent used in Algorithm 9.2, Algorithm 9.1 also involves the residual of the agent itself. Consequently, three cases, rather than two cases as discussed in the proof of Theorem 9.3, are needed to be analyzed in Theorem 9.2.*

Algorithm 9.3 Fully distributed resilient consensus algorithm based on DINIs

---

1: For each agent i with $S_i = 0$ and its neighbors $j \in \mathcal{N}_i$ **do**
2: **for** $j = i^1 : i^{|\mathcal{N}_i|}$ **do**
3:     **if** $S_j = j$ **then**
4:         $a_{ij} = 0$;
5:         $\mathcal{N}_i = \mathcal{N}_i \setminus j$;
6:     **end if**
7: **end for**
8: Design control protocol u_i as in (9.6).

9.4 RESILIENT COOPERATIVE CONTROL UNDER COLLUSIVE ATTACKS

For the high-order NAS (9.1), the consensus control protocol in [166] is reviewed as follows:

$$\begin{cases} u_i = (\rho_i + \varrho_i)K\hat{\varepsilon}_i, \\ \dot{\rho}_i = \hat{\varepsilon}_i^T Q B B^T Q \hat{\varepsilon}_i, \end{cases} \quad (9.6)$$

where $\varrho_i = \hat{\varepsilon}_i^T Q \hat{\varepsilon}_i$, $K = -B^T Q$, Q is a positive matrix satisfying $Q^{-1}A^T + AQ^{-1} - 2BB^T < \mathbf{0}$.

It follows from [166] that for the high-order NAS (9.1) with Assumption 9.1 under the directed communication graph $\mathcal{G}$ containing a spanning tree, consensus cannot be reached by the control protocol (9.6) if there are any attacks in the NAS. To achieve resilient consensus of normal agents, the normal agent i mitigates the effect of its compromised neighbor j by resetting $a_{ij} = 0$. With this idea, the fully distributed resilient consensus algorithm based on DINIs is introduced in Algorithm 9.3.

Remark 9.8 *Algorithm 8.3 only acts on the agents with zero safety indices. However, all agents should implement Algorithm 9.3 against collusive attacks. The reason lies in the mistaken isolation of certain normal agents under Algorithm 9.1. Besides, if the subgraph $\mathcal{G}'$ induced by removing the isolated agents is connected, it can be derived that although the mistakenly isolated normal agents are excluded by other normal ones, the graph consisting of all normal agents contains a directed spanning tree since the mistakenly isolated normal agents still have access to information of other normal agents.*

Then, the fully distributed resilient consensus under F-total FDI attack model is provided as follows.

Corollary 9.1 *Consider a high-order NAS described by (9.1), with a communication topology represented by the graph $\mathcal{G}$. Suppose that Assumptions 9.1–9.2 hold, and the subgraph $\mathcal{G}'$ induced by removing the isolated agents is connected. Under the F-total FDI attack model, fully distributed resilient consensus can be achieved by Algorithms 9.1 and 9.3 if it holds that*

1) Graph $\mathcal{G}$ is $(F, 5)$-isolable with $F = 2$;

2) Graph $\mathcal{G}$ is $(F, 7)$-isolable with $F \geq 3$.

Remark 9.9 *To isolate and exclude the compromised agents, the identities of agents are required to be known. The MSR-based algorithms [49, 51, 157, 160, 194] and the DARC-based algorithms [64, 65, 195] can be used to realize resilient consensus without using agents' identities. In the MSR-based algorithms [49, 51, 157, 160, 194], each normal agent excludes some of its neighbors with extreme values, and derives graph conditions for the achievement of resilient consensus. In the DARC-based algorithms [64, 65, 195], a safe point is selected by each agent from the convex hull of the positions of its normal neighbors, and then each normal agent updates its value based on the convex combination of the selected safe point and its former value. However, the above-mentioned two methods are suitable for integrator-type NASs. As a result, the main advantage of the proposed fully distributed resilient consensus algorithm is its applicability to general high-order NASs, while the price it takes is the requirement of agents' identities and the observer computation.*

The following result establishes the resilient stability of physically coupled NASs under collusive covert attacks.

Corollary 9.2 *Consider the physically coupled NAS under covert attacks described by (8.10), with a communication topology represented by the graph $\mathcal{G}$. Suppose that Assumptions 8.3–8.6 hold. Under the F-total covert attack model, resilient stability of $N - F$ normal agents can be achieved by Algorithms 9.2 and 8.4 if the followings hold simultaneously:*

(i) Graph $\mathcal{G}$ is $(F, 7)$-isolable;

(ii) The subgraph $\mathcal{G}'$ induced by removing the F isolated compromised agents is connected.

9.5 NUMERICAL SIMULATIONS

In this section, two examples are provided to validate the theoretical results in Sections 9.2–9.4, respectively.

Example 1: This example is performed to verify the resilient consensus of high-order NASs under collusive FDI attacks. Consider a NAS consisting of twenty agents described by (9.1), with its communication topology illustrated in Fig. 9.14. By examining the number of neighbors for each agent and the size of each cycle, it can be concluded that Fig. 9.14 is (2,5)-isolable. The elements of the weight matrix satisfy $a_{ij} = 2$ if $a_{ij} > 0$ and $i = 1, 7$, and $a_{ij} = 1$ if $a_{ij} > 0$. The system matrices of each agent is characterized as

$$A = \begin{bmatrix} -2.5 & 0.5 & 0 \\ 0.5 & -0.55 & 0.65 \\ 0 & -4 & 0.5 \end{bmatrix}, \ B = \begin{bmatrix} 1 \\ 2 \\ 1 \end{bmatrix}, \ C = \begin{bmatrix} 1 & 0 & 0 \\ 0 & 0 & 1 \end{bmatrix}.$$

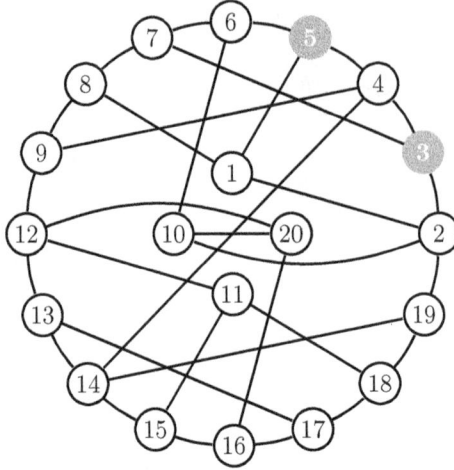

Figure 9.14   The communication topology of the NAS comprising twenty agents.

The preset convergent time instant $\tau = 1$, and the other parameters of the fixed-time observer (9.2) are calculated as

$$H_1 = \begin{bmatrix} 15.8939 & -3.6534 \\ -17.5628 & -10.9562 \\ 4.8665 & 14.5596 \end{bmatrix}, \quad H_2 = \begin{bmatrix} 9.8725 & -0.4240 \\ -2.1768 & -11.5698 \\ 4.6553 & 14.9078 \end{bmatrix}.$$

The control gain Q in (9.6) is given as

$$Q = \begin{bmatrix} 0.6308 & 0.3184 & -0.0977 \\ 0.3184 & 1.5124 & -0.3340 \\ -0.0977 & -0.3340 & 0.1994 \end{bmatrix}.$$

Assume that agents 3 and 5 are compromised with $T_{\eta_3} = T_{\eta_5} = 1.5$. The influence of the FDI attacks are described as $\eta_3(t, T_{\eta_3}) = -\eta_5(t, T_{\eta_5}) = \left(-2 - \cos(t) \quad -2 + \sin(t) \right)^T$. It is obviously that the attacks on these two agents collude by making $\|\tilde{\xi}_4(t)\| = 0$, $t \geq 1$ to disturb attack detection.

The residual trajectories of all agents are depicted in Fig. 9.15, where $\|\tilde{\xi}_i(t)\| \neq 0$ for $i \in \{1,2,3,5,6,7\}$ when $t > 1.5$, then one has $\|\Gamma_i\|_{l_0} = 3$ for $i \in \{1,2,3,5,6,7\}$ and $\|\Gamma_i\|_{l_0} < 3$ for others. Besides, $\|\tilde{\xi}_3\| = \|\tilde{\xi}_5\| > \|\tilde{\xi}_i\|, i \in \mathcal{V}\backslash\{3,5\}$. Consequently, the compromised agents 3 and 5 are isolated by Algorithm 9.1, seen in Fig. 9.16. Finally, the state consensus error trajectories of all agents are demonstrated in Fig. 9.17, in which all normal agents achieve resilient consensus. This illustrates the effectiveness of Corollary 9.1.

Example 2: This example is performed to verify the resilient stability of high-order NASs under collusive covert attacks. For the physically coupled NAS in Fig. 8.10, if the covert attacks satisfy $u_{1a} = -u_{9a}$, one has that the influence of the attacks on agent 10 cancels out since $\|r_{10}(t)\| = 0$ for $t \geq 2$. Consequently, $\|\Upsilon_1\|_{l_0} = \|\Upsilon_4\|_{l_0} = \|\Upsilon_7\|_{l_0} = \|\Upsilon_9\|_{l_0} = 2$, and thus, the compromised agents 1 and 9 may not

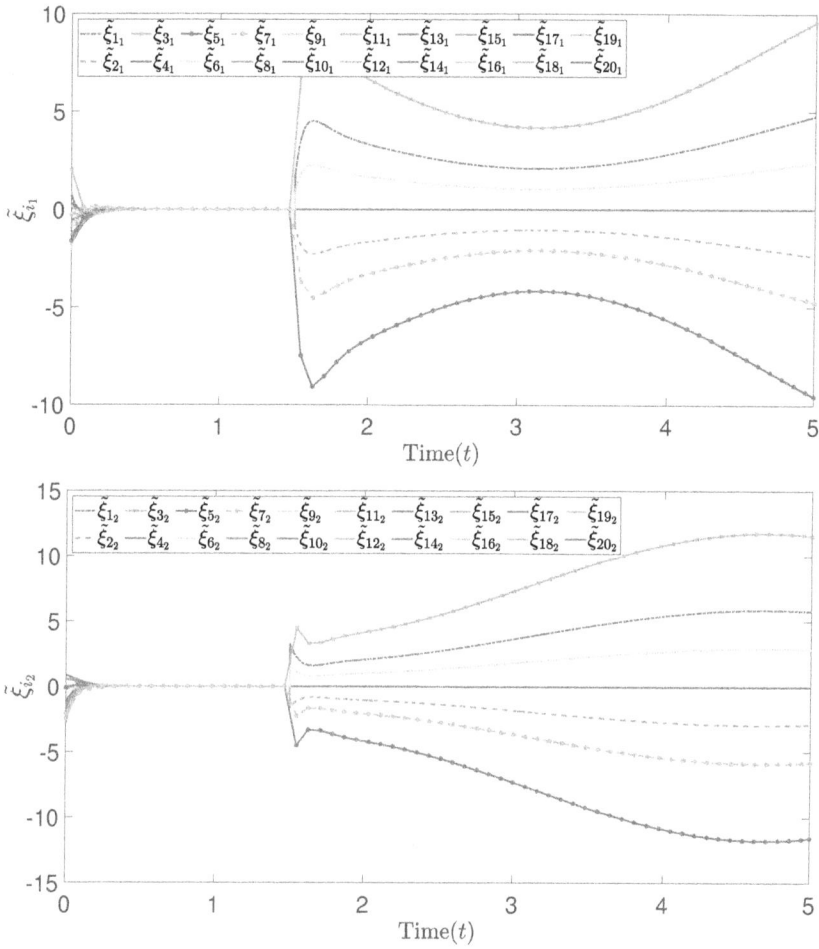

Figure 9.15   The residual trajectory $\tilde{\xi}_i$ of each agent.

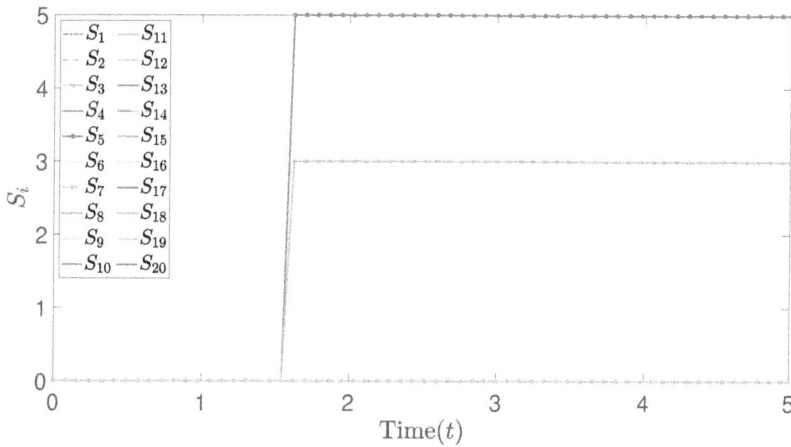

Figure 9.16   The safety index S_i of each agent.

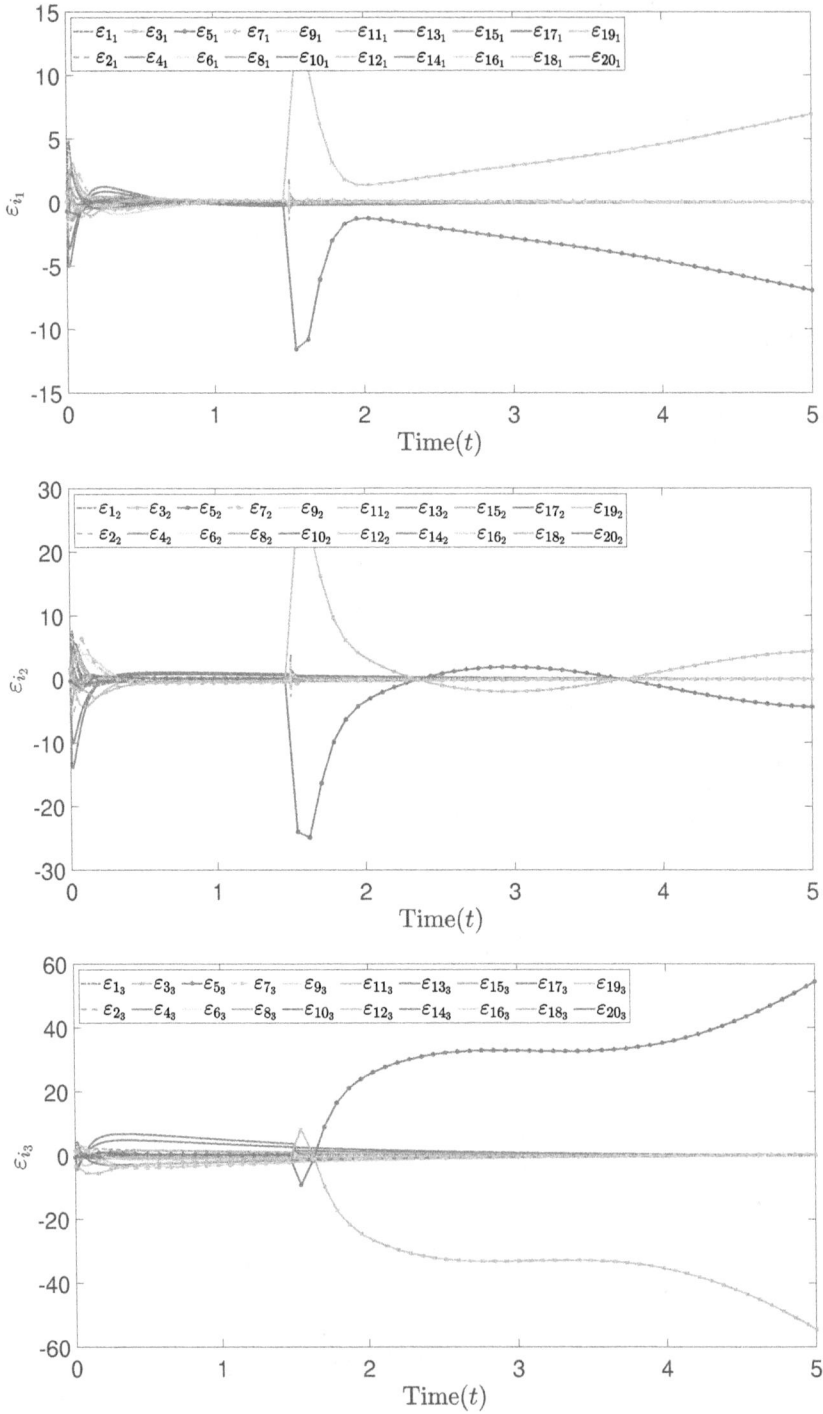

Figure 9.17   The state consensus error trajectory ε_i of each agent.

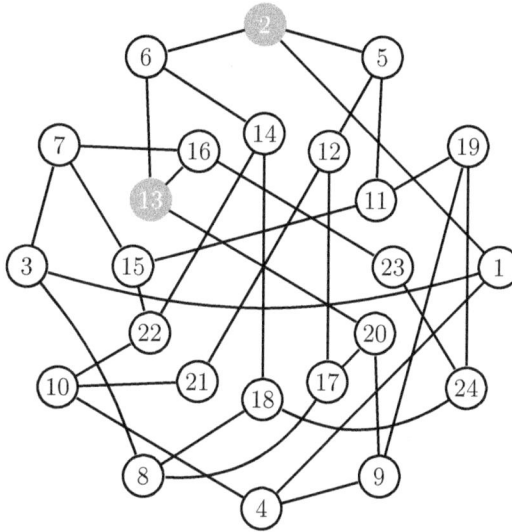

Figure 9.18   The $(2,7)$-isolable graph.

be isolated by Algorithm 9.3. This means that the graph condition r^*-isolability is far from sufficient to achieve attack isolation with collusion. Consider the physically coupled NAS in Fig. 9.18 with (2,7)-isolability, the system matrices of agents 1–5 are the same as the ones in Chapter 8, and the system matrices of agents 6–24 are the same as those of agent 1 in *Example 3* of Chapter 8. The elements of the weight matrix satisfy $a_{ij} = 0.5$ if $a_{ij} > 0$ and $i = 20$, and $a_{ij} = 1$ if $a_{ij} > 0$. Assume that the agents 2 and 13 subject to covert attacks at $t = 2.5$, and the control targets of the covert attacks satisfy $u_{2a} = -u_{13a} = [50 + 5t\ 0\ 50 + 5t]^T$.

It is easy to find that the covert attacks on agents 2 and 13 collude to cancel out their influence on agent 6 such that $\|r_6(t)\| = 0$ for $t \geq 2$, which is depicted in Fig. 9.19. Besides, one derives that $\|r_i(t)\| \neq 0$ for $i \in \{1, 5, 16, 20\}$, $t \geq 2.5$. Then we have $\|\Upsilon_2\|_{l_0} = \|\Upsilon_{13}\|_{l_0} = 2$ while $\|\Upsilon_i\|_{l_0} < 2$ for $i \in \mathcal{V}\backslash\{2, 13\}$. The safety indices of all agents are shown in Fig. 9.20, illustrating that the compromised agents 2 and 13 are isolated sequentially by Algorithm 9.2. The state trajectories of all agents are shown in Fig. 9.21, demonstrating that the resilient stability of twenty-two normal agents is reached. This verifies the correctness of Corollary 9.2.

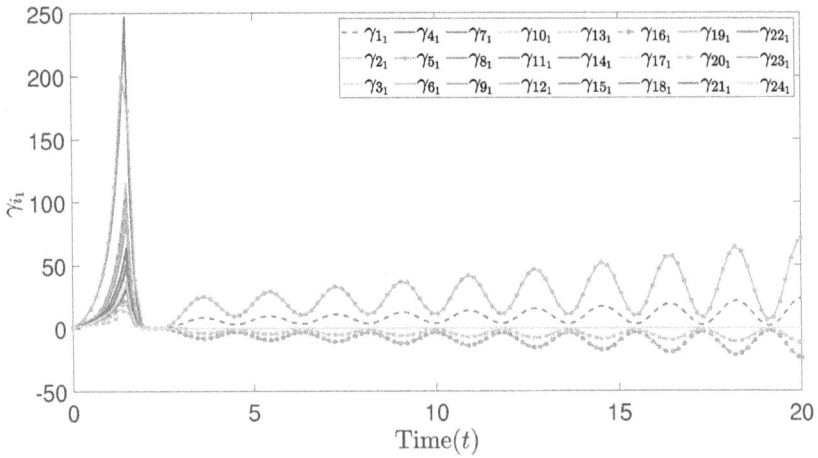

(a)

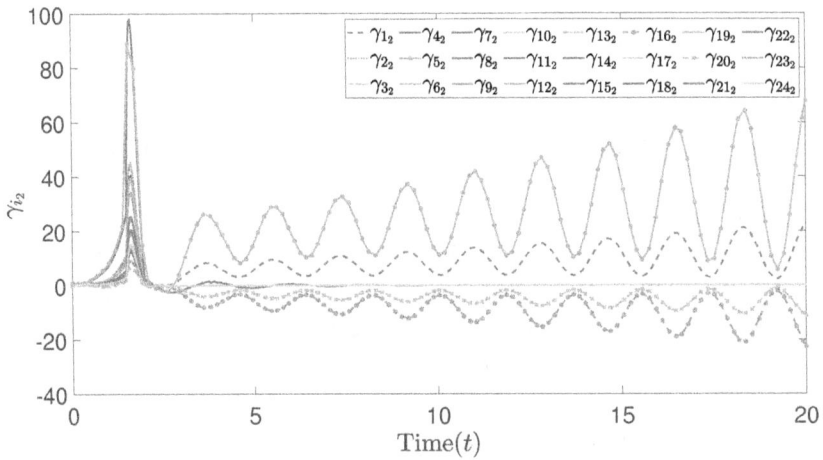

(b)

Figure 9.19 The output residual of each subsystem r_i with collusive covert attacks.

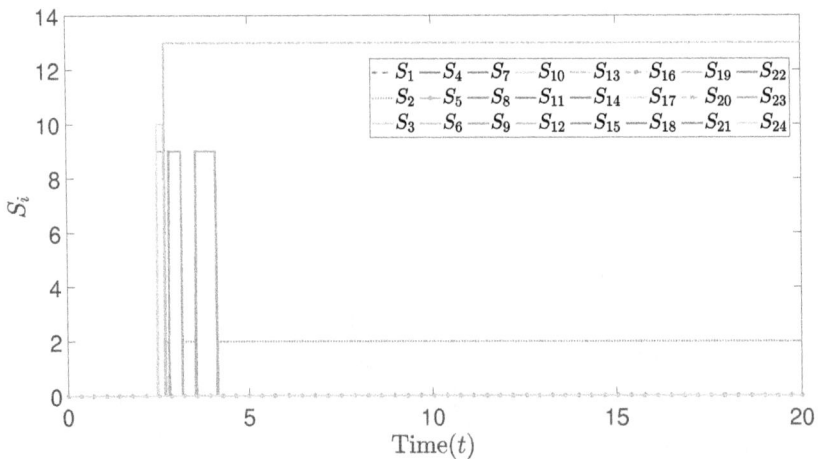

Figure 9.20 The safety index of each subsystem S_i with collusive covert attacks.

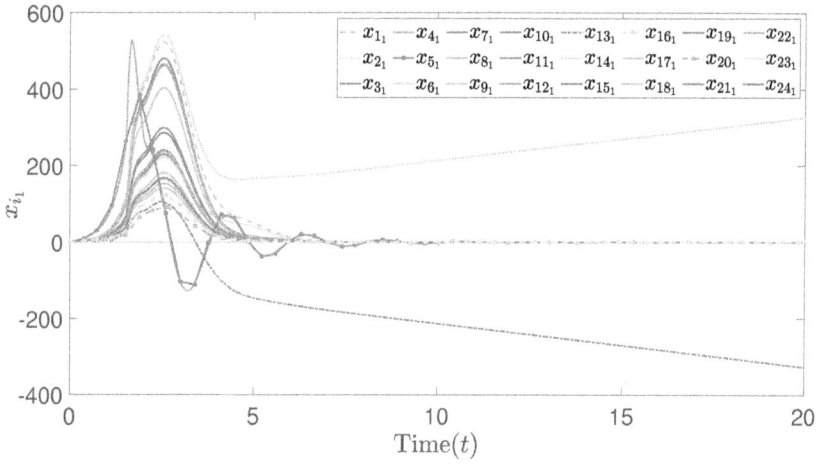

(a)

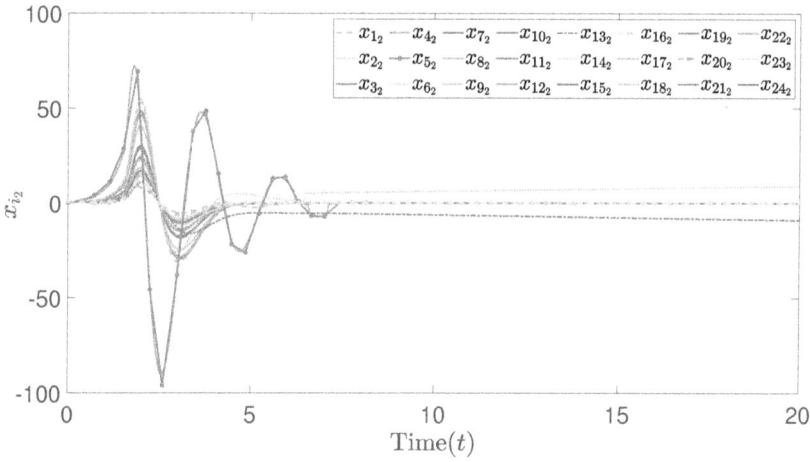

(b)

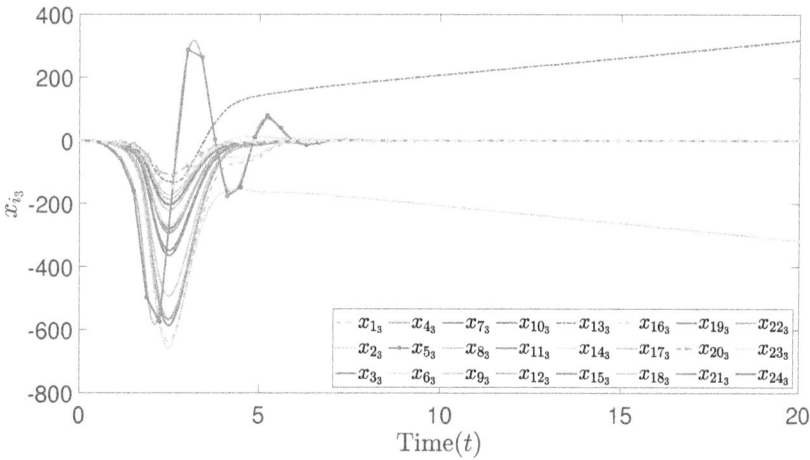

(c)

Figure 9.21   The state trajectory of each subsystem x_i with collusive covert attacks.

9.6 CONCLUSIONS

In this chapter, the resilient cooperative control problem of high-order NASs subject to both collusive FDI attacks and covert attacks has been studied. Two novel isolation algorithms are developed to counter collusive FDI attacks and collusive covert attacks, respectively. A new topological property, termed (r, s)-isolability, is introduced to ensure the effectiveness of the proposed attack isolation strategies. Finally, attack-isolation-based control algorithms are developed to accomplish resilient cooperative control tasks for normal agents, even in the presence of mistakenly isolated normal ones.

CHAPTER 10

Generalized graph-dependent isolation of collusive attacks

This chapter addresses the isolation problem of collusive FDI attacks and covert attacks for high-order NASs, particularly in establishing generalized graph-dependent conditions responsible for attack isolation. Section 10.1 reviews relevant prior work and articulates the motivations that drive the current study. Section 10.2 formulates the problem investigated in this chapter. Note that both the inclusive neighbor-based algorithm and the neighbor-based algorithm in Chapter 9 are capable of isolating both FDI attacks and covert attacks. However, it remains undetermined which algorithm is more suitable for isolating each type of attack, respectively. Section 10.3 investigates the worst-case collusive attack scenarios across different cycle sizes, from both the perspective of the attackers and that of the isolation algorithms, without considering the effect of normal agents. Section 10.4 develops several equivalent graph conditions for isolating collusive FDI attacks under both the inclusive neighbor-based algorithm and the neighbor-based algorithm. Section 10.5 gives the graph conditions for isolating collusive covert attacks under the neighbor-based algorithm. Section 10.6 provides some simulations to illustrate the effectiveness of the theoretical results. Section 10.7 concludes this chapter.

10.1 INTRODUCTION

In the previous chapter, two algorithms involving inclusive neighbor information and neighbor information were developed to isolate collusive FDI attacks and collusive covert attacks, respectively. Both algorithms demonstrate that $(F, 7)$-isolability serves as a sufficient condition for successful attack isolation. It should be noted that the $(F, 7)$-isolability requires that each cycle in the graph consists of at least seven nodes. While, it is preferable to derive generalized graph conditions for attack isolation with any size of cycles. Furthermore, it can be analytically shown that these two algorithms are capable of isolating both collusive FDI attacks and covert attacks. However, it

DOI: 10.1201/9781003669913-10

has yet to be determined that which algorithm is more appropriate for isolating FDI attacks and covert attacks, respectively, if there are cycles consisting of less than seven nodes.

In this chapter, we aim to develop generalized graph conditions for the isolation of collusive FDI attacks and covert attacks, respectively. Firstly, a counterexample is provided to show that the $(F,7)$-isolability is not a unique sufficient condition for attack isolation. By exploring the mechanism of attack collusion, it is found that both the number of agents' one-hop neighbors and the size of cycles in the graph play significant roles in attack isolation. Besides, it is interesting to note that the impact of the smaller size of cycles on attack isolation can be compensated by increasing the number of agents' neighbors. Then, the worst attack collusion cases under different sizes of cycles from the perspective of the attacks and the one of the isolation algorithms without considering the normal agents are developed, respectively. Based on these, several equivalent graph conditions are provided for isolating FDI attacks under these two algorithms. Also, the graph conditions are given for isolating covert attacks under the neighbor-based algorithm. In particular, it is revealed that the graph conditions for isolating FDI attacks under the inclusive neighbor-based algorithm are more relaxed than those under the neighbor-based one, while the neighbor-based algorithm is more suitable for covert attacks.

10.2 PROBLEM FORMULATION

Consider a physically coupled NAS consisting of N agents with heterogeneous dynamics, the dynamics of the ith agent is described as

$$
\begin{cases}
\dot{x}_i(t) = A_{ii}x_i(t) + B_i\bar{u}_i(t) + \sum_{i=1}^N a_{ij}A_{ij}x_j(t), \\
\bar{u}_i(t) = u_i(t) + u_i^a(t), \\
y_i(t) = C_ix_i(t) + y_i^a(t), \ i = 1,2,\ldots,N,
\end{cases} \tag{10.1}
$$

where $x_i(t) \in \mathbb{R}^n$, $\bar{u}_i(t) \in \mathbb{R}^m$, $u_i(t) \in \mathbb{R}^m$, and $y_i(t) \in \mathbb{R}^p$ are, the state, the compromised control input, the nominal control input, and the measurement output, respectively. $\sum_{i=1}^N a_{ij}A_{ij}x_j(t)$ represents the coupling with neighboring agents, the term $(u_i^a(t), y_i^a(t))$ accounts for the attacks on both actuator and sensor channels. $A_{ii} \in \mathbb{R}^{n\times n}$, $B_i \in \mathbb{R}^{n\times m}$, $A_{ij} \in \mathbb{R}^{n\times n_j}$, and $C_i \in \mathbb{R}^{p\times n}$ are the compatible system matrices. Let $A_i = [a_{i1}A_{i1}, a_{i2}A_{i2}, \ldots, a_{iN}A_{iN}]$, $\bar{B}_i = [B_i, A_i]$, the following assumptions are required to derive the main results.

Assumption 10.1 $rank(C_i\bar{B}_i) = rank(\bar{B}_i)$.

Assumption 10.2 $rank\begin{bmatrix} sI_n - A_{ii} & \bar{B}_i \\ C_i & 0 \end{bmatrix} = n + rank(\bar{B}_i), \forall s \in \mathbb{C}.$

Remark 10.1 *The terminology F-total attack model is used in the following if there is no specific distinction between FDI attacks and covert attacks.*

To detect and isolate the compromised agents, a two-stage fixed-time observer is designed in Section 9.3, which is reviewed briefly in the following:

$$
\begin{cases}
\dot{\omega}_i^{\mathbf{u}}(t) = A_{ci}^{\mathbf{u}}\omega_i^{\mathbf{u}}(t) + B_{ci}^{\mathbf{u}}y_i(t), \\
\chi_i(t) = D_{ci}^{\mathbf{u}}\left[\omega_i^{\mathbf{u}}(t) - \exp\left(A_{ci}^{\mathbf{u}}\tau_{1i}\right)\omega_i^{\mathbf{u}}(t - \tau_{1i})\right], \\
\hat{x}_i^{\mathbf{u}}(t) = \chi_i(t) - E_i y_i(t),
\end{cases}
$$

$$
\begin{cases}
\dot{\omega}_i^{\mathbf{l}}(t) = A_{ci}^{\mathbf{l}}\omega_i^{\mathbf{l}}(t) + B_{ci}^{\mathbf{l}}u_i(t) + L_{ci}y_i(t) + C_c\sum_{i=1}^{N}a_{ij}A_{ij}\hat{x}_j^{\mathbf{u}}(t), \\
\hat{x}_i^{\mathbf{l}}(t) = D_{ci}^{\mathbf{l}}\left[\omega_i^{\mathbf{l}}(t) - \exp\left(A_{ci}^{\mathbf{l}}\tau_i\right)\omega_i^{\mathbf{l}}(t - \tau_i)\right], \\
\hat{y}_i(t) = C_i\hat{x}_i^{\mathbf{l}}(t),
\end{cases}
\tag{10.2}
$$

where $\omega_i^{\mathbf{u}}(t) \in \mathbb{R}^{2n}$, $\chi_i(t) \in \mathbb{R}^n$, $\omega_i^{\mathbf{l}}(t) \in \mathbb{R}^{2n}$, $\hat{x}_i^{\mathbf{l}}(t) \in \mathbb{R}^n$ are auxiliary variables satisfying $\omega_i^{\mathbf{u}}(t) = \mathbf{0}$ for $t \in [-\tau_{1i}, 0]$ and $\omega_i^{\mathbf{l}}(t) = \mathbf{0}$ for $t \in [-\tau_i, \tau_{1i}]$. Note that τ_{1i} and τ_i are preset time constants such that $\hat{y}_i(t) = y_i(t)$ for $t \geq \tau_i$ if there is no attacks on agent i and its neighbors. Also, $\hat{x}_i^{\mathbf{u}}(t) \in \mathbb{R}^n$ and $\hat{y}_i(t) \in \mathbb{R}^p$ denote the state and the output of the observer, respectively. $A_{ci}^{\mathbf{u}} = \begin{pmatrix} F_i A_{ii} - H_i^1 C_i & \mathbf{0} \\ \mathbf{0} & F_i A_{ii} - H_i^2 C_i \end{pmatrix}$, $B_{ci}^{\mathbf{u}} = \begin{pmatrix} B_{ci}^{1}{}^T & B_{ci}^{2}{}^T \end{pmatrix}^T$, H_i^k is the gain matrix such that $F_i A_{ii} - H_i^k C_i$ is Hurwitz, $B_{ci}^k = H_i^k(I_p + C_i E_i) - F_i A_{ii} E_i, k = 1, 2$, $E_i = -\bar{B}_i\left(C_i \bar{B}_i\right)^\dagger$, $F_i = I_n + E_i C_i$, $D_{ci}^{\mathbf{u}} = \begin{pmatrix} I_n & \mathbf{0} \end{pmatrix}\begin{pmatrix} C_c & \exp\left(A_{ci}^{\mathbf{u}}\tau_{1i}\right)C_c \end{pmatrix}^{-1}$, $C_c = (I_n \ I_n)^T$, $A_{ci}^{\mathbf{l}} = \begin{pmatrix} A_{ii} - L_i^1 C_i & \mathbf{0} \\ \mathbf{0} & A_{ii} - L_i^2 C_i \end{pmatrix}$, L_i^k makes $A_{ii} - L_i^k C_i$ be Hurwitz, $k = 1, 2$, $B_{ci}^{\mathbf{l}} = (B_i^T \ B_i^T)^T$, $L_{ci} = \left(L_i^{1}{}^T \ L_i^{2}{}^T\right)^T$, $D_{ci}^{\mathbf{l}} = \begin{pmatrix} I_n & \mathbf{0} \end{pmatrix}\begin{pmatrix} C_c & \exp\left(A_{ci}^{\mathbf{l}}\tau_i\right)C_c \end{pmatrix}^{-1}$.

Denote by $\gamma_i(t) = y_i(t) - \hat{y}_i(t)$ the output residual between agent i and its observer, it can be derived from (10.1) and (10.2) that

$$
\begin{cases}
\gamma_i(t) = C_i D_{ci}^{\mathbf{l}}\left[\int_{t-\tau_i}^{t} \exp\left(A_{ci}^{\mathbf{l}}(t - s)\right) C_c \sum_{j=1}^{N}a_{ij}A_{ij}\tilde{x}_j^{\mathbf{u}}(s)ds \right. \\
\left. \qquad + \int_{t-\tau_i}^{t} \exp\left(A_{ci}^{\mathbf{l}}(t - s)\right) f_i(s)ds\right] + y_i^a(t), \\
\tilde{x}_j^{\mathbf{u}}(t) = D_{cj}^{\mathbf{u}}\int_{t-\tau_{j1}}^{t} \exp\left(A_{cj}^{\mathbf{u}}(t - s)\right) y_j^a(s)\,ds + E_j y_j^a(t), \ t \geq \tau_i,
\end{cases}
\tag{10.3}
$$

where $f_i(s) = B_{ci}^{\mathbf{l}}u_{ia}(s) - L_{ci}y_{ia}(s)$ and $\tilde{x}_i^{\mathbf{u}}(t) = x_i(t) - \hat{x}_i^{\mathbf{u}}(t)$.

Particularly, if the attack signal $(u_i^a(t), y_i^a(t))$ satisfies

$$
\begin{cases}
\dot{x}_i^a(t) = A_{ii}x_i^a(t) + B_i u_i^a(t), \\
y_i^a(t) = -C_i x_i^a(t), \ x_i^a(t_a) = \mathbf{0},
\end{cases}
\tag{10.4}
$$

where t_a is the time instant that the attack occurs. Then, the output residual $\gamma_i(t)$ in (10.3) can be modified as

$$
\begin{cases}
\gamma_i(t) = C_i D_{ci}^{\mathbf{l}}\int_{t-\tau_i}^{t} \exp\left(A_{ci}^{\mathbf{l}}(t - s)\right) C_c \sum_{j=1}^{N}a_{ij}A_{ij}\tilde{x}_j^{\mathbf{u}}(s)ds, \\
\tilde{x}_j^{\mathbf{u}}(t) = -x_j^a(t), \ t \geq \tau_i.
\end{cases}
\tag{10.5}
$$

Algorithm 10.1 Simplified DINIs algorithm against collusive attacks

---

Input: $S_i = 0$, γ_i, $i \in \mathcal{V}$;
Output: S_i, $i \in \mathcal{V}$;
  1: **for** $l = 1 : F$ **do**
  2:    $a_i(0) = \|\Gamma_i\|_{l_0}$, $i \in \mathcal{V}$;
  3:    **for** $k = 1 : N - 1$ **do**
  4:      $a_i(k) = \max_{j \in \mathcal{J}_i}\{a_j(k-1)\}$, $i \in \mathcal{V}$;
  5:    **end for**
  6:    Set $S_j = j$ if $\|\Gamma_j\|_{l_0} = a_1(N-1)$;
  7:    Reset $\|\gamma_q\| \neq 0$, $\forall q \in \mathcal{J}_j$;
  8:    Reset $\|\Gamma_j\|_{l_0} = 0$;
  9:    Update $\|\Gamma_i\|_{l_0}$, $i \in \mathcal{V}\backslash\{j\}$.
 10: **end for**

The contrast between (10.3) and (10.5) suggests that the output residual $\gamma_i(t)$ for agent i is not able to detect the attack on itself under the attack signal (10.4). Here, the attack $(u_i^a(t), y_i^a(t))$ is called a covert attack if its dynamics satisfy (10.4) (see [15, 155]), and is called an FDI attack otherwise. Therefore, if there is no attack collusion, the output residual γ_i is able to detect the FDI attacks on its *inclusive neighbors* while detecting the covert attacks on its *neighbors*. Moreover, it follows from the previous chapter that the attacks on multiple agents may collude to avoid being detected and cause serious damage to the whole system's performance. Especially, the FDI attacks in the *inclusive neighbor set* of agent i, i.e., $\bigcup_{j \in \mathcal{J}_i} \left(u_j^a(t), y_j^a(t) \right)$, can collude to make $\gamma_i(t) = \mathbf{0}$ for $t \geq \tau_i$ such that the residual for agent i fails to detect the attacks on its inclusive neighbors. The same conclusion can be obtained for the covert attacks in the *neighbor set* of agent i, i.e., $\bigcup_{j \in \mathcal{N}_i} \left(u_j^a(t), y_j^a(t) \right)$.

Denote $\Gamma_i = \left(\|\gamma_i\|, \|\gamma_{i^1}\|, \cdots, \|\gamma_{i^{|\mathcal{N}_i|}}\| \right) \in \mathbb{R}^{|\mathcal{J}_i|}$ and $\Upsilon_i = \left(\|\gamma_{i^1}\|, \|\gamma_{i^2}\|, \cdots, \|\gamma_{i^{|\mathcal{N}_i|}}\| \right) \in \mathbb{R}^{|\mathcal{N}_i|}$ the inclusive neighbor isolation vector and the neighbor isolation vector of agent i, respectively. Then, a simplified algorithm based on Algorithm 9.1 is given below. For simplicity, assume that the preset time constant $\tau_i = \tau$, $i \in \mathcal{V}$. In the following, consider the time instant t satisfying $t \geq \tau$, and the time index t is omitted to simplify the analysis.

It is easy to derive from Chapter 9 that for the F-total attack model with $F \geq 3$, attack isolation can be achieved under Algorithm 10.1 or Algorithm 9.2 if the graph $\mathcal{G}$ is $(F, 7)$-isolable. Besides, the condition of $(F, 7)$-isolability is sufficient yet sharp since there exists an $(F, 6)$-isolable graph failing to achieve attack isolation. However, it's worth noting that the condition of $(F, 7)$-isolability is not the only sharply sufficient condition. For the 3-total FDI attack model in Fig. 10.1, the neighbor number of each agent satisfies $|\mathcal{N}_i| = 5$, the cycles not shown satisfy $\mathcal{C}_k, k \geq 6$, the residuals of normal agent m, compromised agents 1, 2, 3, and the neighbors of 1, 2, and 3 not shown are non-zero, and the rest of the normal ones not shown are zero. Then one has $\|\Gamma_3\|_{l_0} = 6$, $\|\Gamma_1\|_{l_0} = \|\Gamma_2\|_{l_0} = 5$, $\|\Gamma_k\|_{l_0} \leq 4, k \in \mathcal{V}\backslash\{1,2,3\}$, which suggests

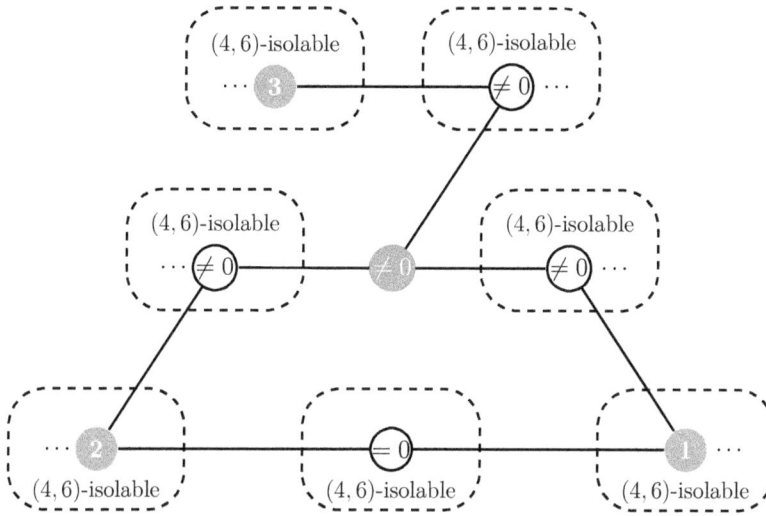

Figure 10.1   An example of a $(4,6)$-isolable graph for FDI attacks, where the neighbor number of each agent satisfies $|\mathcal{N}_i| = 5$, and the cycles not shown satisfy $\mathcal{C}_k, k \geq 6$.

that Algorithm 10.1 is able to achieve attack isolation under the $(4,6)$-isolable graph. A similar conclusion can be derived for the isolation of 3-total covert attack model under Algorithm 9.2 in Fig. 10.2.

In each round of Algorithm 10.1 or Algorithm 9.2, the agent j with $\|\Gamma_j\|_{l_0} = \max_{i \in \mathcal{V}} \{\|\Gamma_i\|_{l_0}\}$ or $\|\Upsilon_j\|_{l_0} = \max_{i \in \mathcal{V}} \{\|\Upsilon\|_{l_0}\}$ is isolated as the compromised one. To ensure the compromised agents own sufficiently large values of $\|\Gamma_j\|_{l_0}$ or $\|\Upsilon_j\|_{l_0}$, the size of cycles in the graph is required to be large enough to reduce the collusion among multiple attacks. However, if there is no collusion among attacks, it follows from (10.3) and (10.5) that for the compromised agent j, the residuals of all its neighbors are non-zero, while for the normal agent m, the residual of its neighbor is non-zero only if the neighbor is compromised. Besides, the attacks are able to collude if they are inclusive neighbors or neighbors of certain agents. These facts inspire us to ask whether the influence of smaller cycles can be resisted by more neighbors of agents. Fortunately, the answer is yes! This chapter will focus on the generalized graph conditions under any size of cycles for isolating collusive FDI attacks and covert attacks, respectively.

10.3   THE WORST CASES OF ATTACK COLLUSION

In this section, the worst attack collusive cases from the perspective of the attacks themselves will be discussed for $\mathcal{C}_{>4}$ and $\mathcal{C}_{\geq 3}$, respectively. Based on this, the worst attack collusive cases from the perspective of the isolation algorithms without considering the effect of $\max_{j \in \mathcal{V}_R} \{\|\Gamma_j\|_{l_0}\}$ and $\max_{j \in \mathcal{V}_R} \{\|\Upsilon_j\|_{l_0}\}$ will be further investigated. For simplicity, this chapter will make no distinction between attacks and compromised agents.

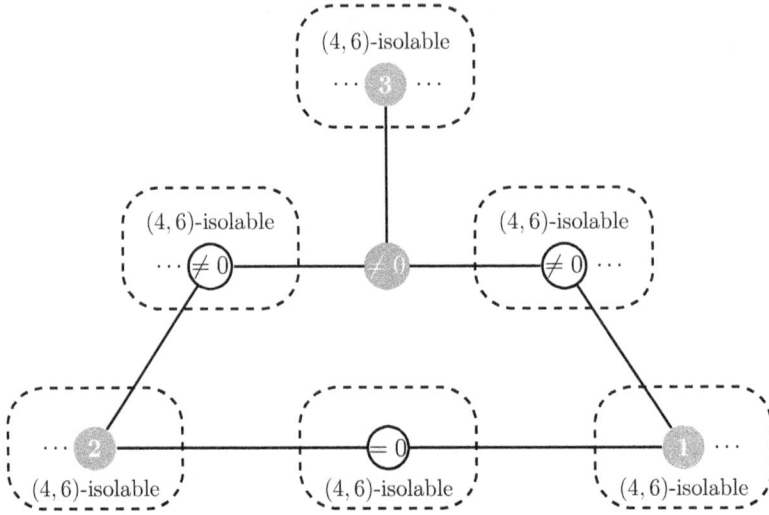

Figure 10.2 An example of a $(4,6)$-isolable graph for covert attacks, where '$\neq 0$' and '$= 0$' respectively denote that $\|\gamma_i\| \neq 0$ and $\|\gamma_i\| = 0$.

10.3.1 The worst attack collusive cases from the perspective of attacks themselves with $\mathcal{C}_{>4}$

Firstly, the worst collusive cases among FDI attacks for $\|\Gamma_i\|_{l_0}$, $i \in \mathcal{V}_A$ with $\mathcal{C}_{>4}$ are addressed.

Obviously, the attacks tend to collude to decrease $\|\Gamma_i\|_{l_0}$ of *all* $i \in \mathcal{V}_A$, rather than *some* $i \in \mathcal{V}_A$, to avoid being isolated. Denote by $\left\{\|\Gamma_i\|_{l_0}\right\}$, $i \in \mathcal{V}_A$ the set consisting of the l_0 norm of the inclusive neighbor isolation vectors of the compromised agents, one has the maximum element of set $\left\{\|\Gamma_i\|_{l_0}\right\}$, $i \in \mathcal{V}_A$ with the smallest mean, denoted by $\max_{i \in \mathcal{V}_A} \left\{\|\Gamma_i\|_{l_0}\right\}^{\mathbb{E}_{\min}}$, characterizes the worst case of attack collusion for $\|\Gamma_i\|_{l_0}$, $i \in \mathcal{V}_A$. Since the attacks cannot be preset, assume $|\mathcal{J}_i| = |\mathcal{J}|$ for simplicity. Then, one has $\|\Gamma_i\|_{l_0} = |\mathcal{J}| - \overline{\|\Gamma_i\|}_{l_0}$, and the derivation of $\max_{i \in \mathcal{V}_A} \left\{\|\Gamma_i\|_{l_0}\right\}^{\mathbb{E}_{\min}}$ is equivalent to solving the minimum value of set $\left\{\overline{\|\Gamma_i\|}_{l_0}\right\}$, $i \in \mathcal{V}_A$ with the largest mean, denoted by $\min_{i \in \mathcal{V}_A} \left\{\overline{\|\Gamma_i\|}_{l_0}\right\}^{\mathbb{E}_{\max}}$. The followings will focus on the solution of $\min_{i \in \mathcal{V}_A} \left\{\overline{\|\Gamma_i\|}_{l_0}\right\}^{\mathbb{E}_{\max}}$.

It follows from Fig. 9.3 that there are two ways for two FDI attacks to collude to increase the value of their $\overline{\|\Gamma_i\|}_{l_0}$ under cycles $\mathcal{C}_{>4}$. The two attacks i and j satisfy $\|\gamma_i\| = 0$ or $\|\gamma_j\| = 0$ in Fig. 9.3(a), and $\tilde{x}_i^{\mathbf{u}} = -\tilde{x}_j^{\mathbf{u}}$ in Fig. 9.3(b). Note that the way in Fig. 9.3(a) is able to increase the value of $\overline{\|\Gamma_i\|}_{l_0}$ of one of these two compromised agents, while the way in Fig. 9.3(b) can increase the value of $\overline{\|\Gamma_i\|}_{l_0}$ of both these two compromised agents. Then, the following result is provided for the worst collusion among FDI attacks for $\|\Gamma_i\|_{l_0}$, $i \in \mathcal{V}_A$ with $\mathcal{C}_{>4}$.

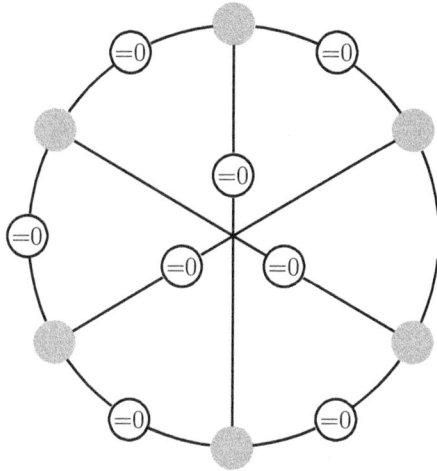

Figure 10.3   The worst collusive way for attacks with $F_0 = 2q$ and $\mathcal{C}_{k>4}$.

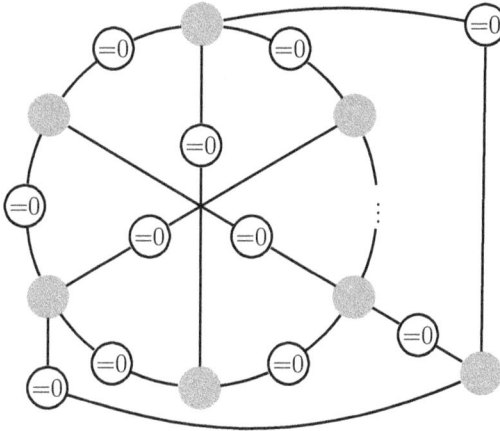

Figure 10.4   The worst collusive way for attacks with $F_0 = 2q+1, \mathcal{C}_{k>4}$ and $k_1 = m$.

Theorem 10.1 *Suppose that there are F_0 ($F_0 \leq F$) agents being compromised by FDI attacks and all cycles in the graph satisfy $\mathcal{C}_k$, $k > 4$. The worst attack collusive ways for $\|\Gamma_i\|_{l_0}$, $i \in \mathcal{V}_A$ are shown in Fig. 10.3 for $F_0 = 2q$ and Fig. 10.4 for $F_0 = 2q+1$, $q \in \mathbb{N}^+$, respectively. Particularly, the minimum value of set $\left\{ \overline{\|\Gamma_i\|_{l_0}} \right\}$, $i \in \mathcal{V}_A$ satisfies*

$$\min_{i \in \mathcal{V}_A} \left\{ \overline{\|\Gamma_i\|_{l_0}} \right\}^{\mathrm{Emax}} = \left\lfloor \frac{F_0}{2} \right\rfloor.$$

Proof 10.1 *Assume that the compromised agent j satisfies $\overline{\|\Gamma_j\|}_{l_0} = \min_{i \in \mathcal{V}_A} \left\{ \overline{\|\Gamma_i\|}_{l_0} \right\}$ $= m$, $m \in \mathbb{N}$. There are k_1 attacks colluding with j under the way in Fig. 9.3(b), k_2 attacks colluding with j and k such that $\|\gamma_j\| = 0$, and $m - k_1 - \mathbf{1}_{k_2>0}$ attacks colluding with j under the way in Figs. 9.3(a) such that $\|\gamma_k\| = 0$, $k \in \mathcal{N}_j \cap \mathcal{V}_A$.*

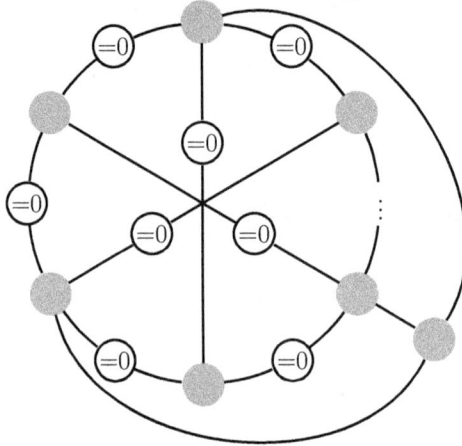

Figure 10.5   The worst collusive way for attacks with $F_0 = 2q + 1, \mathcal{C}_{k>4}$ and $k_2 = m$.

The relationship between the compromised agent j and the other ones is shown in Fig. 10.6. Then, one has $\left|\Phi_j^1\right| = k_1$, $\left|\Phi_j^2\right| = k_2$, $\left|\Phi_j^3\right| = m - k_1 - \mathbf{1}_{k_2>0}$, and $\left|\Phi_j^4\right| = F_0 - m - k_2 + \mathbf{1}_{k_2>0} - 1$. It follows from (10.3) that 1) any two attacks in $\Phi_j^1 \cup \Phi_j^2 \cup \Phi_j^3$ are not able to collude; 2) each agent in $\Phi_j^1 \cup \Phi_j^3$ can provide one zero-element for the $\overline{\|\Gamma_j\|}_{l_0}$ of agent j, while all agents in Φ_j^2 can only provide one zero-element for the $\overline{\|\Gamma_j\|}_{l_0}$ of agent j; 3) the attacks in Φ_j^4 is only able to collude with the attacks in one of the sets $\Phi_j^1 \cup \Phi_j^2 \cup \Phi_j^3$.

Then, the value of m with respect to different k_1 and k_2 is analyzed as follows.

- $k_1 = m$. One has $|\Phi_j^2| = 0$ and $|\Phi_j^3| = 0$. To make each agent in $\Phi_j^1 \cup \Phi_j^4$ satisfy $\overline{\|\Gamma_i\|}_{l_0} \geq m$, $\forall i \in \Phi_j^1 \cup \Phi_j^4$, the cardinality of set Φ_j^4 should satisfy $\left|\Phi_j^4\right| \geq m-1$, and any two attacks in Φ_j^1 and Φ_j^4 collude under the way in Fig. 9.3(b). Then, one has $F_0 - m - 1 \geq m - 1$, and thus, $m \leq \lfloor F_0/2 \rfloor$.

- $0 < k_1 < m$, $k_2 = 0$. One has $|\Phi_j^2| = 0$. First, it takes at least $(m-1) + (m-2)$ attacks in Φ_j^4 to make each agent in $\Phi_j^1 \cup \Phi_j^3$ satisfy $\overline{\|\Gamma_i\|}_{l_0} \geq m$, $\forall i \in \Phi_j^1 \cup \Phi_j^3$. Note that any two attacks $h_1, h_2 \in \Phi_j^4$ colluding with the ones in Φ_j^1 and Φ_j^3 cannot collude with each other. Then, extra $(m - k_1) + k_1$ attacks are needed to ensure h_1, h_2 to satisfy $\overline{\|\Gamma_{h_1}\|}_{l_0} \geq m$ and $\overline{\|\Gamma_{h_2}\|}_{l_0} \geq m$, and so on. Based on the above analysis, it can be concluded that the cardinality of set Φ_j^4 should satisfy $\left|\Phi_j^4\right| > (m-1) + (m-2) + (m - k_1) + k_1$, and thus, $m < \lfloor (F_0 + 2)/4 \rfloor$.

- $0 < k_1 < m$, $k_2 > 0$. The analysis is similar to that of the case $0 < k_1 < m$ and $k_2 = 0$, then one has $m < \lfloor (F_0 + 1)/6 \rfloor$.

- $k_1 = 0, k_2 = 0$. One has $|\Phi_j^3| = m$, $|\Phi_j^4| \geq m$, and any two attacks in Φ_j^3 and

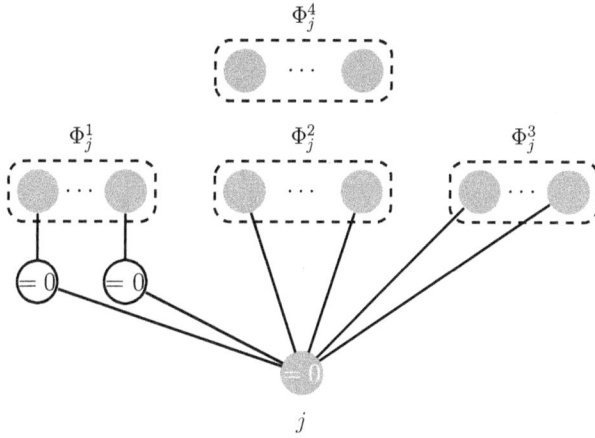

Figure 10.6   The collusion among the FDI attack on agent j and the other ones with $C_{>4}$.

Φ_j^4 collude under the way in Fig. 9.3(b). This suggests $F_0 - m - 1 \geq m$, and thus, $m \leq \lfloor (F_0 - 1)/2 \rfloor$.

- $k_1 = 0$, $k_2 > 0$. The analysis is similar to that of the case $0 < k_1 < m$ and $k_2 > 0$, then one has $m < \lfloor F_0/4 \rfloor$.

Now, the following conclusions can be derived from the above analysis.

1) $\min_{i \in \mathcal{V}_A} \left\{ \overline{\|\Gamma_i\|_{l_0}} \right\}^{\mathbb{E}_{\max}} = \lfloor F_0/2 \rfloor$ holds under the worst collusion among attacks.

2) For $F_0 = 2q$, the worst attack collusive way can be obtained with $k_1 = m$, as seen in Fig. 10.3. Then, one has that all compromised agents satisfy $\overline{\|\Gamma_i\|_{l_0}} = F_0/2$, $\forall i \in \mathcal{V}_A$.

3) For $F_0 = 2q+1$, there are two ways to achieve $\min_{i \in \mathcal{V}_A} \left\{ \overline{\|\Gamma_i\|_{l_0}} \right\}^{\mathbb{E}_{\max}} = \lfloor F_0/2 \rfloor$, including 1) $k_1 = m$; 2) $k_1 = 0$ and $k_2 = 0$, as seen in Figs. 10.4–10.5. For the collusive way in Fig. 10.4, $(F_0 + 1)/2$ compromised agents satisfy $\overline{\|\Gamma_i\|_{l_0}} = \lceil F_0/2 \rceil$, and the others satisfy $\overline{\|\Gamma_i\|_{l_0}} = \lfloor F_0/2 \rfloor$. For the collusive way in Fig. 10.5, all compromised agents satisfy $\overline{\|\Gamma_i\|_{l_0}} = \lfloor F_0/2 \rfloor$, $\forall i \in \mathcal{V}_A$. Since the $\overline{\|\Gamma_i\|_{l_0}}$ of compromised agents in Fig. 10.4 are no greater than those in Fig. 10.5, one can obtain that the worst attack collusive way for $F_0 = 2q+1$ is the one in Fig. 10.4.

Then, the worst collusive cases among FDI attacks for $\|\Upsilon\|_{l_0}$, $i \in \mathcal{V}_A$ with $C_{k>4}$ is analyzed.

Denote by $\max_{i \in \mathcal{V}_A} \left\{ \|\Upsilon_i\|_{l_0} \right\}^{\mathbb{E}_{\min}}$, the maximum value of set $\left\{ \|\Upsilon\|_{l_0} \right\}$, $i \in \mathcal{V}_A$ with the smallest mean, and $\min_{i \in \mathcal{V}_A} \left\{ \overline{\|\Upsilon_i\|_{l_0}} \right\}^{\mathbb{E}_{\max}}$ the minimum value of set $\left\{ \overline{\|\Upsilon\|_{l_0}} \right\}$, $i \in \mathcal{V}_A$ with the largest mean. Assume that $|\mathcal{N}_i| = |\mathcal{N}|$ for simplicity. Then, one has

$\|\Upsilon\|_{l_0} = |\mathcal{N}| - \overline{\|\Upsilon\|}_{l_0}$. The following corollary illustrates the worst collusive cases among FDI attacks for $\|\Upsilon\|_{l_0}$, $i \in \mathcal{V}_A$ with $\mathcal{C}_{k>4}$.

Corollary 10.1 *Suppose that there are F_0 ($F_0 \leq F$) agents being compromised by FDI attacks and all cycles in the graph satisfy $\mathcal{C}_k$, $k > 4$. The worst attack collusive ways for $\|\Upsilon\|_{l_0}$, $i \in \mathcal{V}_A$ are shown in Fig. 10.3 for $F_0 = 2q$ and Fig. 10.4 for $F_0 = 2q+ 1$, $q \in \mathbb{N}^+$, respectively. Particularly, the minimum value of the set $\left\{\overline{\|\Upsilon_i\|}_{l_0}\right\}$, $i \in \mathcal{V}_A$ satisfies*

$$\min_{i \in \mathcal{V}_A}\left\{\overline{\|\Upsilon_i\|}_{l_0}\right\}^{\mathbb{E}_{\max}} = \left\lfloor\frac{F_0}{2}\right\rfloor.$$

Proof 10.2 *It can be derived from Figs. 10.3 and 10.4 that $\|\gamma_i\| \neq 0, \forall i \in \mathcal{V}_A$ under the worst attack collusion for $\|\Gamma_i\|_{l_0}$, $i \in \mathcal{V}_A$. With $\overline{\|\Gamma_i\|}_{l_0} = \overline{\|\Upsilon\|}_{l_0} + \overline{\|\gamma_i\|}_{l_0}$, it concludes that $\overline{\|\Upsilon\|}_{l_0} = \overline{\|\Gamma_i\|}_{l_0}$ under the worst attack collusion for $\|\Gamma_i\|_{l_0}$, $i \in \mathcal{V}_A$. The above analysis suggests that the worst collusive ways for $\|\Upsilon\|_{l_0}$, $i \in \mathcal{V}_A$ are the same as those for $\|\Gamma_i\|_{l_0}$, $i \in \mathcal{V}_A$.*

Now, the worst collusive cases among covert attacks for $\|\Upsilon\|_{l_0}$, $i \in \mathcal{V}_A$ with $\mathcal{C}_{>4}$ are provided.

Lemma 10.1 *Suppose that there are F_0 ($F_0 \leq F$) agents being compromised by covert attacks and all cycles in the graph satisfy $\mathcal{C}_k$, $k > 4$. The worst attack collusive ways for $\|\Upsilon\|_{l_0}$, $i \in \mathcal{V}_A$ are shown in Fig. 10.3 for $F_0 = 2q$ and Fig. 10.4 for $F_0 = 2q+ 1$, $q \in \mathbb{N}^+$, respectively. Particularly, the minimum value of the set $\left\{\overline{\|\Upsilon_i\|}_{l_0}\right\}$, $i \in \mathcal{V}_A$ satisfies*

$$\min_{i \in \mathcal{V}_A}\left\{\overline{\|\Upsilon_i\|}_{l_0}\right\}^{\mathbb{E}_{\max}} = \left\lfloor\frac{F_0}{2}\right\rfloor.$$

Proof 10.3 *Compared with FDI attacks, there is only one way for two covert attacks to collude, as seen in Fig. 9.10(b). Assume that agent j is compromised by a covert attack, satisfying $\overline{\|\Upsilon_j\|}_{l_0} = \min_{i \in \mathcal{V}_A}\left\{\overline{\|\Upsilon\|}_{l_0}\right\} = m$, $m \in \mathbb{N}$. There are k_1 attacks colluding with j under the way in Fig. 9.10(b), $m - k_1$ compromised neighbors of agent j satisfying $\|\gamma_k\| = 0, k \in \mathcal{N}_j \cap \mathcal{V}_A$.*

The relationship between compromised agent j and other ones is shown in Fig. 10.7. One has $\left|\Omega_j^1\right| = k_1$, $\left|\Omega_j^2\right| = m - k_1$, and $\left|\Omega_j^3\right| = F_0 - m - 1$. It follows from (10.5) that 1) any two attacks in $\Omega_j^1 \cup \Omega_j^2$ are not able to collude; 2) there is at least one attack in Ω_j^3 being the one-hop neighbor of each attack in Ω_j^2.

Similar to the analysis in Theorem 10.1, the worst attack collusive way can be obtained with $k_1 = m$ for both $F = 2q$ and $F = 2q + 1$, as seen in Figs. 10.3 and 10.4. Then, one has $\min_{i \in \mathcal{V}_A}\left\{\overline{\|\Upsilon_i\|}_{l_0}\right\}^{\mathbb{E}_{\max}} = \lfloor F_0/2 \rfloor$.

Remark 10.2 *It follows from Theorem 10.1, Corollary 10.1, and Lemma 10.1 that the worst collusive ways for covert attacks are the same as those for FDI attacks. The difference between them is that all FDI attacks satisfy $\|\gamma_i\| \neq 0$, while the covert attacks satisfy $\|\gamma_i\| = 0$.*

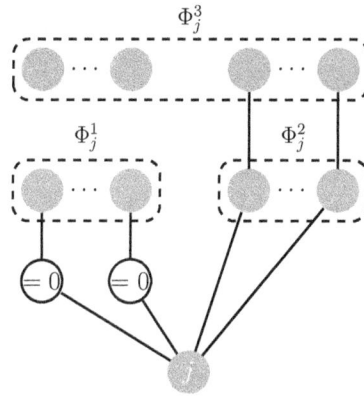

Figure 10.7   The collusion between the covert attack j and the other ones with $\mathcal{C}_{k>4}$.

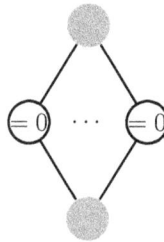

Figure 10.8   The worst collusive way for two attacks with $\mathcal{C}_k$, $k = 4$.

10.3.2   The worst attack collusive cases from the perspective of attacks themselves with $\mathcal{C}_{\geq 3}$

If any two attacks collude under the way in Fig. 10.8 in a cycle $\mathcal{C}_4$, they can provide two zero elements for the $\overline{\|\Gamma_i\|}_{l_0}$ of each other. Then, the relationship between the value of $\min_{i \in \mathcal{V}_A}\left\{\overline{\|\Gamma_i\|}_{l_0}\right\}^{\mathbb{E}_{\max}}$ and the number of cycles $\mathcal{C}_4$ can be derived as follows.

Lemma 10.2 *Suppose that there are F_0 ($F_0 \leq F$) agents being compromised by attacks. The minimum value of the set $\left\{\overline{\|\Gamma_i\|}_{l_0}\right\}$, $i \in \mathcal{V}_A$ under the worst collusive ways satisfies*

$$\min_{i \in \mathcal{V}_A}\left\{\overline{\|\Gamma_i\|}_{l_0}\right\}^{\mathbb{E}_{\max}} = \left\lfloor \frac{F_0}{2} \right\rfloor + p, p \left\lfloor \frac{F_0}{2} \right\rfloor \leq |\mathcal{C}_4| < 2p \left\lfloor \frac{F_0}{2} \right\rfloor, p \in \mathbb{N}^+.$$

Proof 10.4 *The worst attack collusive ways with $\mathcal{C}_{\geq 3}$ can be obtained by appropriately adding cycles $\mathcal{C}_4$ on the ones with $\mathcal{C}_{k>4}$. For simplicity, assume that the attack signals satisfy either $\tilde{x}^u$ or $-\tilde{x}^u$ in Figs. 10.3 and 10.4. Then, it can be easily obtained from Figs. 10.3 and 10.8 that the value of $\min_{i \in \mathcal{V}_A}\left\{\overline{\|\Gamma_i\|}_{l_0}\right\}^{\mathbb{E}_{\max}}$ increases by one if the following two conditions hold. One is that the number of cycles $\mathcal{C}_4$ satisfies $\lfloor F_0/2 \rfloor \leq |\mathcal{C}_4| < 2\lfloor F_0/2 \rfloor$. The other is that the cycles $\mathcal{C}_4$ are set to ensure that there is an extra normal neighbor between each attack i with $\tilde{x}_i^u = \tilde{x}^u$ and any one j with $\tilde{x}_j^u = -\tilde{x}^u$. Thus, the conclusion can be proved in the same manner for $p > 1$.*

Remark 10.3 *Note that the worst attack collusive ways for* $\left\{\overline{\|\Upsilon_i\|}_{l_0}\right\}$, $i \in \mathcal{V}_A$ *with* $\mathcal{C}_{\geq 3}$ *are the same as those for* $\left\{\overline{\|\Gamma_i\|}_{l_0}\right\}$, $i \in \mathcal{V}_A$. *Thus, one has* $\min_{i\in\mathcal{V}_A}\left\{\overline{\|\Upsilon\|}_{l_0}\right\}^{\mathbb{E}_{\max}} = \min_{i\in\mathcal{V}_A}\left\{\overline{\|\Gamma_i\|}_{l_0}\right\}^{\mathbb{E}_{\max}}$ *under cycles* $\mathcal{C}_{\geq 3}$.

10.3.3 The worst attack collusive cases from the perspective of the isolation algorithms without considering the effect of normal agents

The worst attack collusive ways for the attack isolation algorithms without considering $\max_{j\in\mathcal{V}_R}\left\{\|\Gamma_j\|_{l_0}\right\}$ are investigated in the following.

Consider the case that the relationship between the normal agent m with $\|\Gamma_m\|_{l_0} = \max_{j\in\mathcal{V}_R}\left\{\|\Gamma_j\|_{l_0}\right\}$ and the attacks have no effect on the way of attack collusion. Denote by $\overline{\|\Gamma\|}_{l_0}$, the key value of set $\left\{\overline{\|\Gamma_i\|}_{l_0}\right\}$, $i \in \mathcal{V}_A$, in which the compromised agent s with $\overline{\|\Gamma_s\|}_{l_0} = \overline{\|\Gamma\|}_{l_0}$ satisfies that 1) the isolation of compromised agents r with $\|\Gamma_r\|_{l_0} > \|\Gamma_s\|_{l_0}$ cannot lead to the isolation of agent s; 2) the isolation of agent s will lead to the sequential isolation of the rest compromised agents. Note that Algorithm 10.1 is able to uncover the collusion among attacks. Compared with the worst attack collusive cases from the perspective of attacks, the one from the perspective of Algorithm 10.1 only focuses on the value of $\overline{\|\Gamma\|}_{l_0}$. The derivation of $\overline{\|\Gamma\|}_{l_0}$ with $\mathcal{C}_{>4}$ is first presented as follows.

It follows from Figs. 10.3–10.5 that the isolation of any attack j with $\overline{\|\Gamma_j\|}_{l_0} = \min_{i\in\mathcal{V}_A}\left\{\overline{\|\Gamma_i\|}_{l_0}\right\}^{\mathbb{E}_{\max}} = \lfloor F_0/2 \rfloor$ will lead to the sequential isolation of the rest ones. Therefore, all collusive ways with $\overline{\|\Gamma\|}_{l_0} = \lfloor F_0/2 \rfloor$ are considered as the worst ones for Algorithm 10.1. It can be easily derived that the worst collusive way for Algorithm 10.1 is the same as those for the attacks with $F_0 = 2q$, as seen in Fig. 10.3. In contrast, the collusive ways of attacks with $F_0 = 2q + 1$ in Figs. 10.4–10.5 are two of the worst ones for Algorithm 10.1. Inspired by Fig. 10.3, one has that the worst attack collusive way for Algorithm 10.1 with $F_0 = 2q + 1$ is that there are $2\lfloor F_0/2 \rfloor$ attacks colluding under the way in Fig. 10.3, and no constraint is imposed on the last attack. Then, the following conclusion is provided.

Lemma 10.3 *Suppose that there are $F_0 (F_0 \leq F)$ agents being compromised by attacks, and all cycles in the graph satisfy $\mathcal{C}_k$, $k > 4$. The worst attack collusive way for Algorithm 10.1 without considering $\max_{j\in\mathcal{V}_R}\left\{\|\Gamma_j\|_{l_0}\right\}$ is that there are $2\lfloor F_0/2 \rfloor$ attacks colluding under the way in Fig. 10.3, under which, $\overline{\|\Gamma\|}_{l_0} = \lfloor F_0/2 \rfloor$.*

The main result of the worst attack collusive ways for the isolation algorithms without considering $\max_{j\in\mathcal{V}_R}\left\{\|\Gamma_j\|_{l_0}\right\}$ with $\mathcal{C}_{\geq 3}$ can be directly derived as follows.

Lemma 10.4 *Suppose that there are $F_0 (F_0 \leq F)$ agents being compromised by attacks, and all cycles in the graph satisfy $\mathcal{C}_k$, $k \geq 3$. The worst attack collusive way for Algorithm 10.1 without considering $\max_{j\in\mathcal{V}_R}\left\{\|\Gamma_j\|_{l_0}\right\}$ is that there are $2\lfloor F_0/2 \rfloor$ attacks colluding under the way in Fig. 10.3, and all cycles $\mathcal{C}_4$ are averagely set to increase the normal neighbors between the normal agent m with*

$\|\Gamma_m\|_{l_0} = \max_{j \in \mathcal{V}_R} \left\{ \|\Gamma_j\|_{l_0} \right\}$ *and every one of these* $2\lfloor F_0/2 \rfloor$ *attacks, under which,* $\overline{\|\Gamma\|}_{l_0} = \lfloor F_0/2 \rfloor + p$, $p \lfloor F_0/2 \rfloor \leq |\mathcal{C}_4| < 2p \lfloor F_0/2 \rfloor$.

Remark 10.4 *Note that Lemmas 10.3 and 10.4 hold for FDI attacks under both Algorithm 10.1 and Algorithm 9.2. Similar conclusions can be derived for covert attacks under Algorithm 9.2. The reason lies in that the worst collusive ways for both FDI attacks and covert attacks are the same, and* $\min_{i \in \mathcal{V}_A} \left\{ \overline{\|\Gamma_i\|}_{l_0} \right\}^{\mathbb{E}_{\max}} = \min_{i \in \mathcal{V}_A} \left\{ \overline{\|\Upsilon_i\|}_{l_0} \right\}^{\mathbb{E}_{\max}}$.

10.4 THE GRAPH CONDITIONS FOR ISOLATING FDI ATTACKS

In this section, the graph conditions for isolating FDI attacks under Algorithms 10.1 and 9.2 will be studied, respectively.

10.4.1 The graph conditions for isolating FDI attacks under Algorithm 10.1

Suppose that there are F_0 attacks in the F-total FDI attack model with $F_0 \leq F$. It follows from Algorithm 10.1 that attack isolation can be achieved if $\max_{i \in \mathcal{V}_A} \left\{ \|\Gamma_i\|_{l_0} \right\} > \max_{j \in \mathcal{V}_R} \left\{ \|\Gamma_j\|_{l_0} \right\}$ hold in the first F_0 rounds of isolation. This suggests that the achievement of attack isolation depends on both $\max_{i \in \mathcal{V}_A} \left\{ \|\Gamma_i\|_{l_0} \right\}$ and $\max_{j \in \mathcal{V}_R} \left\{ \|\Gamma_j\|_{l_0} \right\}$. The worst attack collusive ways for Algorithm 10.1 without considering of $\max_{j \in \mathcal{V}_R} \left\{ \|\Gamma_j\|_{l_0} \right\}$ has been discussed in Section 10.3. The following will analyze the value of $\max_{j \in \mathcal{V}_R} \left\{ \|\Gamma_j\|_{l_0} \right\}$ under different cycle sizes.

First, the effect of one attack on the $\|\Gamma_m\|_{l_0}$ of normal agent m is analyzed under the cycle $\mathcal{C}_k$ containing the attack and no other attacks. If the cycle $\mathcal{C}_k$ satisfies $k > 4$, it can be seen from Fig. 10.9(a) that the attack can provide two non-zero elements for $\|\Gamma_m\|_{l_0}$ when it is the one-hop neighbor of normal agent m. In contrast, the attack can only provide one non-zero element for $\|\Gamma_m\|_{l_0}$ when it is the two-hop neighbor of m, as seen in Fig. 10.9(b). If the cycle $\mathcal{C}_k$ satisfies $k = 4$, it is shown in Fig. 10.9(c) that the attack is able to provide two non-zero elements for $\|\Gamma_m\|_{l_0}$. If the cycle $\mathcal{C}_k$ satisfies $k = 3$, Fig. 10.9(d) shows that the attack is able to provide three non-zero elements for $\|\Gamma_m\|_{l_0}$. Then, the value of $\max_{j \in \mathcal{V}_R} \left\{ \|\Gamma_j\|_{l_0} \right\}$ under different cycle sizes is provided as follows.

Denote by $\left| \mathcal{C}'_{\leq 4} \right|$ the number of cycles $\mathcal{C}_{\leq 4}$ satisfying that 1) there is only a single attack in each cycle $\mathcal{C}_{\leq 4}$; 2) there is at least one non-common edge between each cycle $\mathcal{C}_{\leq 4}$ and all the other ones; 3) the cycles $\mathcal{C}_4$ are not counted if the cycles containing the same normal agent and the attack satisfy $|\mathcal{C}_3| \neq 0$. Then, one has that 1) $\max_{j \in \mathcal{V}_R} \left\{ \|\Gamma_j\|_{l_0} \right\} = F_0 + 1$ under cycles $\mathcal{C}_{>4}$ if there is a normal agent m being the one-hop neighbor of at least one attack, and being the two-hop neighbor of the others; 2) $\max_{j \in \mathcal{V}_R} \left\{ \|\Gamma_j\|_{l_0} \right\} = F_0 + 1 + \left| \mathcal{C}'_{\leq 4} \right|$ under cycles $\mathcal{C}_{\geq 3}$ if there is a normal agent m being the one-hop neighbor of at least one attack, and being the two-hop

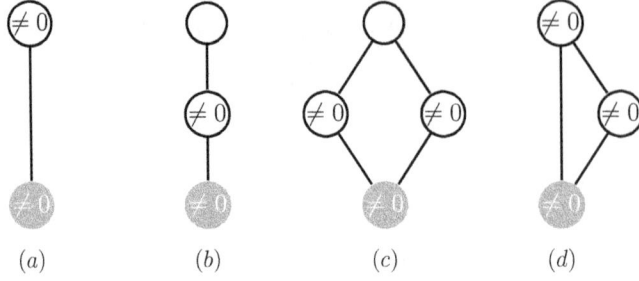

Figure 10.9 The effect of a single FDI attack on the $\|\Gamma_m\|_{l_0}$ of normal agent m under different cycles $\mathcal{C}_k$ containing the attack and m , where (a)-(b) $k > 4$, (c) $k = 4$, and (d) $k = 3$.

neighbor of the others, and all cycles $\mathcal{C}_{\leq 4}$ are set to increase the number of common normal neighbors between agent m and any attacks.

The following result is developed to provide some equivalent graph conditions for isolating FDI attacks under Algorithm 10.1.

Theorem 10.2 *Consider a heterogeneous, physically coupled NAS under attacks described by (10.1), with a communication topology represented by the graph $\mathcal{G}$. Suppose that Assumptions 10.1–10.2 hold. In the F-total FDI attack model, attack isolation can be achieved under Algorithm 10.1 if one of the following conditions holds:*

i) $\mathcal{G}$ is $(F,7)$-isolable;

ii) $\mathcal{G}$ is $\left(\left\lceil \frac{3F}{2} \right\rceil - 1, 6\right)$-isolable;

iii) $\mathcal{G}$ is $\left(\left\lfloor \frac{3F}{2} \right\rfloor, 5\right)$-isolable;

iv) $\mathcal{G}$ is $\left(\left\lfloor \frac{3F}{2} \right\rfloor + \left|\mathcal{C}'_{\leq 4}\right|, 3\right)$-isolable.

Proof 10.5 *Assume that the normal agent m satisfies $\|\Gamma_m\|_{l_0} = \max_{j \in \mathcal{V}_R}\left\{\|\Gamma_j\|_{l_0}\right\} = \alpha$, $\alpha \in \mathbb{N}$. Obviously, $\alpha \leq F_0 + 1$ under cycles $\mathcal{C}_{>4}$, and $\alpha \leq F_0 + 1 + \left|\mathcal{C}'_{\leq 4}\right|$ under cycles $\mathcal{C}_{\geq 3}$.*

Define $\{\Xi_1^\alpha, \Xi_2^\alpha, \ldots\}$ as the attack collusive ways with respect to a given α. Denote by $\|\Gamma\|_{l_0}^{\Xi_z^\alpha}$ the key value of the set $\left\{\|\Gamma_i\|_{l_0}\right\}$, $i \in \mathcal{V}_A$, under the zth attack collusive way. The compromised agent s with $\|\Gamma_s\|_{l_0} = \|\Gamma\|_{l_0}^{\Xi_z^\alpha}$ satisfies that 1) the compromised agent s still cannot be isolated after the isolation of all compromised agents r with $\|\Gamma_r\|_{l_0} > \|\Gamma_s\|_{l_0}$, 2) the rest compromised agents r' with $\|\Gamma_{r'}\|_{l_0} \leq \|\Gamma_s\|_{l_0}$ will be sequentially isolated after the isolation of agent s. Note that the collusion among attacks will lead to an increase of $\overline{\|\Gamma_i\|}_{l_0}$, and consequently, a decrease of $\|\Gamma_i\|_{l_0}$, $i \in \mathcal{V}_A$. Denote by $\|\Gamma\|_{l_0}{}^\alpha$ and $\overline{\|\Gamma\|}_{l_0}{}^\alpha$, the key values of sets $\left\{\|\Gamma_i\|_{l_0}\right\}$ and $\left\{\overline{\|\Gamma_i\|}_{l_0}\right\}$, under the worst attack collusive way with respect to the given α. Then, one has $\|\Gamma\|_{l_0}{}^\alpha = \min\left\{\|\Gamma\|_{l_0}^{\Xi_z^\alpha}\right\}$. This means that attack isolation can be surely achieved *under any cases if it can be achieved under the worst attack collusive way with respect to any given α. Therefore, it only needs to prove that $\|\Gamma\|_{l_0}{}^{\alpha} > \alpha$ holds for any given α.*

Assume that there are m_1 compromised agents being the one-hop neighbors of agent m. It follows from Figs. 10.9(a)–(b) that $|\mathcal{N}_m^1 \cap \mathcal{V}_A| = m_1$, $|\mathcal{N}_m^2 \cap \mathcal{V}_A| = \alpha - m_1 - \mathbf{1}_{m_1 \geq 1}$, $|\mathcal{N}_m^{>2} \cap \mathcal{V}_A| = F_0 - \alpha + \mathbf{1}_{m_1 \geq 1}$. Note that $m_1 \geq 1$ for $\alpha = F_0 + 1$ and $m_1 \geq 0$ for $\alpha \leq F_0$. For $\mathcal{C}_{\geq 7}$, the proof can be divided into the following steps.

Step 1: Analyze whether the attacks in sets $\mathcal{N}_m^1 \cap \mathcal{V}_A$, $\mathcal{N}_m^2 \cap \mathcal{V}_A$, and $\mathcal{N}_m^{>2} \cap \mathcal{V}_A$ can collude or not under $\mathcal{C}_{\geq 7}$.

  (a) *Any two attacks in $\mathcal{N}_m^1 \cup \mathcal{N}_m^2 \cap \mathcal{V}_A$ cannot collude with each other.*

  (b) *The attacks in $\mathcal{N}_m^{>2} \cap \mathcal{V}_A$ can collude with only one attack in $\mathcal{N}_m^1 \cap \mathcal{V}_A$, and all the ones in $\mathcal{N}_m^2 \cap \mathcal{V}_A$.*

Step 2: Derive the worst attack collusive ways with respect to any given α by Lemma 10.3.

  (a) *For the case $\mathbf{1}_{m_1 \geq 1} + |\mathcal{N}_m^2 \cap \mathcal{V}_A| \leq |\mathcal{N}_m^{>2} \cap \mathcal{V}_A|$, i.e., $\alpha \leq \lfloor (F_0 + m_1 + \mathbf{1}_{m_1 \geq 1})/2 \rfloor$, there are $2\lfloor (F_0 - m_1 + \mathbf{1}_{m_1 \geq 1})/2 \rfloor$ attacks colluding under the way in Fig. 10.3. Here, the signals of one of the attacks in $\mathcal{N}_m^1 \cap \mathcal{V}_A$ and all attacks in $\mathcal{N}_m^2 \cap \mathcal{V}_A$ are the same.*

  (b) *For the case $\mathbf{1}_{m_1 \geq 1} + |\mathcal{N}_m^2 \cap \mathcal{V}_A| > |\mathcal{N}_m^{>2} \cap \mathcal{V}_A|$, i.e., $\alpha > \lfloor (F_0 + m_1 + \mathbf{1}_{m_1 \geq 1})/2 \rfloor$, there are $2|\mathcal{N}_m^{>2} \cap \mathcal{V}_A|$ attacks colluding under the way in Fig. 9.3(b), under which, the signals of $|\mathcal{N}_m^{>2} \cap \mathcal{V}_A|$ of the attacks in $\mathcal{N}_m^1 \cup \mathcal{N}_m^2 \cap \mathcal{V}_A$ are the same. Note that there are at most one attack belonging to $\mathcal{N}_m^1 \cap \mathcal{V}_A$ for these $|\mathcal{N}_m^{>2} \cap \mathcal{V}_A|$ attacks in $\mathcal{N}_m^1 \cup \mathcal{N}_m^2 \cap \mathcal{V}_A$.*

Step 3: Solve $\|\Gamma\|_{l_0}{}^{\alpha}$ according to Step 2.

  (a) *For $\alpha \leq \lfloor (F_0 + m_1 + \mathbf{1}_{m_1 \geq 1})/2 \rfloor$, one has $\overline{\|\Gamma\|_{l_0}}^{\alpha} = \lfloor (F_0 - m_1 + \mathbf{1}_{m_1 \geq 1})/2 \rfloor$.*

  (b) *For $\alpha > \lfloor (F_0 + m_1 + \mathbf{1}_{m_1 \geq 1})/2 \rfloor$, one has $\overline{\|\Gamma\|_{l_0}}^{\alpha} = F_0 - \alpha + \mathbf{1}_{m_1 \geq 1}$.*

Step 4: Judge whether $\|\Gamma\|_{l_0}{}^{\alpha} > \alpha$ holds for any given α based on condition i).
Since $|\mathcal{J}| \geq F + 2$, $F \geq F_0$, and $\|\Gamma\|_{l_0}^{\alpha} = |\mathcal{J}| - \overline{\|\Gamma\|_{l_0}}^{\alpha}$, it can be concluded that $\|\Gamma\|_{l_0}{}^{\alpha} > \alpha$ for any given α.

The proving procedures for $\mathcal{C}_{\geq 6}$, $\mathcal{C}_{\geq 5}$, and $\mathcal{C}_{\geq 3}$ are similar to those for $\mathcal{C}_{\geq 7}$. However, the attack collusion in sets $\mathcal{N}_m^1 \cap \mathcal{V}_A$, $\mathcal{N}_m^2 \cap \mathcal{V}_A$, and $\mathcal{N}_m^{>2} \cap \mathcal{V}_A$ can vary across different cycle sizes, resulting significant different worst attack collusive ways with respect to the given α. In the following, we only focus on the difference in each detailed step between $\mathcal{C}_{\geq 7}$ and other cycles.

The case of $\mathcal{C}_{\geq 6}$ is as follows.

Step 1: (a) *The attacks $i \in \mathcal{N}_m^1 \cap \mathcal{V}_A$ can collude neither with each other nor with the ones in $\mathcal{N}_m^2 \cap \mathcal{V}_A$.*

(b) *Any two attacks in $\mathcal{N}_m^2 \cap \mathcal{V}_A$ can collude with each other.*

(c) *The attacks in $\mathcal{N}_m^{>2} \cap \mathcal{V}_A$ can collude with all the ones in $\mathcal{N}_m^1 \cup \mathcal{N}_m^2 \cap \mathcal{V}_A$.*

Step 2: (a) *For the case $|\mathcal{N}_m^1 \cap \mathcal{V}_A| + |\mathcal{N}_m^2 \cap \mathcal{V}_A| \leq |\mathcal{N}_m^{>2} \cap \mathcal{V}_A|$, i.e., $\alpha \leq \lfloor F_0/2 \rfloor + \mathbf{1}_{m_1 \geq 1}$, there are $2\lfloor F_0/2 \rfloor$ attacks colluding under the way in Fig. 10.3, with the signals of the attacks in $\mathcal{N}_m^1 \cup \mathcal{N}_m^2 \cap \mathcal{V}_A$ being the same.*

(b) *For the case $|\mathcal{N}_m^1 \cap \mathcal{V}_A| + |\mathcal{N}_m^2 \cap \mathcal{V}_A| > |\mathcal{N}_m^{>2} \cap \mathcal{V}_A|$ and $|\mathcal{N}_m^1 \cap \mathcal{V}_A| \leq \lfloor |\mathcal{N}_m^2 \cup \mathcal{N}_m^{>2} \cap \mathcal{V}_A|/2 \rfloor$, i.e., $\alpha > \lfloor F_0/2 \rfloor + \mathbf{1}_{m_1 \geq 1}$ and $m_1 \leq \lfloor (F_0 - m_1)/2 \rfloor$, there are $2\lfloor (F_0 - m_1)/2 \rfloor$ attacks in $\mathcal{N}_m^2 \cup \mathcal{N}_m^{>2} \cap \mathcal{V}_A$ colluding under the way in Fig. 10.3.*

(c) *For the case $|\mathcal{N}_m^1 \cap \mathcal{V}_A| + |\mathcal{N}_m^2 \cap \mathcal{V}_A| > |\mathcal{N}_m^{>2} \cap \mathcal{V}_A|$ and $|\mathcal{N}_m^1 \cap \mathcal{V}_A| > \lfloor |\mathcal{N}_m^2 \cup \mathcal{N}_m^{>2} \cap \mathcal{V}_A|/2 \rfloor$, i.e., $\alpha > \lfloor F_0/2 \rfloor + \mathbf{1}_{m_1 \geq 1}$ and $m_1 > \lfloor (F_0 - m_1)/2 \rfloor$, there are $2|\mathcal{N}_m^{>2}|$ attacks colluding under the way in Fig. 9.3(b), with at least $|\mathcal{N}_m^{>2}|$ of the attacks in $\mathcal{N}_m^1 \cup \mathcal{N}_m^2 \cap \mathcal{V}_A$ being the same.*

Step 3: (a) *For $\alpha \leq \lfloor F_0/2 \rfloor + \mathbf{1}_{m_1 \geq 1}$, one has $\overline{\|\Gamma\|}_{l_0}{}^\alpha = \lfloor F_0/2 \rfloor$.*

(b) *For $\alpha > \lfloor F_0/2 \rfloor + \mathbf{1}_{m_1 \geq 1}$ and $m_1 \leq \lfloor (F_0 - m_1)/2 \rfloor$, one has $\overline{\|\Gamma\|}_{l_0}{}^\alpha = \lfloor (F_0 - m_1)/2 \rfloor$.*

(c) *For $\alpha > \lfloor F_0/2 \rfloor + \mathbf{1}_{m_1 \geq 1}$ and $m_1 > \lfloor (F_0 - m_1)/2 \rfloor$, one has $\overline{\|\Gamma\|}_{l_0}{}^\alpha = F_0 - \alpha + \mathbf{1}_{m_1 \geq 1}$.*

The case of $\mathcal{C}_{\geq 5}$ is as follows.

Step 1: Any two attacks except for the ones in $\mathcal{N}_m^1 \cap \mathcal{V}_A$ can collude with each other.

Step 2: (a) *For the case $|\mathcal{N}_m^1 \cap \mathcal{V}_A| \leq |\mathcal{N}_m^2 \cap \mathcal{V}_A| + |\mathcal{N}_m^{>2}|$, i.e., $m_1 \leq \lfloor F_0/2 \rfloor$, there are $2\lfloor F_0/2 \rfloor$ attacks colluding under the way in Fig. 10.3, with the signals of the attacks in $\mathcal{N}_m^1 \cap \mathcal{V}_A$ being the same.*

(b) *For the case $|\mathcal{N}_m^1 \cap \mathcal{V}_A| > |\mathcal{N}_m^2 \cap \mathcal{V}_A| + |\mathcal{N}_m^{>2}|$, i.e., $m_1 > \lfloor F_0/2 \rfloor$, there are $2(F_0 - m_1)$ attacks colluding under the way in Fig. 9.3(b), with the signals of at least $F_0 - m_1$ of the attacks in $\mathcal{N}_m^1 \cap \mathcal{V}_A$ being the same.*

Step 3: (a) *For $m_1 \leq \lfloor F_0/2 \rfloor$, one has $\overline{\|\Gamma\|}_{l_0}{}^\alpha = \lfloor F_0/2 \rfloor$.*

(b) *For $m_1 > \lfloor F_0/2 \rfloor$, one has $\overline{\|\Gamma\|}_{l_0}{}^\alpha = F_0 - m_1$.*

The case of $\mathcal{C}_{\geq 3}$ is as follows.

Step 1: Any two attacks can collude with each other.

Step 2: It follows from Lemmas 10.2 and 10.4, and Figs. 10.9(c)–(d) that it needs at least $\lfloor F_0/2 \rfloor$ cycles $\mathcal{C}_4$ to increase $\overline{\|\Gamma\|}_{l_0}{}^\alpha$, while one cycle $\mathcal{C}_4$ is enough to increase $\max_{m \in \mathcal{V}_R} \{ \|\Gamma_m\|_{l_0} \}$. This means that the situation is much worse for Algorithm 10.1 if all cycles $\mathcal{C}_{\leq 4}$ are set to increase the $\|\Gamma_m\|_{l_0}$ of normal agent m, rather than that of the attacks.

Step 3: For any α, one has $\overline{\|\Gamma\|}_{l_0}{}^{\alpha} = \lfloor F_0/2 \rfloor$.

Remark 10.5 *Note that Theorem 1 provides generalized sufficient conditions for any size of cycles, improving the application of the isolation algorithm. Besides, it can be seen from those conditions in i)-iv) that the number of agents' neighbors increases as the cycle size decreases. This fact demonstrates that the impact of smaller cycle sizes on attack isolation can be compensated by increasing the number of neighboring agents.*

Remark 10.6 *It can be directly derived from conditions i) and ii) that for the 2-total FDI attack model, attack isolation can be achieved under Algorithm 10.1 if the graph is $(2,6)$-isolable. Compared to Algorithm 9.1, which can isolate 2-total FDI attacks under $(2,5)$-isolable graphs, Algorithm 10.1 does not incorporate the value of $\|\gamma_i\|$ into attack isolation. The price it takes is a more rigorous graph condition.*

Remark 10.7 *Define $\min\{\|\Gamma\|_{l_0}{}^{\alpha} - \alpha\}$ as the worst case for Algorithm 10.1. $\overline{\|\Gamma\|}_{l_0}{}^{\alpha^*}$, $\|\Gamma\|_{l_0}{}^{\alpha^*}$, and α^* are, respectively, called the worst values of $\overline{\|\Gamma\|}_{l_0}{}^{\alpha}$, $\|\Gamma\|_{l_0}{}^{\alpha}$ and α for Algorithm 10.1. The way to achieve $\overline{\|\Gamma\|}_{l_0}{}^{\alpha^*}$ is called the worst attack collusive one for Algorithm 10.1. It follows from Step 2 in the proof that the worst attack collusive ways for Algorithm 10.1 with $\mathcal{C}_{\geq 7}$ and $\mathcal{C}_{\geq 6}$ depend on the relationship between the normal agent m with $\|\Gamma_m\|_{l_0} = \alpha^*$ and the compromised ones. However, the worst attack collusive ways for Algorithm 10.1 with $\mathcal{C}_{\geq 5}$ and $\mathcal{C}_{\geq 3}$ are irrelevant to the relationship between agent m and the compromised ones.*

Remark 10.8 *It follows from conditions ii) and iii) that for the F-total FDI attack model with $F = 2q$, attack isolation can be achieved under Algorithm 10.1 if the graph is $(3F/2 - 1, 6)$-isolable or $(3F/2, 5)$-isolable. However, attack isolation can be achieved whether the graph is $(\lfloor 3F/2 \rfloor, 6)$-isolable or $(\lfloor 3F/2 \rfloor, 5)$-isolable for $F = 2q + 1$. The reason lies in the different worst cases for $F = 2q$ with $\mathcal{C}_{\geq 6}$ and $\mathcal{C}_{\geq 5}$. It can be obtained from the above analysis that the worst case for $F = 2q$ with $\mathcal{C}_{\geq 6}$ can be $\overline{\|\Gamma\|}_{l_0}{}^{\alpha^*} = F/2$ and $\alpha^* = F$, or $\overline{\|\Gamma\|}_{l_0}{}^{\alpha^*} = (F-2)/2$ and $\alpha^* = F+1$, while the one for $F = 2q$ with $\mathcal{C}_{\geq 5}$ is $\overline{\|\Gamma\|}_{l_0}{}^{\alpha^*} = F/2$ and $\alpha^* = F+1$. In contrast, the worst cases for $F = 2q+1$ with both $\mathcal{C}_{\geq 6}$ and $\mathcal{C}_{\geq 5}$ are $\overline{\|\Gamma\|}_{l_0}{}^{\alpha^*} = \lfloor F/2 \rfloor$ and $\alpha^* = F+1$.*

It follows from Proposition 9.2 that the condition i) in Theorem 10.2 is sufficient yet sharp, the sharpness of the rest conditions in Theorem 10.2 is provided as follows.

Proposition 10.1 *For a heterogeneous, physically coupled NAS (10.1) under F-total FDI attack model with $F = 2q$, there exists a $(3F/2 - 1, 5)$-isolable graph containing only one cycle $\mathcal{C}_5$ such that attack isolation fails under Algorithm 10.1.*

Proof 10.6 *Consider the case that $|\mathcal{N}_i| = 3F/2$, $i \in \mathcal{V}$, all attacks collude under the way in Fig. 10.3, and there exists a normal agent m being the one-hop neighbor of a compromised agent, and being the two-hop neighbor of the others. Then, one has $\|\Gamma\|_{l_0}{}^{\alpha^*} = |\mathcal{J}| - F/2 = F+1$ and $\alpha^* = F+1$, which suggests the failure of attack isolation.*

Similar to the above analysis, the following conclusion can be obtained.

Proposition 10.2 *For a heterogeneous, physically coupled NAS* (10.1) *under F-total FDI attack model, there exists a $(\lfloor 3F/2 \rfloor, 4)$-isolable graph containing only one cycle $\mathcal{C}_4$ such that attack isolation fails under Algorithm 10.1.*

10.4.2 The graph conditions for isolating FDI attacks under Algorithm 9.2

Theorem 10.3 *Consider a heterogeneous, physically coupled NAS under attacks described by* (10.1), *with a communication topology represented by the graph $\mathcal{G}$. Suppose that Assumptions 10.1–10.2 hold. In the F-total FDI attack model, attack isolation can be achieved under Algorithm 9.2 if one of the following conditions holds:*

   i) $\mathcal{G}$ is $(F, 7)$-isolable;

   ii) $\mathcal{G}$ is $\left(\left\lfloor \frac{3F}{2} \right\rfloor, 5 \right)$-isolable;

   iii) $\mathcal{G}$ is $\left(\left\lfloor \frac{3F}{2} \right\rfloor + \left| \mathcal{C}'_{\leq 4} \right|, 3 \right)$-isolable.

Proof 10.7 *Assume that the normal agent m satisfies $\|\Upsilon_m\|_{l_0} = \max_{j \in \mathcal{V}_R} \left\{ \|\Upsilon_j\|_{l_0} \right\} = \beta$, $\beta \in \mathbb{N}$. It can be easily derived from Fig. 10.9 that $\beta \leq F_0$ under cycles $\mathcal{C}_{>4}$, and $\beta \leq F_0 + \left| \mathcal{C}'_{\leq 4} \right|$ under cycles $\mathcal{C}_{\geq 3}$. Denote by $\|\Upsilon\|_{l_0}^{\beta}$, the key value of set $\left\{ \|\Upsilon_i\|_{l_0} \right\}$, $i \in \mathcal{V}_A$, under the worst attack collusive way with respect to the given β. Similar to the analysis in Theorem 10.2, it only needs to prove that $\|\Upsilon\|_{l_0}^{\beta} > \beta$ holds for any given β.*

   Compared with Algorithm 10.1, which focuses on the non-zero $\|\gamma_j\|$ of both the agent i itself and its neighbors, i.e., $\|\gamma_j\|$, $j \in \{i\} \cup \mathcal{N}_i$, Algorithm 9.2 only involves the non-zero $\|\gamma_j\|$ of the neighbors of agent i, i.e., $\|\gamma_j\|$, $j \in \mathcal{N}_i$. Assume that there are m_1 compromised agents being the one-hop neighbors of agent m, then one has $|\mathcal{N}_m^1 \cap \mathcal{V}_A| = m_1$, $|\mathcal{N}_m^2 \cap \mathcal{V}_A| = \beta - m_1$, $|\mathcal{N}_m^{>2} \cap \mathcal{V}_A| = F_0 - \beta$.

   Note that Step 1 in the proof of Theorem 10.3 is the same as that of Theorem 10.2, and the main difference between Theorems 10.2 and 10.3 lies in the $\|\Upsilon\|_{l_0}^{\beta}$ for different β. In the following, we only point out the difference in Step 3 under cycles $\mathcal{C}_k$.

   The case of $\mathcal{C}_{\geq 7}$ is as follows.

Step 3: *(a) For $\beta \leq \lfloor (F_0 + m_1 - \mathbf{1}_{m_1 \geq 1})/2 \rfloor$, one has $\overline{\|\Upsilon\|}_{l_0}^{\beta} = \lfloor (F_0 - m_1 + \mathbf{1}_{m_1 \geq 1})/2 \rfloor$.*

     (b) For $\beta > \lfloor (F_0 + m_1 - \mathbf{1}_{m_1 \geq 1})/2 \rfloor$, one has $\overline{\|\Upsilon\|}_{l_0}^{\beta} = F - \beta$.

   The case of $\mathcal{C}_{\geq 6}$ is as follows.

Step 3: *(a) For $\beta \leq \lfloor F_0/2 \rfloor$, one has $\overline{\|\Upsilon\|}_{l_0}^{\beta} = \lfloor F_0/2 \rfloor$.*

     (b) For $\beta > \lfloor F_0/2 \rfloor$ and $m_1 \leq \lfloor (F_0 - m_1)/2 \rfloor$, one has $\overline{\|\Upsilon\|}_{l_0}^{\beta} = \lfloor (F_0 - m_1)/2 \rfloor$.

(c) For $\beta > \lfloor F_0/2 \rfloor$ and $m_1 > \lfloor (F_0 - m_1)/2 \rfloor$, one has $\overline{\|\Upsilon\|}_{l_0}^{\beta} = F_0 - \beta$.

The case of $\mathcal{C}_{\geq 5}$ is as follows.

Step 3: (a) For $m_1 \leq \lfloor F_0/2 \rfloor$, one has $\overline{\|\Upsilon\|}_{l_0}^{\beta} = \lfloor F_0/2 \rfloor$.

(b) For $m_1 > \lfloor F_0/2 \rfloor$, one has $\overline{\|\Upsilon\|}_{l_0}^{\beta} = F_0 - m_1$.

The case of $\mathcal{C}_{\geq 3}$ is as follows.

Step 3: For any β, one has $\overline{\|\Upsilon\|}_{l_0}^{\beta} = \lfloor F_0/2 \rfloor$.

Remark 10.9 Define $\min \left\{ \overline{\|\Upsilon\|}_{l_0}^{\beta} - \beta \right\}$ as the worst case for Algorithm 9.2. $\overline{\|\Upsilon\|}_{l_0}^{\beta^*}$, $\|\Upsilon\|_{l_0}^{\beta^*}$, and β^* are, respectively, called the worst values of $\overline{\|\Upsilon\|}_{l_0}^{\beta}$, $\|\Upsilon\|_{l_0}^{\beta}$ and β for Algorithm 9.2. The way to achieve $\overline{\|\Upsilon\|}_{l_0}^{\beta^*}$ is called the worst attack collusive way for Algorithm 9.2. It follows from the proof that the worst attack collusive ways for Algorithm 9.2 with $\mathcal{C}_{\geq 7}$ depend on the relationship between the normal agent m with $\|\Upsilon_m\|_{l_0} = \beta^*$ and the compromised ones. However, the worst attack collusive ways with $\mathcal{C}_{\geq 6}$, $\mathcal{C}_{\geq 5}$, and $\mathcal{C}_{\geq 3}$ are irrelevant to the relationship between agent m and the compromised ones.

Remark 10.10 It can be derived from Theorems 10.2 and 10.3 that under the F-total FDI attack model with $F = 2q$ and $\mathcal{C}_{\geq 6}$, the graph condition for Algorithm 9.2 is more rigorous than that for Algorithm 10.1. The main reason is that when $F = 2q$ and $\mathcal{C}_{\geq 6}$, the worst attack collusive way for Algorithm 10.1 is the same as that for Algorithm 9.2, i.e., $\overline{\|\Gamma\|}_{l_0}^{\alpha^*} = \overline{\|\Upsilon\|}_{l_0}^{\beta^*}$, $\alpha^* = \beta^*$. Thus, one has that $\|\Gamma\|_{l_0}^{\alpha^*} = |\mathcal{N}| + 1 - \overline{\|\Gamma\|}_{l_0}^{\alpha^*} > \|\Upsilon\|_{l_0}^{\beta^*}$. Besides, for cycles $\mathcal{C}_{\geq 7}$, $\mathcal{C}_{\geq 5}$, and $\mathcal{C}_{\geq 3}$, the graph conditions for Algorithm 9.2 are the same as those for Algorithm 10.1. Then, it can be concluded that Algorithm 10.1 is more effective for isolating FDI attacks.

Remark 10.11 Similar to Propositions 10.1 and 10.2, the conditions in Theorem 10.3 are sufficient yet sharp since there exist an $(F, 6)$-isolable graph containing only one cycle $\mathcal{C}_6$ and a $(\lfloor 3F/2 \rfloor, 4)$-isolable graph containing only one cycle $\mathcal{C}_4$ such that attack isolation fails under Algorithm 9.2.

10.5 THE GRAPH CONDITIONS FOR ISOLATING COVERT ATTACKS

The graph conditions for isolating covert attacks under Algorithm 9.2 will be provided in this section.

The relationship between $\max_{j \in \mathcal{V}_R} \left\{ \|\Upsilon_j\|_{l_0} \right\}$ and the size of cycles is first provided. The effect of one covert attack on the $\|\Upsilon_m\|_{l_0}$ of normal agent m under different cycle sizes is shown in Fig. 10.10. Compared with the FDI attack in Fig. 10.9, one has that the covert attack cannot increase $\|\Upsilon_m\|_{l_0}$ if it is the one-hop neighbor of agent m under the cycle $\mathcal{C}_{>4}$. Besides, the covert attack can only provide one non-zero element for $\|\Upsilon_m\|_{l_0}$ under the cycle $\mathcal{C}_3$. Denote by $|\mathcal{C}_4''|$ the number of cycles $\mathcal{C}_4$ satisfying that 1)

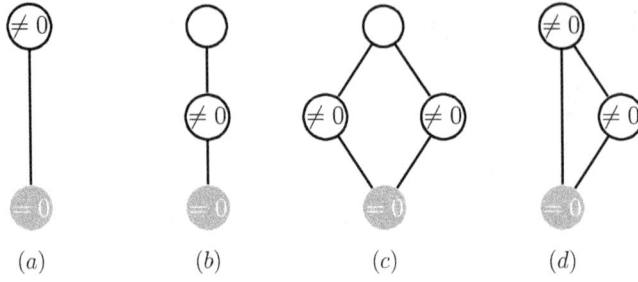

(a)   (b)   (c)   (d)

Figure 10.10   The effect of a single covert attack on the $\|\Upsilon_m\|_{l_0}$ of normal agent m under different cycles $\mathcal{C}_k$ containing the attack and m , where (a–b) $k > 4$, (c) $k = 4$, and (d) $k = 3$.

there is only a single attack in each cycle $\mathcal{C}_4$; 2) there is at least one non-common edge between each $\mathcal{C}_4$ and other cycles. Suppose that there are F_0 covert attacks in the F-total covert attack model. Then, one has that 1) $\max_{j \in V_R} \left\{ \|\Upsilon_j\|_{l_0} \right\} = F_0$ under cycles $\mathcal{C}_{>4}$ if there is a normal agent m being the two-hop neighbor of each attack; 2) $\max_{j \in V_R} \left\{ \|\Upsilon_j\|_{l_0} \right\} = F_0 + |\mathcal{C}_4''|$ under cycles $\mathcal{C}_{\geq 3}$ if there is a normal agent m being the two-hop neighbor of each attack, and all cycles $\mathcal{C}_4$ are set to increase the number of common normal neighbors between agent m and any attacks.

Theorem 10.4 *Consider a heterogeneous, physically coupled NAS under attacks described by (10.1), with a communication topology represented by the graph $\mathcal{G}$. Suppose that Assumptions 10.1–10.2 hold. In the F-total covert attack model, attack isolation can be achieved under Algorithm 9.2 if one of the following conditions holds:*

i) $\mathcal{G}$ is $(F, 7)$-isolable;

ii) $\mathcal{G}$ is $\left(\left\lfloor \frac{3F}{2} \right\rfloor, 5 \right)$-isolable;

iii) $\mathcal{G}$ is $\left(\left\lfloor \frac{3F}{2} \right\rfloor + |\mathcal{C}_4''|, 3 \right)$-isolable.

Proof 10.8 *Assume that the normal agent m satisfies $\|\Upsilon_m\|_{l_0} = \max_{j \in V_R} \left\{ \|\Upsilon_j\|_{l_0} \right\} = \beta$, $\beta \in \mathbb{N}$. Then, one has $\beta \leq F_0$ under cycles $\mathcal{C}_{>4}$, and $\beta \leq F_0 + |\mathcal{C}_4''|$ under cycles $\mathcal{C}_{\geq 3}$. Obviously, attack isolation can be achieved if $\|\Upsilon\|_{l_0}^{\beta} > \beta$ holds for any given β.*

Different from the FDI attack with $\|\gamma_i\| \neq 0$, the agent i compromised by a covert attack satisfies $\|\gamma_i\| = 0$. Assume that there are m_1 attacks being the one-hop neighbors of agent m, then one has $|\mathcal{N}_m^1 \cap V_A| = m_1$, $|\mathcal{N}_m^2 \cap V_A| = \beta$, $|\mathcal{N}_m^{\geq 2} \cap V_A| = F_0 - m_1 - \beta$.

The main difference between Theorems 10.3 and 10.4 lies in the $\|\Upsilon\|_{l_0}^{\beta}$ for different β. In the following, we only point out the difference in Step 3 under cycles $\mathcal{C}_k$.

The case of $\mathcal{C}_{\geq 7}$ is as follows.

Step 3:   (a) for $\beta \leq \lfloor (F_0 - m_1 - \mathbf{1}_{m_1 \geq 1})/2 \rfloor$, one has $\overline{\|\Upsilon\|}_{l_0}^{\beta} = \lfloor (F_0 - m_1 + \mathbf{1}_{m_1 \geq 1})/2 \rfloor$.

(b) for $\beta > \lfloor (F_0 - m_1 - \mathbf{1}_{m_1 \geq 1})/2 \rfloor$, one has $\overline{\overline{\|\Upsilon\|}}_{l_0}^{\beta} = F_0 - \beta$.

The case of $\mathcal{C}_{\geq 6}$ is as follows.

Step 3: (a) for $\beta \leq \lfloor (F_0 - 2m_1)/2 \rfloor$, one has $\overline{\overline{\|\Upsilon\|}}_{l_0}^{\beta} = \lfloor F_0/2 \rfloor$.

(b) for $\beta > \lfloor (F_0 - 2m_1)/2 \rfloor$ and $m_1 \leq \lfloor (F_0 - m_1)/2 \rfloor$, one has $\overline{\overline{\|\Upsilon\|}}_{l_0}^{\beta} = \lfloor (F_0 - m_1)/2 \rfloor$.

(c) for $\beta > \lfloor (F_0 - 2m_1)/2 \rfloor$ and $m_1 > \lfloor (F_0 - m_1)/2 \rfloor$, one has $\overline{\overline{\|\Upsilon\|}}_{l_0}^{\beta} = F_0 - \beta$.

The cases of $\mathcal{C}_{\geq 5}$ and $\mathcal{C}_{\geq 3}$ are the same as those in Theorem 10.3, and hence are omitted.

Remark 10.12 *Compared with condition iii) in Theorem 10.3, the number of cycles $\mathcal{C}_3$ does not affect the isolation of covert attacks under Algorithm 9.2. The reason lies in that the FDI attack itself can provide two non-zero elements, while the covert attack can only provide one non-zero element for the $\|\Upsilon_m\|_{l_0}$ under each cycle $\mathcal{C}_3$ containing only the attack.*

Remark 10.13 *Note that the conditions in Theorem 10.4 are sufficient yet sharp since there exists an $(F, 6)$-isolable graph containing only one cycle $\mathcal{C}_6$ and a $(\lfloor 3F/2 \rfloor, 4)$-isolable graph containing only one cycle $\mathcal{C}_4$ such that Algorithm 9.2 fails to isolate covert attacks.*

Remark 10.14 *From a brief analysis, one can derive that $\alpha^* = \beta^*$ and $\overline{\overline{\|\Gamma\|}}_{l_0}^{\alpha^*} = \overline{\overline{\|\Upsilon\|}}_{l_0}^{\beta^*}$ hold for $F = 2q$ and $\mathcal{C}_{\geq 6}$, otherwise, $\alpha^* = \beta^* + 1$ and $\overline{\overline{\|\Gamma\|}}_{l_0}^{\alpha^*} = \overline{\overline{\|\Upsilon\|}}_{l_0}^{\beta^*}$ hold. This indicates that the graph condition for isolating covert attacks with $F = 2q$ and $\mathcal{C}_{\geq 6}$ under Algorithm 10.1 is more rigorous than that under Algorithm 9.2. Consequently, it can be concluded that Algorithm 9.2 is more effective for isolating covert attacks.*

For hybrid attacks consisting of FDI attacks and covert attacks, it follows from (10.3) and (10.5) that multiple attacks in $\bigcup_{j \in \mathcal{J}_i} \left(u_j^a(t), y_j^a(t) \right)$ can collude only if they belong to the same type of attacks. Therefore, the worst collusive cases for hybrid attacks depend on the type of attacks with larger numbers. Furthermore, it follows from Figs. 10.9–10.10 that the ways to derive $\max_{j \in \mathcal{V}_R} \left\{ \|\Gamma_j\|_{l_0} \right\}$ and $\max_{j \in \mathcal{V}_R} \left\{ \|\Upsilon_j\|_{l_0} \right\}$ under hybrid attacks are the same as those under FDI attacks. Thus, it can be concluded from Theorems 10.2–10.4 and Remarks 10.10 and 10.14 that Algorithm 10.1 is more effective for isolating hybrid attacks if the number of FDI attacks is no smaller than the number of covert attacks; otherwise, Algorithm 9.2 is more suitable.

10.6 NUMERICAL SIMULATIONS

In this section, two examples are provided to validate the theoretical results in Sections 10.3–10.5, respectively.

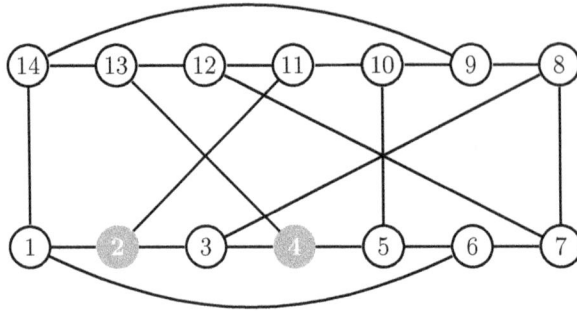

Figure 10.11   The $(2, 6)$-isolable graph.

The system matrices in (10.1) are given as follows:

$$A_i = \begin{pmatrix} -2.25 & 9 & 0 \\ 1 & -1 & 1 \\ 0 & -a_i & 0 \end{pmatrix}, B_i = \begin{pmatrix} b_i \\ 0 \\ 0 \end{pmatrix},$$

$$A_{ij} = \begin{pmatrix} f_{ij} & f_{ij} & 0 \\ 0 & 0 & 0 \\ 0 & 0 & 0 \end{pmatrix}, C_i = \begin{pmatrix} 1 & 0 & 0 \\ 0 & 1 & 0 \end{pmatrix}.$$

The system parameters a_i, b_i, and f_{ij} will be provided in the following. The preset time constants are set as $\tau_{1i} = 0.6, \tau_i = 1$.

Example 1: This example is performed to illustrate that the graph condition for isolating FDI attacks under Algorithm 10.1 is more relaxed than that under Algorithm 9.2. Consider the physically coupled NAS consisting of fourteen agents, the communication topology among the agents is shown in Fig. 10.11, which is $(2, 6)$-isolable. The system parameters satisfy $a_i = 15$, $b_i = 1$ for $i \in \mathcal{V}\backslash\{4, 5\}$, $a_i = 18, b_i = 3$ for $i \in \{4, 5\}$, the couplings among the agents satisfy $f_{ij} = 2$ if $a_{ij} = 1$ with $i = 11$ or $i = 14$, and $f_{ij} = 0$ otherwise. Assume that agents 2 and 4 are compromised by FDI attacks with $t_a = 1.5$, and the attacks on them satisfy $u_2^a = u_4^a = \mathbf{0}$, $y_2^a = -y_4^a = 10\,(\sin(0.2t)\;\cos(1/(t+1)))^T$.

It is shown in Fig. 10.12 that the attacks on agents 2 and 4 collude to make $\|\gamma_3\| = 0$, besides, $\|\gamma_i\| \neq 0$ for $i \in \{1, 2, 5, 11, 13\}$. Then, one has $\|\Gamma_2\|_{l_0} = \|\Gamma_4\|_{l_0} = 3$, and $\|\Gamma_i\|_{l_0} < 3$ for $i \in \mathcal{V}\backslash\{2, 4\}$. By contrast, one gets $\|\Upsilon_2\|_{l_0} = \|\Upsilon_3\|_{l_0} = \|\Upsilon_4\|_{l_0} = \|\Upsilon_{12}\|_{l_0} = \|\Upsilon_{14}\|_{l_0} = 2$, and $\|\Gamma_i\|_{l_0} < 2$ for $i \in \mathcal{V}\backslash\{2, 3, 4, 12, 14\}$. The safety indices of the agents are shown in Figs. 10.13–10.14. It can see from Fig. 10.13 that the compromised agents 2 and 4 are correctly isolated under Algorithm 10.1. However, Fig. 10.14 shows that the compromised agent 4 is not isolated, indicating the failure of attack isolation under Algorithm 9.2. This validates the correctness of Remark 10.10.

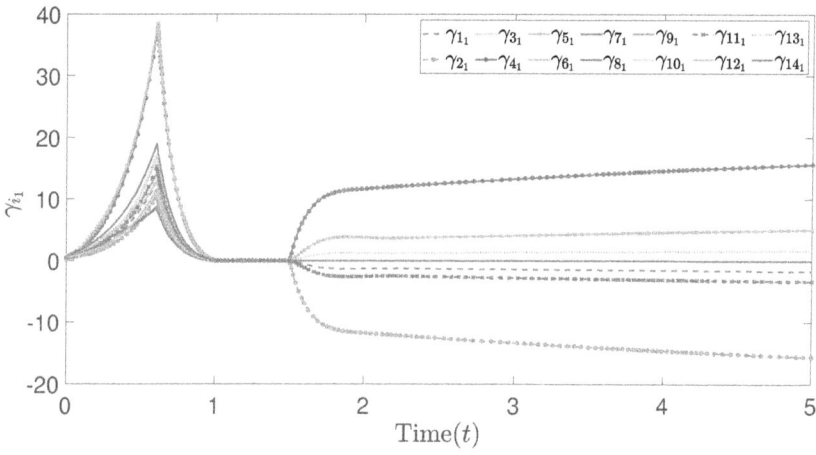

(a) γ_{i1}.

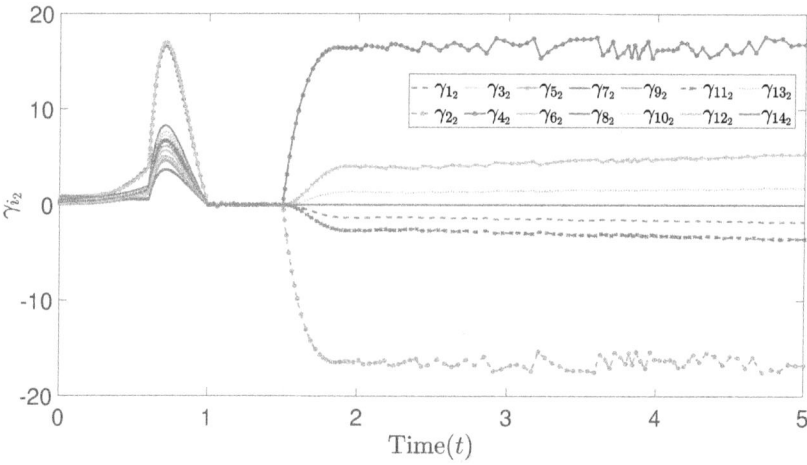

(b) γ_{i2}.

Figure 10.12   The output residual γ_i for each agent.

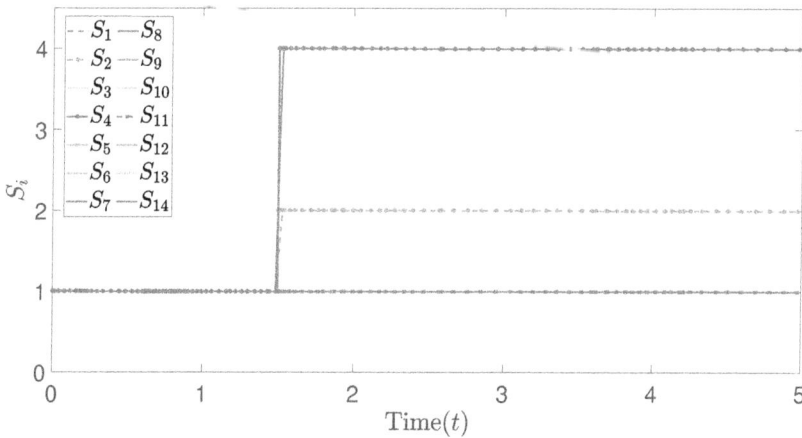

Figure 10.13   The safety index S_i for each agent under Algorithm 10.1.

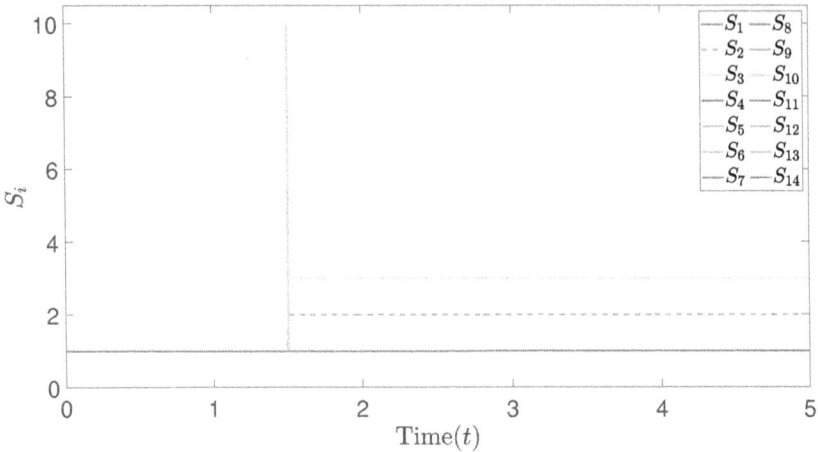

Figure 10.14   The safety index S_i for each agent under Algorithm 9.2.

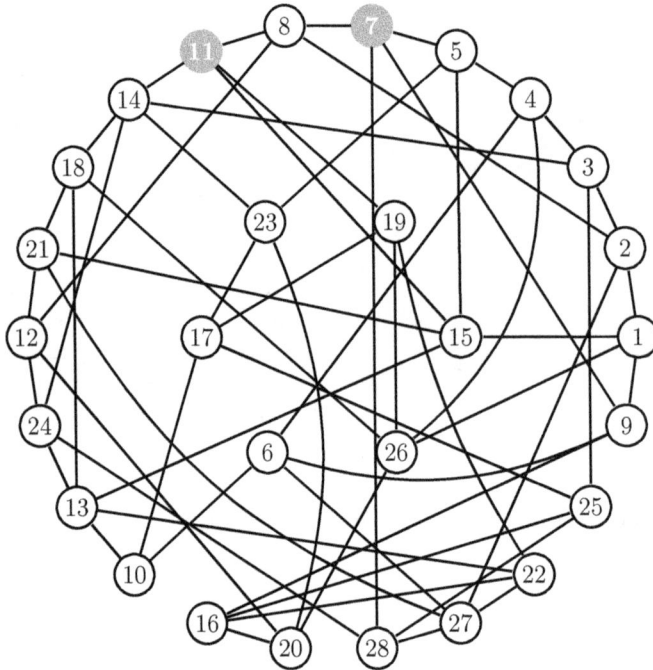

Figure 10.15   The $(3,5)$-isolable graph.

Example 2: This example verifies the equivalent graph condition for isolating covert attacks under Algorithm 9.2. Consider the physically coupled NAS consisting of twenty-eight agents, the communication topology among the agents is shown in Fig. 10.15, which is $(3,5)$-isolable. The system parameters satisfy $a_i = 15$, $b_i = 1$ for $i \in \mathcal{V}\backslash\{4,5\}$, $a_i = 18, b_i = 3$ for $i \in \{4,5\}$, the couplings among the agents satisfy $f_{ij} = 2$ if $a_{ij} = 1$ with $i = 5$ or $i = 19$, $f_{ij} = 0.5$ if $a_{ij} = 1$ with $i = 9$ or $i = 14$, and $f_{ij} = 0$ otherwise. Assume that agents 7 and 11 are compromised by covert attacks with $T_a = 2$, and the attacks on them satisfy $u_7^a = -u_{11}^a = (200 + 200\cos(t)\ 0\ 200 + 200\sin(t))^T$.

The example in Chapter 9 shows that the above covert attacks can be isolated if the graph is $(2,7)$-isolable. For the $(3,5)$-isolable graph in Fig. 10.15, it can be seen from Fig. 10.16 that the attacks on agents 7 and 11 collude to make $\|\gamma_8\| = 0$. Then, one has $\|\Upsilon_7\|_{l_0} = \|\Upsilon_{11}\|_{l_0} = 3$, and $\|\Upsilon_i\|_{l_0} < 3$ for $i \in \mathcal{V}\backslash\{7, 11\}$. The safety indices of the agents are depicted in Fig. 10.17, showing that the compromised agents 7 and 11 are correctly isolated. This validates the correctness of the condition ii) in Theorem 10.4.

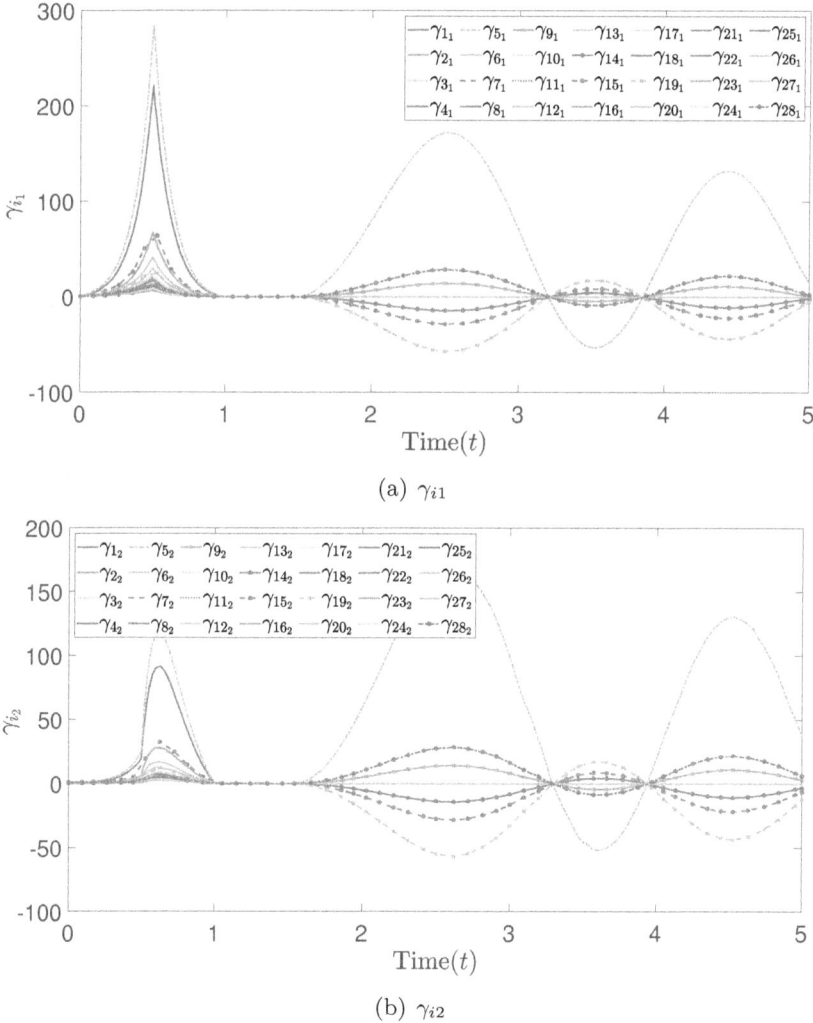

(a) γ_{i1}

(b) γ_{i2}

Figure 10.16 The output residual γ_i for each agent under the $(3,5)$-isolable graph.

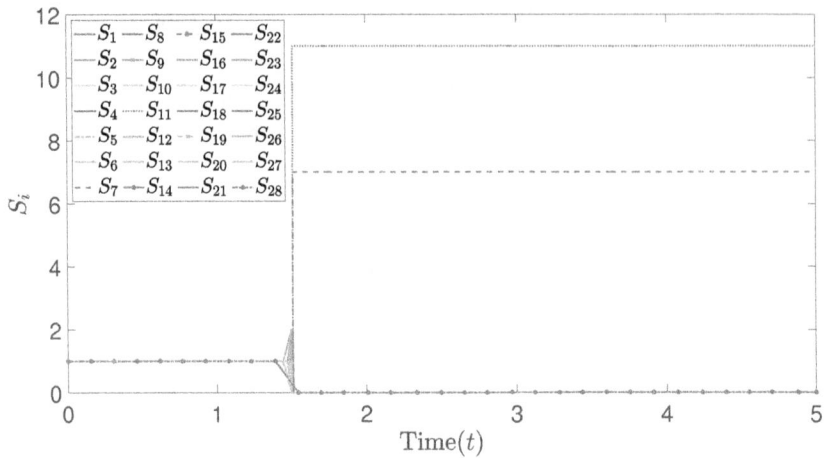

Figure 10.17　The safety index S_i for each agent under the $(3,5)$-isolable graph.

10.7　CONCLUSIONS

In this chapter, the isolation problem of collusive FDI attacks and covert attacks has been studied under Algorithm 10.1 and Algorithm 9.2, respectively. To generalize the application of these algorithms, the worst attack collusive cases from the perspective of the algorithms under different cycle sizes have been derived. Then, several equivalent graph conditions have been developed such that the attack isolation could be achieved under any cycle size. The superior algorithms have been provided for FDI attacks and covert attacks, respectively.

CHAPTER 11

Resilient consensus of NASs under link attacks

This chapter investigates the resilient consensus problem of NASs under cyber attacks on communication links, where the attacks on different links may collude to remain stealthy and undetectable. Section 11.1 reviews relevant prior work and articulates the motivations that drive the current study. Section 11.2 formulates the resilient consensus problem under link attacks. Section 11.3 develops two attack isolation algorithms tailored to non-collusive and collusive link attacks, respectively, and provides theoretical analyses to validate their effectiveness. Section 11.4 provides the resilient consensus algorithm based on attack isolation. Section 11.5 provides some simulations to illustrate the effectiveness of the theoretical results. Section 11.6 concludes this chapter.

11.1 INTRODUCTION

Link attacks were addressed in [196] by introducing random network topologies, wherein two distributed consensus algorithms were proposed. In [197, 198], the synchronization problem of complex networks under switching topologies induced by recoverable link attacks was investigated, and the relationship between attack frequency and average recovery time was established. Building upon the concept of switching topologies, [199] derived mean-square consensus tracking criteria for NASs under both connectivity-maintained and connectivity-broken attacks. In [200], the resilient consensus problem of NASs was studied under the assumption that link attacks followed a Bernoulli distribution. In [54], both malicious agents and stochastic link failures were considered, and a discrete-time consensus protocol was developed to ensure almost sure consensus among the normal agents. Denial-of-Service (DoS) attacks that disrupt communication across the entire network were considered in [120, 148, 201], while [202] further investigated secure consensus under DoS attacks affecting partial links at different time instants. The authors in [59, 60] developed resilient consensus algorithms under the F-local link attack model, wherein each node is assumed to have at most F compromised links incident to it. It is worth noting that the core idea of these resilient consensus algorithms lies in discarding the extreme values received from neighboring nodes, which limits their applicability to general high-order NASs. A potential approach to address this limitation is to first detect and isolate the compromised links, and subsequently discard the isolated links to ensure the successful execution of the cooperative task. A novel estimator based on unknown input observers was proposed in [203] to enable the detection and isolation of link attacks. Furthermore, [204] enhanced the detectability of certain stealthy attacks by introducing additional Luenberger observers.

In this chapter, we aim to incorporate attack detection and isolation into the resilient consensus framework for general high-order NASs, with particular consideration given to collusive link attacks. The main challenges lie in decoupling the collusion among link attacks and mitigating their impact on resilient consensus. First, the distributed fixed-time observer in [166] is introduced, and the consensus estimation residual is calculated for each agent. The attack isolation algorithm against non-collusive link attacks is then developed by checking the non-zero residuals of the corresponding two agents on each link. For the case that there exists collusion among attacks on different links, the residual of the agent cannot detect whether there are link attacks involving itself. Then, another attack isolation algorithm is designed by checking the difference of the residual of the associated agent before and after disconnecting this link. Finally, a resilient consensus control algorithm is then presented by discarding the isolated compromised links.

11.2 PROBLEM FORMULATION

Consider a high-order NAS consisting of N agents, the dynamics of the ith agent are given as

$$\begin{cases} \dot{x}_i(t) = Ax_i(t) + Bu_i(t), \\ y_i(t) = Cx_i(t), \ i = 1, 2, \ldots, N, \end{cases} \tag{11.1}$$

where $x_i(t) \in \mathbb{R}^n$, $u_i(t) \in \mathbb{R}^m$, $y_i(t) \in \mathbb{R}^p$ are, respectively, the state, the control input, and the measurement output of the ith agent, and A, B and C are constant system matrices, satisfying the following assumptions.

Assumption 11.1 *rank(CB)=rank(B)= m.*

Assumption 11.2 $rank\begin{pmatrix} sI_n - A & B \\ C & \mathbf{0} \end{pmatrix} = n + m, \ \forall s \in \mathbb{C}.$

The fixed-time observer (8.5) in Chapter 8 and the observer-based control protocol (9.6) in Chapter 9 can be used to achieve resilient consensus of the NAS (11.1). Note that the observer (8.5) for the ith agent relies on the output information of its neighbors, i.e., $y_j(t)$, $j \in \mathcal{N}_i$. When the hackers launch attacks on the communication links, the information agent i receives via link (j, i) may not be exactly $y_j(t)$. Denote $y_j^i(t)$ the information that agent i receives from its neighbor j through link (j, i), and one has

$$y_j^i(t) = \begin{cases} y_j(t), & (j, i) \text{ is normal}, \\ p_j^i(t), & (j, i) \text{ is compromised}, \end{cases} \tag{11.2}$$

where $p_j^i(t)$ is the false information that agent i receives from its neighbor j.

Remark 11.1 *With appropriate definition of $p_j^i(t)$, (11.2) can capture different types of attacks, such as: 1) a FDI attack with $p_j^i(t)$ being the injection signal; 2) a replay attack with $p_j^i(t) = y_j(t - mT)$, where $m \in \mathbb{N}^+$ and T is the periodicity of the attack; 3) a DoS attack with $p_j^i(t) = y_j(t)$ if $(q - 1)T \le t < (q - 1)T + t_{off}$ and $p_j^i(t) = 0$ if $(q - 1)T + t_{off} \le t < qT$ with $t_{off} < T$ and $q \in \mathbb{N}^+$.*

Remark 11.2 *In [196–199], the link attacks are recoverable, and thus, time-varying graphs are introduced to characterize the influence of link attacks. In this chapter, we consider the worst-case, i.e., the link attacks cannot be recovered. This makes the resilient consensus problem more challenging.*

Then, the problem addressed in this chapter can be formulated as follows.

Problem. Given a high-order NAS (11.1) with Assumptions 11.1 and 11.2 under link attacks (11.2), design algorithms to achieve attack isolation and to realize consensus in the sense that $\lim_{t \to \infty} \|x_i(t) - x_j(t)\| = 0$, $\forall i, j \in \mathcal{V}$.

11.3   ISOLATION ALGORITHMS FOR LINK ATTACKS

This section will present two attack isolation algorithms against non-collusive and collusive link attacks, respectively. First, the attack detection strategy based on the observer (8.5) is introduced as follows:

The distributed fixed-time observer (8.5) under link attacks (11.2) can be rewritten as follows:

$$\begin{cases} \dot{\hat{z}}_i(t) = A_c \hat{z}_i(t) + B_c \sum_{j \in \mathcal{N}_i} \left(y_i(t) - y_j^i(t) \right), \\ \hat{\chi}_i(t) = D_c[\hat{z}_i(t) - \exp(A_c \tau)\hat{z}_i(t - \tau)], \\ \hat{\varepsilon}_i(t) = \hat{\chi}_i(t) - E \sum_{j \in \mathcal{N}_i} \left(y_i(t) - y_j^i(t) \right). \end{cases} \quad (11.3)$$

Denote the consensus estimation residual between the output consensus error of agent i and its estimation as $\tilde{\gamma}_i(t) = \sum_{j \in \mathcal{N}_i} (y_i(t) - y_j^i(t)) - C\hat{\varepsilon}_i(t)$. With (11.1) and (11.3), one has

$$\tilde{\varepsilon}_i(t) = -CD_c \int_{t-\tau}^t \exp\left(A_c \left(t - s \right) \right) B_c \sum_{j \in \mathcal{N}_i} \left(y_j^i \left(s \right) - y_j \left(s \right) \right) ds, \ t \ge \tau. \quad (11.4)$$

It can be concluded from (11.4) that 1) Suppose that there is at most one link in $\mathcal{K}_i$ being compromised, the attack occurs *if and only if* $\exists t \ge \tau$ s.t. $\|\tilde{\gamma}_i(t)\| \ne 0$ holds; 2) There exist several links in $\mathcal{K}_i$ being compromised *if* $\exists t \ge \tau$ s.t. $\|\tilde{\gamma}_i(t)\| \ne 0$ holds. Note that if there are multiple links in $\mathcal{K}_i$ being compromised, $\|\tilde{\gamma}_i(t)\| = 0$ for some $t \ge \tau$ does not mean that there is no link attacks, since multiple attacks may collude to make $\sum_{j \in \mathcal{N}_i} \left(y_j^i(t) - y_j(t) \right) = \mathbf{0}$ to avoid being detected. Take the attacks in Fig. 11.1(a) for example. The links $(2, 3)$, $(2, 4)$, $(3, 5)$, and $(4, 5)$ are compromised with $p_4^2 - y_4 = -(p_3^2 - y_3)$, $p_5^3 - y_5 = -(p_2^3 - y_2)$, $p_5^4 - y_5 = -(p_2^4 - y_2)$, $p_4^5 - y_4 = -(p_3^5 - y_3)$. Clearly, one has $\|\tilde{\gamma}_2(t)\| = 0$, $\|\tilde{\gamma}_3(t)\| = 0$, $\|\tilde{\gamma}_4(t)\| = 0$, and $\|\tilde{\gamma}_5(t)\| = 0$ for $t \ge \tau$. That is, the collusion among the attacks on links $(2, 3)$, $(2, 4)$, $(3, 5)$, and $(4, 5)$ makes

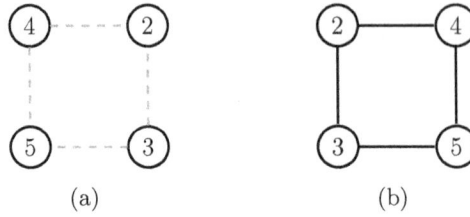

Figure 11.1   A comparison between the compromised and normal cases. (a) The case where link attacks collude. (b) The case where there are no link attacks.

Algorithm 11.1 DIs algorithm against non-collusive link attacks

---

Input: $S_{ij} = 0$, $(j, i) \in \mathcal{E}$, $\tilde{\gamma}_i, i \in \mathcal{V}$;
Output: S_{ij}, $(j, i) \in \mathcal{E}$;
  1: **if** $t \geq \tau$ **then**
  2:    **if** $\|\tilde{\gamma}_i\| \neq 0$ and $\|\tilde{\gamma}_j\| \neq 0$ **then**
  3:       $S_{ij} = 1$.
  4:    **end if**
  5: **end if**

---

the residual of each agent $\tilde{\gamma}_i, i \in \mathcal{V}$ in Fig. 11.1(a) perform the same as the one in Fig. 11.1(b). This means that the link attacks in Fig. 11.1(a) cannot be detected by the residual in (11.4). Then, the definition of collusive link attacks is introduced as follows.

Definition 11.1 (Collusive link attacks) *The link attacks $\bigcup_{(j,i)\in\mathcal{E}_A} p_j^i(t)$ are said to be collusive if for any agent i with $\mathcal{K}_i \cap \mathcal{E}_A \neq \varnothing$, the errors between the information received from its neighbors $j \in \mathcal{N}_i$ and their actual outputs satisfy $\left\| \sum_{j\in\mathcal{N}_i} \left(y_j^i(t) - y_j(t) \right) \right\| = 0$.*

11.3.1   Attack isolation for non-collusive link attacks

For the undirected graph $\mathcal{G}$, it follows from (11.4) that $\|\tilde{\gamma}_i(t)\| \neq 0$ and $\|\tilde{\gamma}_j(t)\| \neq 0$ hold simultaneously for $t \geq \max\{T_{ij}, \tau\}$ if the link (j, i) is compromised, where T_{ij} is the time instant when the attack occurs on link (j, i). In view of this, the distributed isolation (DIs) algorithm against non-collusive link attacks is developed in Algorithm 11.1, where S_{ij} is the safety index of link (j, i) with $S_{ij} = 1$ being isolated as compromised, and $S_{ij} = 0$ otherwise.

The following result gives a sufficient condition for the attack isolation.

Lemma 11.1 *Consider a high-order NAS (11.1) with an undirected graph $\mathcal{G}$ subject to non-collusive link attacks (11.2). Suppose that Assumptions 11.1 and 11.2 hold. Attack isolation can be achieved by Algorithm 11.1 if the path length between any two compromised links (h, g) and (j, i) satisfies $|l_{gh\to ij}| > 1$.*

Proof 11.1 *It follows from (11.4) that $\|\tilde{\gamma}_i\| \neq 0$ and $\|\tilde{\gamma}_j\| \neq 0$ if the link (j,i) is compromised. Obviously, the compromised link (j,i) can be isolated by Algorithm 11.1. It remains to prove that no normal link will be isolated mistakenly.*

Suppose that the normal link (i,h) is mistakenly isolated as a compromised one. According to Algorithm 11.1, one has $\|\tilde{\gamma}_h\| \neq 0$ and $\|\tilde{\gamma}_i\| \neq 0$. Then, there are nodes $g \in \mathcal{N}_h$ and $j \in \mathcal{N}_i$ such that $(h,g) \in \mathcal{E}_A$ and $(j,i) \in \mathcal{E}_A$. Clearly, $|l_{gh \to ij}| \leq 1$.

Then, the main result of attack isolation against non-collusive link attacks is provided.

Theorem 11.1 *Consider a high-order NAS (11.1) with an undirected graph $\mathcal{G}$ subject to non-collusive link attacks (11.2). Suppose that Assumptions 11.1 and 11.2 hold. Attack isolation can be achieved by Algorithm 11.1 if and only if the followings hold simultaneously:*

i) The path length between any two compromised links (h,g) and (j,i) satisfies $|l_{gh \to ij}| \neq 1$;

ii) For any two compromised links (h,g) and (g,i) with $|l_{gh \to gi}| = 0$, $(i,h) \notin \mathcal{E}_R$ holds.

Proof 11.2 *Similarly to the proof of Lemma 11.1, we shall show that no mistaken isolation of normal links happens if and only if conditions (i) and (ii) satisfy.*

(Necessity). The necessity will be proved by contradiction. Consider two cases: 1) There are two compromised links (h,g) and (j,i) satisfying $|l_{gh \to ij}| = 1$ with $(i,h) \in \mathcal{E}_R$; 2) There are two compromised links (h,g) and (g,i) satisfying $|l_{gh \to gi}| = 0$ with $(i,h) \in \mathcal{E}_R$. Then, one has that the normal link (i,h) is mistakenly isolated as compromised by Algorithm 11.1 for both 1) and 2).

(Sufficiency). Condition ii) suggests that either $(i,h) \notin \mathcal{E}$ or $(i,h) \in \mathcal{E}_A$ holds for any two compromised links $(h,g) \in \mathcal{E}_A$ and $(g,i) \in \mathcal{E}_A$. The following proof is similar to that of Lemma 11.1, and is hence omitted.

11.3.2 Attack isolation for collusive link attacks

It follows from Lemma 11.1 and Theorem 11.1 that the constraint on the path length for any two compromised links is required if the attack isolation is achieved by Algorithm 11.1. However, the positions of link attacks cannot be preset in reality. Thus, it is necessary to design another attack isolation algorithm to relax this constraint. Besides, if there are multiple compromised links in $\mathcal{K}_i$, they may collude to make $\|\sum_{j \in \mathcal{N}_i} (y_j^i - y_j)\| = 0$, and thus, $\|\tilde{\gamma}_i\| = 0$. Such collusive attacks are known as covert attacks [206]. It is derived from [186–188] that the covert attack can be detected by changing the structure of the system. Partially inspired by this, the main idea of the distributed isolation (DIs) algorithm against collusive link attacks is to change the communication topology. The details of DIs algorithm against collusive link attacks are stated as follows.

Algorithm 11.2 DIs algorithm against collusive link attacks

Input: $S_{ij} = 0$, $(j, i) \in \mathcal{E}$, $\tilde{\gamma}_i, i \in \mathcal{V}$;
Output: S_{ij}, $(j, i) \in \mathcal{E}$;
  1: **if** $t \geq \tau$ **then**
  2:     **for** each link $(j, i) \in \mathcal{E}$ **do**
  3:         Calculate $\hat{\varphi}_i^j$ by (11.5);
  4:         **if** $\|\tilde{\eta}_i^j\| \neq \|\tilde{\gamma}_i\|$ **then**
  5:            $S_{ij} = 1$.
  6:         **end if**
  7:     **end for**
  8: **end if**

For each directed link (j, i), by removing the output information $y_j^i(t)$ received from agent j, a novel distributed fixed-time observer for agent i is designed as follows:

$$\begin{cases} \dot{\hat{z}}_i^j(t) = A_c \hat{z}_i^j(t) + B_c \sum_{k \in \{\mathcal{N}_i \setminus j\}} (y_i(t) - y_k^i(t)), \\ \hat{\chi}_i^j(t) = D_c \left[\hat{z}_i^j(t) - \exp(A_c \tau) \hat{z}_i^j(t - \tau) \right], \\ \hat{\varphi}_i^j(t) = \hat{\chi}_i^j(t) - E \sum_{k \in \{\mathcal{N}_i \setminus j\}} (y_i(t) - y_k^i(t)), \end{cases} \tag{11.5}$$

where $\hat{z}_i^j(t)$ and $\hat{\chi}_i^j(t)$ are auxiliary variables with $\hat{z}_i^j(t) = \mathbf{0}$ for $t \in [-\tau, 0]$, $\hat{\varphi}_i^j(t)$ is the estimation of agent i's new consensus error $\varphi_i^j(t) = \sum_{k \in \{\mathcal{N}_i \setminus j\}} (x_i(t) - x_k(t))$.

Then, the novel residual $\tilde{\eta}_i^j(t) = \sum_{k \in \{\mathcal{N}_i \setminus j\}} (y_i(t) - y_k^i(t)) - C\hat{\varphi}_i^j(t)$ can be calculated for agent i. The attack isolation algorithm is proposed by checking the difference between the original residual $\tilde{\gamma}_i$ and the new residual $\tilde{\eta}_i^j$, which is described in Algorithm 11.2. The following result of attack isolation against collusive link attacks is presented.

Theorem 11.2 *Consider a high-order NAS (11.1) with a directed graph $\mathcal{G}$ subject to link attacks (11.2). Suppose that Assumptions 11.1 and 11.2 hold. Attack isolation can be achieved by Algorithm 11.2.*

Proof 11.3 *It derives from (11.4) that $\sum_{k \in \{\mathcal{N}_i \setminus j\}} (y_k^i - y_k) \neq \sum_{k \in \mathcal{N}_i} (y_k^i - y_k)$ if and only if (j, i) is compromised. Therefore, all compromised links (j, i) can be isolated by Algorithm 11.2 and no normal links would be mistakenly isolated.*

Remark 11.3 *Compared with Algorithm 11.1 against non-collusive link attacks under undirected graphs, Algorithm 11.2 can isolate collusive link attacks under directed graphs. Moreover, the restriction on the positions of compromised links as presented in Lemma 11.1 and Theorem 11.1 is no longer needed. The price it takes is the extra calculation of the residuals $\tilde{\eta}_i^j$ for each link (j, i).*

11.4 RESILIENT CONSENSUS UNDER LINK ATTACKS

This section will provide an attack-isolation-based resilient control algorithm and derive some conditions to ensure the consensus of the whole system under link attacks. The resilient consensus algorithm against link attacks is designed in Algorithm 11.3.

Algorithm 11.3 Resilient consensus algorithm against link attacks

---

Input: S_{ij}, $(j,i) \in \mathcal{E}$;

  1: **for** each link $(j,i) \in \mathcal{E}$ **do**

  2:     **if** $S_{ij} = 1$ **then**

  3:         $\mathcal{N}_i = \mathcal{N}_i \backslash j$;

  4:         Design control protocol u_i as in (9.6).

  5:     **end if**

  6: **end for**

Denote $\mathcal{G}' = (\mathcal{V}, \mathcal{E}')$ with $\mathcal{E}' = \{(j,i)|(j,i) \in \mathcal{E}_R\}$ the subgraph induced by removing the isolated links. Then, the following result of resilient consensus is presented based on Theorem 11.1.

Corollary 11.1 *Consider a high-order NAS (11.1) with an undirected graph $\mathcal{G}$ subject to non-collusive link attacks (11.2). Suppose that Assumptions 11.1 and 11.2 hold. Resilient consensus can be reached by Algorithms 11.1 and 11.3 if and only if the followings hold simultaneously:*

  i) The path length between any two compromised links (h,g) and (j,i) satisfies $|l_{gh \to ij}| \neq 1$;

  ii) For any two compromised links (h,g) and (g,i) with $|l_{gh \to gi}| = 0$, $(i,h) \notin \mathcal{E}_R$ holds;

  iii) $\mathcal{G}'$ is connected.

In light of Theorem 11.2, the following result can be derived.

Corollary 11.2 *Consider a high-order NAS (11.1) with a directed graph $\mathcal{G}$ subject to link attacks (11.2). Suppose that Assumptions 11.1 and 11.2 hold. Resilient consensus can be reached by Algorithms 11.2 and 11.3 if and only if $\mathcal{G}'$ contains a spanning tree.*

In the following, the comparison between Algorithm 11.3 presented in this chapter and the algorithm in [60] is made.

Theorem 11.3 *Consider a high-order NAS (11.1) with a directed graph $\mathcal{G}$ subject to link attacks (11.2). Suppose that Assumptions 11.1 and 11.2 hold. Under F-local link attack model, resilient consensus can be reached by Algorithms 11.2 and 11.3 if and only if $\mathcal{G}$ is $(F+1)$-robust.*

Proof 11.4 (Necessity). *Suppose that $\mathcal{G}$ is not $(F+1)$-robust, then there exist at least two subsets $\mathcal{S}_1, \mathcal{S}_2 \subseteq \mathcal{V}$ such that every node in $\mathcal{S}_1$ and $\mathcal{S}_2$ has at most F neighbors outside the subset it belongs to. That is, $\forall i \in \mathcal{S}_1$, $|\{(j,i) \in \mathcal{K}_i | j \in \mathcal{V} \backslash \mathcal{S}_1\}| \leq F$ and $\forall i \in \mathcal{S}_2$, $|\{(j,i) \in \mathcal{K}_i | j \in \mathcal{V} \backslash \mathcal{S}_2\}| \leq F$. For the F-local link attack model, it is possible to compromise all links in $\{(j,i) \in \mathcal{K}_i \mid i \in \mathcal{S}_1, j \in \mathcal{V} \backslash \mathcal{S}_1\}$ and $\{(j,i) \in \mathcal{K}_i \mid i \in \mathcal{S}_2, j \in \mathcal{V} \backslash \mathcal{S}_2\}$. Then, for graph $\mathcal{G}'$, none of the nodes in $\mathcal{S}_1$ (or $\mathcal{S}_2$) has path to any node outside $\mathcal{S}_1$ (or $\mathcal{S}_2$), and the consensus can never be realized.*

(Sufficiency). *Since the communication topology is $(F + 1)$-robust and there is at most F compromised links in $\mathcal{K}_i$ for each node $i \in \mathcal{V}$, one has that for every pair of disjoint subsets $\mathcal{S}_1$ and $\mathcal{S}_2$, there exists a node $i \in \mathcal{S}_k$, $k \in \{1, 2\}$ such that $|\{(j, i) \in \mathcal{E}_R | j \in \mathcal{N}_i \backslash \mathcal{S}_k\}| \geq 1$ after attack isolation. Therefore, the graph $\mathcal{G}'$ contains a spanning tree, and the resilient consensus of the whole system can be reached.*

Remark 11.4 *It follows from Algorithm 11.2 that attack isolation can be achieved if there are any number of compromised links in $\mathcal{K}_i$, $i \in \mathcal{V}$. However, the resilient consensus may not be achieved in this case. The main reason is that the $(F+1)$-robust graph cannot ensure that the graph $\mathcal{G}'$ contains a spanning tree after deleting the F_0 compromised links in $\mathcal{K}_i$ with $F_0 > F$.*

Remark 11.5 *Note that resilient consensus can be reached under the algorithm proposed in [60] if $\mathcal{G}$ is $(2F + 1)$-robust, which is much stricter than the condition in Theorem 11.3. The main reason lies in the discarding of normal links by the algorithm in [60]. Specifically, there are at most $2F$ extreme values transmitted via the links in $\mathcal{K}_i$ being discarded by agent i, meaning that at least F values transmitted via the normal links are deleted. In this chapter, only the values received from the compromised links are discarded by each agent under the proposed Algorithm 11.3, which benefits from the operation of attack isolation in Algorithm 11.2. Moreover, the index for discarding links is the value transmitted via each link in [60], which makes the resilient consensus algorithm applicable to only integrator-type NASs. This drawback is also overcome by designing the attack isolation algorithm in this chapter.*

Inspired by [59], the notion of *trusted links* is introduced.

Definition 11.2 *The link (j, i) is trusted if it cannot be compromised by any attacks, i.e., $y_j^i(t) \equiv y_j(t)$ with a trusted link (j, i).*

It is practical to assume that there are trusted links in the communication topology. Denote $\mathcal{E}_T$ as the set of trusted links with $\mathcal{E}_T \subseteq \mathcal{E}_R$. Then, the following result is provided.

Theorem 11.4 *Consider a high-order NAS (11.1) with a directed graph $\mathcal{G}$ subject to link attacks (11.2). Suppose that Assumptions 11.1 and 11.2 hold. Resilient consensus can be reached by Algorithms 11.2 and 11.3 if for the trusted link subset $\mathcal{E}_T \subseteq \mathcal{E}$ and any nonempty agent subset $\mathcal{U} \subseteq \mathcal{V}$, at least one of the followings holds:*

 i) There exists a agent $i \in \mathcal{U}$ such that $|\mathcal{N}_i \backslash \mathcal{U}| \geq F + 1$;

 ii) There exists a agent $i \in \mathcal{U}$ such that $(j, i) \in \mathcal{E}_T$, $i \in \mathcal{U}, j \in \mathcal{N}_i \backslash \mathcal{U}$.

Proof 11.5 *It only requires to show that $\mathcal{G}'$ contains a directed spanning tree, which will be demonstrated by contradiction.*

If $\mathcal{G}'$ does not contain a directed spanning tree, there must be a nonempty subset $\mathcal{V}_0 \subseteq \mathcal{V}$ such that $(j, i) \in \mathcal{E}_A$, $\forall i \in \mathcal{V}_0$, $j \in \mathcal{V} \backslash \mathcal{V}_0$. Clearly, $\mathcal{V}_0$ satisfies neither condition i) nor condition ii).

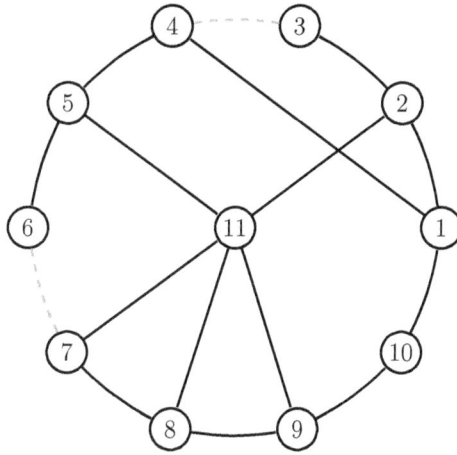

Figure 11.2 The communication topology of the NASs comprising eleven agents.

11.5 NUMERICAL SIMULATIONS

In this section, two examples are provided to validate the theoretical results in Sections 11.2–11.4, respectively. The system matrices of a three-order NAS are given as follows:

$$A = \begin{bmatrix} -2 & 6.8 & 1 \\ 1.1 & -1.6 & 1.2 \\ 0 & -5.6 & 0 \end{bmatrix}, \quad B = \begin{bmatrix} 1 \\ 0 \\ 1 \end{bmatrix}, \quad C = \begin{bmatrix} 1 & 0 & 0 \\ 0 & 1 & 1 \end{bmatrix}.$$

The parameters of the observer (11.3) and the parameters of the control protocol (9.6) are calculated as:

$$H_1 = \begin{bmatrix} 6.2598 & 4.9061 \\ 5.3081 & 3.0674 \\ 0.2984 & 0.8612 \end{bmatrix}, \quad H_2 = \begin{bmatrix} 3.9834 & 2.6845 \\ 0.3207 & -0.5412 \\ -0.4702 & 4.2530 \end{bmatrix},$$

$$Q = \begin{bmatrix} 0.4851 & 0.1186 & 0.3075 \\ 0.1186 & 1.7495 & 0.0211 \\ 0.3075 & 0.0211 & 0.5212 \end{bmatrix}.$$

Example 1: This example demonstrates the resilient consensus of NASs under an undirected graph in the presence of non-collusive link attacks. Consider a NAS of eleven agents with the undirected communication topology is given in Fig. 11.2. The preset convergent time is given as $\tau = 1.5$. The attacks occurring on links $(3,4)$ and $(6,7)$ are described in the following:

$$\begin{cases} y_4^3(t) = y_4(t) + 3(\sin t, \cos t)^T \\ y_3^4(t) = y_3(t) + 2(\sin t, \cos t)^T \end{cases}, \quad (3,4) \in \mathcal{E}_A \text{ with } T_{34} = 1.5,$$

$$\begin{cases} y_7^6(t) = -2y_7(t) \\ y_6^7(t) = -3y_6(t) \end{cases}, \quad (6,7) \in \mathcal{E}_A \text{ with } T_{67} = 2.5.$$

(a)

(b)

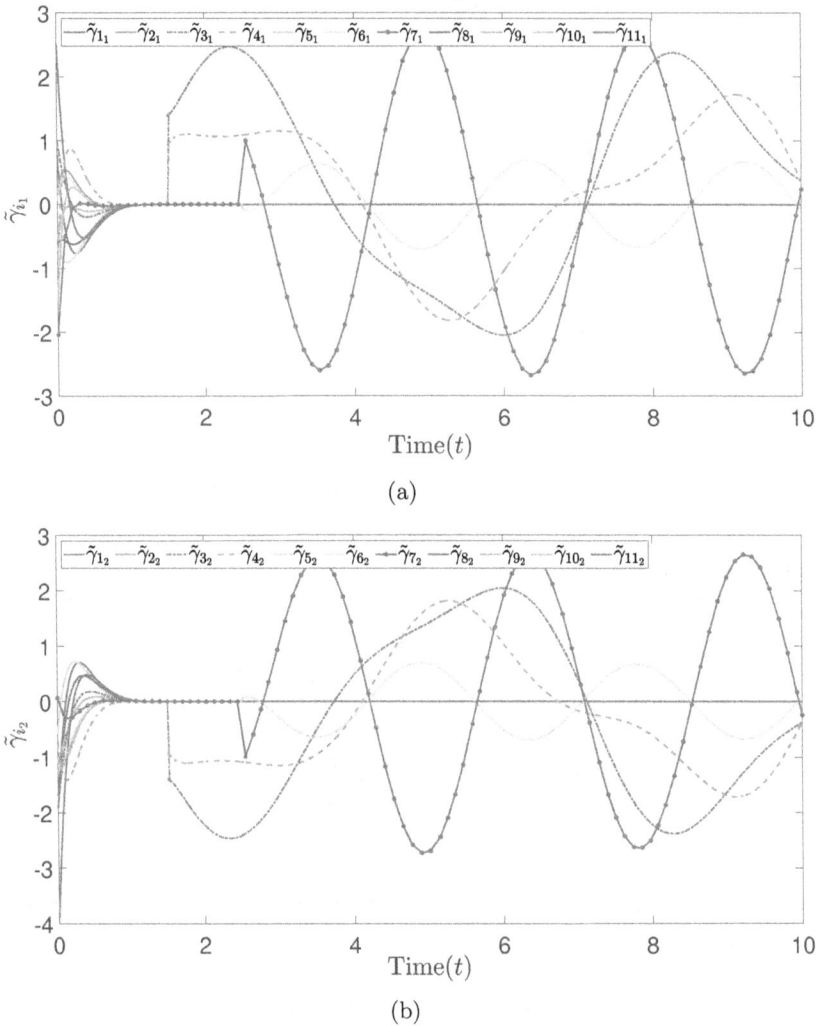

Figure 11.3   The consensus estimation residual trajectory of each agent $\tilde{\gamma}_i$.

The consensus estimation residual of each agent and the safety index of each link are depicted in Figs. 11.3 and 11.4, respectively. It shows in Fig. 11.3 that $\|\tilde{\gamma}_3(t)\| \neq 0$, and $\|\tilde{\gamma}_4(t)\| \neq 0$ after $t \geq 1.5$, which implies that the link $(3,4)$ is compromised, and S_{3-4} turns to be 1 at $t = 1.5$, corroborating the occurrence of attack on link $(3,4)$. Similarly, one has $\|\tilde{\gamma}_6(t)\| \neq 0$ and $\|\tilde{\gamma}_7(t)\| \neq 0$ after $t \geq 2.5$, and the safety index $S_{6-7} = 1$ suggests that link $(6,7)$ is attacked. The above results demonstrate that any compromised link (i,j) can be isolated at the time instant $t = \max\{\tau, T_{ij}\}$. Finally, Fig. 11.5 illustrates the state trajectory of each agent, demonstrating the achievement of resilient consensus under Algorithms 11.1 and 11.3.

Example 2: This example compares the Algorithm 1 in [60] and Algorithms 11.2 and 11.3 proposed in this work. Consider a group of eight sensors measuring the temperature cooperatively. The communication topology of the sensors is described

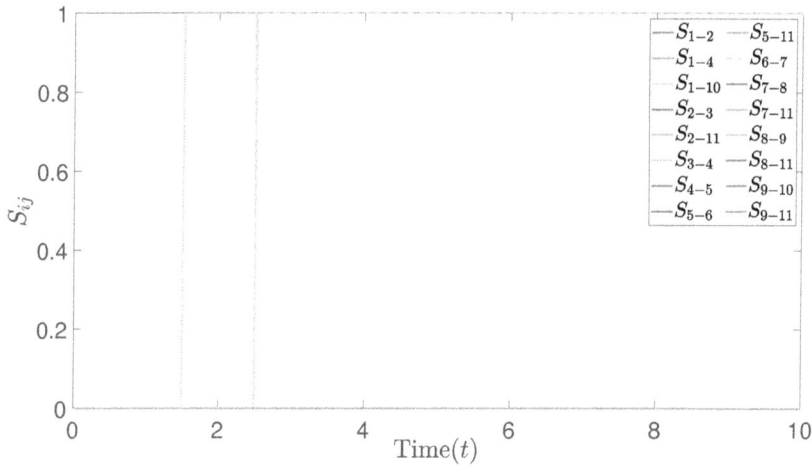

Figure 11.4   The safety index for each link S_{ij} under Algorithm 11.1.

by Fig. 11.6, where the communication links $(2,1)$ and $(6,7)$ subject to attacks with $y_2^1(t) = \cos(t)$ and $y_4^5(t) = y_5^4(t) = 3\sin(t)$. The initial value measured by each sensor is given as $x_i(0) \in [0,1]$. Besides, the preset convergent time is given as $\tau = 0.5$.

It follows from Definition 2.18 that the NAS is 1-local link attacked. Clearly, the sensor network in Fig. 11.6 is 2-robust but not 3-robust. It shows in Fig. 11.7 that the resilient consensus cannot be realized under the algorithm proposed in [60]. However, Fig. 11.8 demonstrates that the resilient consensus is achieved under Algorithms 11.2 and 11.3 since only the compromised links are isolated and the graph induced by deleting the isolated compromised links contains a spanning tree. This suggests that the resilient consensus algorithm based on attack isolation indeed relaxes the graph condition of resilient consensus.

11.6   CONCLUSIONS

In this chapter, the resilient consensus problem of high-order NASs subject to link attacks has been studied. By introducing a distributed fixed-time observer for each agent, the consensus estimation residual can be derived. Based on this, an attack isolation algorithm against non-collusive link attacks has been developed, and a necessary and sufficient condition has been derived to ensure the accurate isolation of compromised links and no mistaken isolation of normal ones. For the case that link attacks collude to maintain undetectable, a novel attack isolation algorithm has been designed by checking the difference of the residual of the corresponding agent before and after disconnecting each link. Finally, a control algorithm has been designed by discarding the isolated links to achieve resilient consensus of high-order NASs.

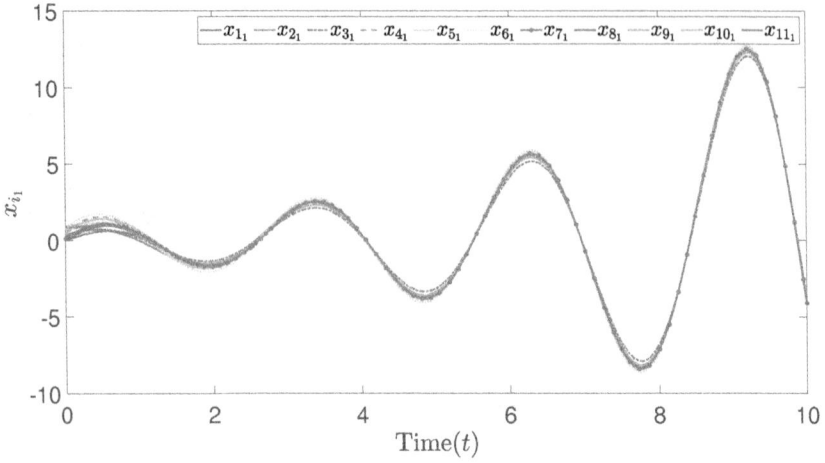

(a)

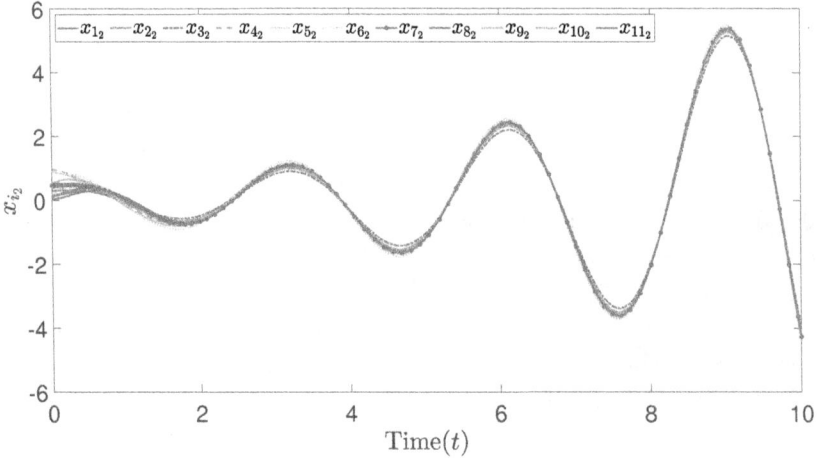

(b)

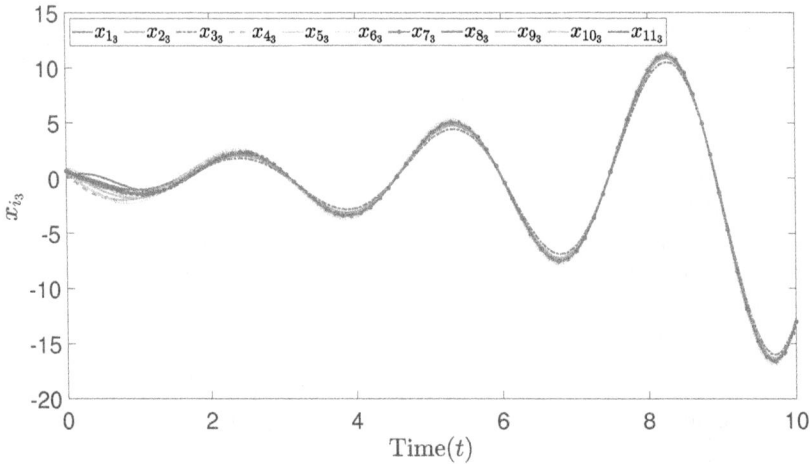

(c)

Figure 11.5 The state trajectory of each agent x_i under Algorithms 11.1 and 11.3.

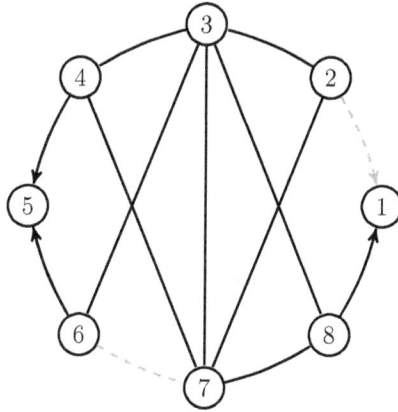

Figure 11.6   The communication topology of the NASs comprising eight agents.

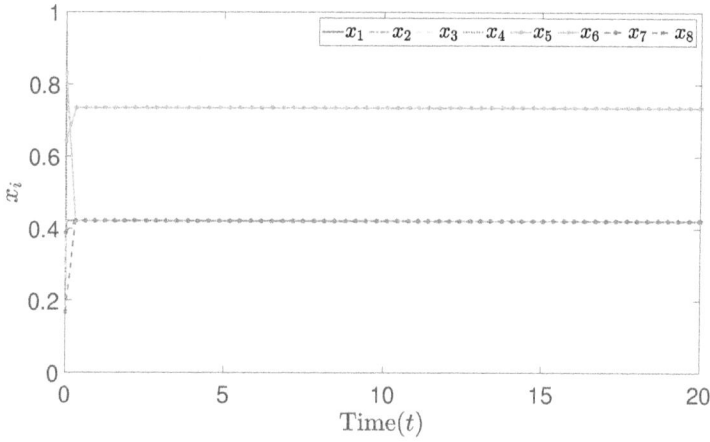

Figure 11.7   The measured value of each sensor under the algorithm proposed in [60].

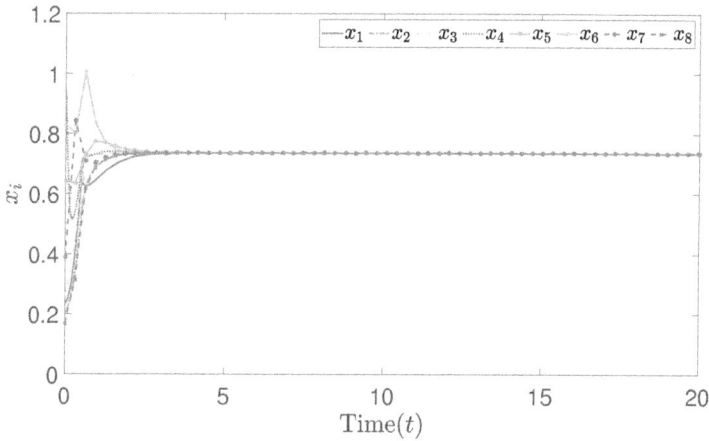

Figure 11.8   The measured value of each sensor under Algorithms 2 and 3 proposed in this paper.

Bibliography

[1] P. Tabuada, S. Y. Caliskan, M. Rungger, and R. Majumdar, "Towards robustness for cyber–physical systems," *IEEE Transactions on Automatic Control*, vol. 59, no. 12, pp. 3151–3163, 2014.

[2] R. Olfati-Saber, J. A. Fax, and R. M. Murray, "Consensus and cooperation in networked multi–agent systems," *Proceedings of the IEEE*, vol. 95, no. 1, pp. 215–233, 2007.

[3] A. Bemporad, M. Heemels, M. Johansson, *Networked control systems.* New York: Springer, 2010, vol. 406.

[4] L. Lombardo, S. Corbellini, M. Parvis, A. Elsayed, E. Angelini, and S. Grassini, "Wireless sensor network for distributed environmental monitoring," *IEEE Transactions on Instrumentation and Measurement*, vol. 67, no. 5, pp. 1214–1222, 2017.

[5] M. S. Rahman, M. A. Mahmud, A. M. T. Oo, and H. R. Pota, "Multi–agent approach for enhancing security of protection schemes in cyber–physical energy systems," *IEEE Transactions on Industrial Informatics*, vol. 13, no. 2, pp. 436–447, 2016.

[6] D. An, F. Zhang, Q. Yang, and C. Zhang, "Data integrity attack in dynamic state estimation of smart grid: Attack model and countermeasures," *IEEE Transactions on Automation Science and Engineering*, vol. 19, no. 3, pp. 1631–1644, 2022.

[7] J. Wang, J. Liu, and N. Kato, "Networking and communications in autonomous driving: A survey," *IEEE Communications Surveys & Tutorials*, vol. 21, no. 2, pp. 1243–1274, 2018.

[8] L. Yang, C. Ding, M. Wu, and K. Wang, "Robust detection of false data injection attacks for data aggregation in an internet of things–based environmental surveillance," *Computer Networks*, vol. 129, pp. 410–428, 2017.

[9] A. M. El-Sayed, P. Scarborough, L. Seemann, and S. Galea, "Social network analysis and agent–based modeling in social epidemiology," *Epidemiologic Perspectives & Innovations*, vol. 9, no. 1, pp. 1–9, 2012.

[10] W. Ren, R. Beard, and E. Atkins, "Information consensus in multivehicle cooperative control," *IEEE Control Systems Magazine*, vol. 27, no. 2, pp. 71–82, 2007.

[11] W. Ren, N. Sorensen, "Distributed coordination architecture for multi-robot formation control," *Robotics and Autonomous Systems*, vol. 56, no. 4, pp. 324–333, 2008.

[12] H. Wu, B. Zhou, and C. Zhang, "Secure distributed estimation against data integrity attacks in internet–of–things systems," *IEEE Transactions on Automation Science and Engineering*, vol. 19, no. 3, pp. 2552–2565, 2022.

[13] Z. Li, G. Wen, Z. Duan, and W. Ren, "Designing fully distributed consensus protocols for linear multi–agent systems with directed graphs," *IEEE Transactions on Automatic Control*, vol. 60, no. 4, pp. 1152–1157, 2015.

[14] M. Andreasson, D. V. Dimarogonas, H. Sandberg, and K. H. Johansson, "Distributed control of networked dynamical systems: Static feedback, integral action and consensus," *IEEE Transactions on Automatic Control*, vol. 59, no. 7, pp. 1750–1764, 2014.

[15] F. Pasqualetti, F. Dörfler, and F. Bullo, "Attack detection and identification in cyber–physical systems," *IEEE Transactions on Automatic Control*, vol. 58, no. 11, pp. 2715–2729, 2013.

[16] G. Wen, W. Yu, X. Yu, and J. Lv, "Complex cyber–physical networks: From cybersecurity to security control," *Journal of Systems Science and Complexity*, vol. 30, no. 1, pp. 46–67, 2017.

[17] Y. Shoukry and P. Tabuada, "Event–triggered state observers for sparse sensor noise/attacks," *IEEE Transactions on Automatic Control*, vol. 61, no. 8, pp. 2079–2091, 2016.

[18] J. Li and G.-H. Yang, "Disturbance decoupled secure state estimation: An orthogonal projection–based method," *Automatica*, vol. 147, p. 110740, 2023.

[19] M. Showkatbakhsh, Y. Shoukry, S. N. Diggavi, and P. Tabuada, "Securing state reconstruction under sensor and actuator attacks: Theory and design," *Automatica*, vol. 116, p. 108920, 2020.

[20] H. Fawzi, P. Tabuada, and S. Diggavi, "Secure estimation and control for cyber–physical systems under adversarial attacks," *IEEE Transactions on Automatic control*, vol. 59, no. 6, pp. 1454–1467, 2014.

[21] Y. H. Chang, Q. Hu, and C. J. Tomlin, "Secure estimation based kalman filter for cyber–physical systems against sensor attacks," *Automatica*, vol. 95, pp. 399–412, 2018.

[22] A.-Y. Lu and G.-H. Yang, "A polynomial–time algorithm for the secure state estimation problem under sparse sensor attacks via state decomposition technique," *IEEE Transactions on Automatic Control*, vol. 68, no. 12, pp. 7451–7465, 2023.

[23] A.-Y. Lu and G.-H. Yang, "Secure switched observers for cyber–physical systems under sparse sensor attacks: A set cover approach," *IEEE Transactions on Automatic Control*, vol. 64, no. 9, pp. 3949–3955, 2019.

[24] Y. Shoukry, P. Nuzzo, A. Puggelli, A. Sangiovannivincentelli, S. A. Seshia, and P. Tabuada, "Secure state estimation for cyber–physical systems under sensor attacks: A satisfiability modulo theory approach," *IEEE Transactions on Automatic Control*, vol. 62, no. 10, pp. 4917–4932, 2017.

[25] Z. Zhao, Y. Li, Y. Yang, L. Li, Y. Xu, and J. Zhou, "Sparse undetectable sensor attacks against cyber–physical systems: A subspace approach," *IEEE Transactions on Circuits and Systems II: Express Briefs*, vol. 67, no. 11, pp. 2517–2521, 2019.

[26] L. An and G.-H. Yang, "Fast state estimation under sensor attacks: A sensor categorization approach," *Automatica*, vol. 142, p. 110395, 2022.

[27] C. Liu, J. Wu, C. Long, and Y. Wang, "Dynamic state recovery for cyber–physical systems under switching location attacks," *IEEE Transactions on Control of Network Systems*, vol. 4, no. 1, pp. 14–22, 2017.

[28] M. Pajic, I. Lee, and G. J. Pappas, "Attack–resilient state estimation for noisy dynamical systems," *IEEE Transactions on Control of Network Systems*, vol. 4, no. 1, pp. 82–92, 2017.

[29] N. R. Chowdhury, J. Belikov, D. Baimel, and Y. Levron, "Observer–based detection and identification of sensor attacks in networked CPSs," *Automatica*, vol. 121, p. 109166, 2020.

[30] Y. Mao and P. Tabuada, "Decentralized secure state–tracking in multiagent systems," *IEEE Transactions on Automatic Control*, vol. 68, no. 7, pp. 4053–4064, 2023.

[31] D. Han, Y. Mo, and L. Xie, "Convex optimization based state estimation against sparse integrity attacks," *IEEE Transactions on Automatic Control*, vol. 64, no. 6, pp. 2383–2395, 2019.

[32] L. An and G.-H. Yang, "State estimation under sparse sensor attacks: A constrained set partitioning approach," *IEEE Transactions on Automatic Control*, vol. 64, no. 9, pp. 3861–3868, 2019.

[33] L. An and G.-H. Yang, "Distributed secure state estimation for cyber–physical systems under sensor attacks," *Automatica*, vol. 107, pp. 526–538, 2019.

[34] X. Luo, M. Pajic, and M. M. Zavlanos, "An optimal graph–search method for secure state estimation," *Automatica*, vol. 123, p. 109323, 2021.

[35] T. Shinohara, T. Namerikawa, and Z. Qu, "Resilient reinforcement in secure state estimation against sensor attacks with a priori information," *IEEE Transactions on Automatic Control*, vol. 64, no. 12, pp. 5024–5038, 2019.

[36] A.-Y. Lu and G.-H. Yang, "Detection and identification of sparse sensor attacks in cyber–physical systems with side information," *IEEE Transactions on Automatic Control*, vol. 68, no. 9, pp. 5349–5364, 2023.

[37] Y. Chen, S. Kar, and J. M. Moura, "Resilient distributed estimation: Sensor attacks," *IEEE Transactions on Automatic Control*, vol. 64, no. 9, pp. 3772–3779, 2019.

[38] L. An and G.-H. Yang, "Supervisory nonlinear state observers for adversarial sparse attacks," *IEEE Transactions on Cybernetics*, vol. 52, no. 3, pp. 1575–1587, 2022.

[39] Y. Chen, S. Kar, and J. M. Moura, "Attack resilient distributed estimation: A consensus + innovations approach," in *2018 Annual American Control Conference (ACC)*. IEEE, 2018, pp. 1015–1020.

[40] X. He, X. Ren, H. Sandberg, and K. H. Johansson, "How to secure distributed filters under sensor attacks," *IEEE Transactions on Automatic Control*, vol. 67, no. 6, pp. 2843–2856, 2022.

[41] X. He, E. Hashemi, and K. H. Johansson, "Distributed control under compromised measurements: Resilient estimation, attack detection, and vehicle platooning," *Automatica*, vol. 134, p. 109953, 2021.

[42] Y. Chen, S. Kar, and J. M. Moura, "Resilient distributed parameter estimation with heterogeneous data," *IEEE Transactions on Signal Processing*, vol. 67, no. 19, pp. 4918–4933, 2019.

[43] X. Lei, G. Wen, W. X. Zheng, and J. Fu, "Security strategy against location–varying sparse attack on distributed state monitoring," *IEEE Transactions on Automatic Control*, vol. 69, no. 4, pp. 2514–2521, 2024.

[44] S. Mishra, Y. Shoukry, N. Karamchandani, S. N. Diggavi, and P. Tabuada, "Secure state estimation against sensor attacks in the presence of noise," *IEEE Transactions on Control of Network Systems*, vol. 4, no. 1, pp. 49–59, 2017.

[45] X. Liu, Y. Mo, and E. Garone, "Local decomposition of kalman filters and its application for secure state estimation," *IEEE Transactions on Automatic Control*, vol. 66, no. 10, pp. 5037–5044, 2021.

[46] X. He, C. Hu, Y. Hong, L. Shi, and H.-T. Fang, "Distributed kalman filters with state equality constraints: Time–based and event–triggered communications," *IEEE Transactions on Automatic Control*, vol. 65, no. 1, pp. 28–43, 2020.

[47] D. Dolev, N. A. Lynch, S. S. Pinter, E. W. Stark, and W. E. Weihl, "Reaching approximate agreement in the presence of faults," *Journal of the ACM (JACM)*, vol. 33, no. 3, pp. 499–516, 1986.

[48] N. A. Lynch, *Distributed algorithms*. San Francisco, CA, USA: Morgan Kaufmann, 1996.

[49] H. J. Leblanc, H. Zhang, X. Koutsoukos, and S. Sundaram, "Resilient asymptotic consensus in robust networks," *IEEE Journal on Selected Areas in Communications*, vol. 31, no. 4, pp. 766–781, 2013.

[50] S. M. Dibaji and H. Ishii, "Consensus of second–order multi–agent systems in the presence of locally bounded faults," *System & Control Letters*, vol. 79, pp. 23–29, 2015.

[51] S. M. Dibaji and H. Ishii, "Resilient consensus of second–order agent networks: Asynchronous update rules with delays," *Automatica*, vol. 81, pp. 123–132, 2017.

[52] W. Abbas, A. Laszka, and X. Koutsoukos, "Improving network connectivity and robustness using trusted nodes with application to resilient consensus," *IEEE Transactions on Control of Network Systems*, vol. 5, no. 4, pp. 2036–2048, 2018.

[53] H. J. LeBlanc and X. Koutsoukos, "Resilient first–order consensus and weakly stable, higher order synchronization of continuous–time networked multiagent systems," *IEEE Transactions on Control of Network Systems*, vol. 5, no. 3, pp. 1219–1231, 2018.

[54] H. Rezaee, T. Parisini, and M. M. Polycarpou, "Almost sure resilient consensus under stochastic interaction: Links failure and noisy channels," *IEEE Transactions on Automatic Control*, vol. 66, no. 12, pp. 5727–5741, 2021.

[55] D. Saldana, A. Prorok, S. Sundaram, M. F. M. Campos, and V. Kumar, "Resilient consensus for time–varying networks of dynamic agents," in *2017 American Control Conference (ACC)*. IEEE, 2017, pp. 252–258.

[56] J. Huang, Y. Wu, L. Chang, X. He, and S. Li, "Resilient consensus with switching networks and double–integrator agents," in *7th Data Driven Control and Learning Systems Conference*. IEEE, 2018, pp. 802–807.

[57] J. Usevitch and D. Panagou, "Resilient leader–follower consensus to arbitrary reference values in time–varying graphs," *IEEE Transactions on Automatic Control*, vol. 65, no. 4, pp. 1755–1762, 2020.

[58] G. Wen, Y. Lv, W. X. Zheng, J. Zhou, and J. Fu, "Joint robustness of time–varying networks and its applications to resilient consensus," *IEEE Transactions on Automatic Control*, vol. 68, no. 11, pp. 6466–6480, 2023.

[59] W. Fu, J. Qin, Y. Shi, W. X. Zheng, and Y. Kang, "Resilient consensus of discrete–time complex cyber–physical networks under deception attacks," *IEEE Transactions on Industrial Informatics*, vol. 16, no. 7, pp. 4868–4877, 2019.

[60] W. Fu, J. Qin, W. X. Zheng, Y. Chen, and Y. Kang, "Resilient cooperative source seeking of double–integrator multi–robot systems under deception attacks," *IEEE Transactions on Industrial Electronics*, vol. 68, no. 5, pp. 4218–4227, 2020.

[61] N. H. Vaidya and V. K. Garg, "Byzantine vector consensus in complete graphs," in *Proceedings of the 2013 ACM Symposium on Principles of Distributed Computing*, 2013, pp. 65–73.

[62] N. H. Vaidya, "Iterative Byzantine vector consensus in incomplete graphs," in *Distributed Computing and Networking: 15th International Conference*, 2014, pp. 14–28.

[63] H. Mendes, M. Herlihy, N. Vaidya, and V. K. Garg, "Multidimensional agreement in Byzantine systems," *Distributed Computing*, vol. 28, no. 6, pp. 423–441, 2015.

[64] H. Park and S. A. Hutchinson, "Fault–tolerant rendezvous of multirobot systems," *IEEE Transactions on Robotics*, vol. 33, no. 3, pp. 565–582, 2017.

[65] M. Shabbir, J. Li, W. Abbas, and X. Koutsoukos, "Resilient vector consensus in multi–agent networks using centerpoints," in *2020 American Control Conference (ACC)*. IEEE, 2020, pp. 4387–4392.

[66] W. Abbas, M. Shabbir, J. Li, and X. Koutsoukos, "Resilient distributed vector consensus using centerpoint," *Automatica*, vol. 136, p.110046, 2022.

[67] F. Pasqualetti, F. Dörfler, and F. Bullo, "A divide–and–conquer approach to distributed attack identification," in *54th IEEE Conference on Decision and Control (CDC)*, 2015, pp. 5801–5807.

[68] D. Zhao, Y. Lv, X. Yu, G. Wen, and G. Chen, "Resilient consensus of higher order multiagent networks: An attack isolation–based approach," *IEEE Transactions on Automatic Control*, vol.67, no.2, pp. 1001–1007, 2022.

[69] D. Zhao, Y. Lv, J. Zhou, G. Wen, and T. Huang, "Attack–isolation–based resilient control of large–scale systems against collusive attacks," *IEEE Transactions on Network Science and Engineering*, vol. 9, no. 4, pp. 2857–2869, 2022.

[70] D. Zhao, Y. Lv, G. Wen, and Z. Gao, "Resilient consensus of high–order networks against collusive attacks," *Automatica*, vol. 151, p. 110934, 2023.

[71] D. Zhao, G. Wen, Z.-G. Wu, Y. Lv, and J. Zhou, "Resilient consensus of multiagent systems under collusive attacks on communication links," *IEEE Transactions on Cybernetics*, vol. 54, no. 4, pp. 2076–2085, 2024.

[72] C. R. Johnson and R. A. Horn, *Matrix analysis*. Cambridge, UK: Cambridge University Press, 1985.

[73] R. B. Bapat, *Linear algebra and linear models*. New York, NY, USA: Springer Science & Business Media, 2012.

[74] A. J. Laub, *Matrix analysis for scientists and engineers*. Philadelphia, PA: Society for Industrial and Applied Mathematics, 2005.

[75] S. Boyd and L. Vandenberghe, *Convex optimization*. Cambridge, UK: Cambridge University Press, 2004.

[76] Z. Sun and S. S. Ge, *Stability theory of switched dynamical systems*. London, UK: Springer, 2011.

[77] H. Zhang and S. Sundaram, "Robustness of information diffusion algorithms to locally bounded adversaries," in *2012 American Control Conference (ACC)*. IEEE, 2012, pp. 5855–5861.

[78] W. Ren and Y. Cao, *Distributed coordination of multi–agent networks: Emergent problems, models, and issues*. New York, NY, USA: Springer Science & Business Media, 2010.

[79] C. Godsil and G. F. Royle, *Algebraic graph theory*. New York, NY, USA: Springer Science & Business Media, 2001, vol. 207.

[80] G. Lixin, Z. Yan, and W. Liyong, "Consensus of discrete multi–agent systems under switching topologies," in *Proceedings of the 31st Chinese Control Conference*. IEEE, 2012, pp. 6135–6140.

[81] L. An and G.-H. Yang, "Byzantine–resilient distributed state estimation: A min–switching approach," *Automatica*, vol. 129, p. 109664, 2021.

[82] H. Rezaee, T. Parisini, and M. M. Polycarpou, "Resiliency in dynamic leader–follower multiagent systems," *Automatica*, vol. 125, p. 109384, 2021.

[83] A. Mitra and S. Sundaram, "Byzantine–resilient distributed observers for LTI systems," *Automatica*, vol. 108, p. 108487, 2019.

[84] S. Sundaram and B. Gharesifard, "Distributed optimization under adversarial nodes," *IEEE Transactions on Automatic Control*, vol. 64, no. 3, pp. 1063–1076, 2019.

[85] A. Mitra, F. Ghawash, S. Sundaram, and W. Abbas, "On the impacts of redundancy, diversity, and trust in resilient distributed state estimation," *IEEE Transactions on Control of Network Systems*, vol. 8, no. 2, pp. 713–724, 2021.

[86] L. An and G.-H. Yang, "Mean–square exponential convergence for Byzantine–resilient distributed state estimation," *Automatica*, vol. 163, p. 111592, 2024.

[87] Z. Zuo, Q.-L. Han, and B. Ning, *Fixed–time cooperative control of multi–agent systems*. Berlin, Germany: Springer, 2019.

[88] X. Liu, D. W. C. Ho, Q. Song, and W. Xu, "Finite/fixed–time pinning synchronization of complex networks with stochastic disturbances," *IEEE Transactions on Cybernetics*, vol. 49, no. 6, pp. 2398–2403, 2019.

[89] B. Ning, Q.-L. Han, Z. Zuo, L. Ding, Q. Lu, and X. Ge, "Fixed–time and prescribed–time consensus control of multiagent systems and its applications: A survey of recent trends and methodologies," *IEEE Transactions on Industrial Informatics*, vol. 19, no. 2, pp. 1121–1135, 2022.

[90] M. Goldberg, "Equivalence constants for l_p norms of matrices," *Linear Multilinear Algebra*, vol. 21, no. 2, pp. 173–179, 1987.

[91] S. Yu, X. Yu, B. Shirinzadeh, and Z. Man, "Continuous finite–time control for robotic manipulators with terminal sliding mode," *Automatica*, vol. 41, no. 11, pp. 1957–1964, 2005.

[92] A. Polyakov, "Nonlinear feedback design for fixed–time stabilization of linear control systems," *IEEE Transactions on Automatic Control*, vol. 57, no. 8, pp. 2106–2110, 2012.

[93] X. Lei, G. Wen, and G. Chen, "Resilient distributed parameter estimation for sensor networks against sparse–varying attacks," *IEEE Transactions on Systems, Man, and Cybernetics: Systems*, vol. 54, no. 12, pp. 7331–7340, 2024.

[94] Y. Mo and B. Sinopoli, "On the performance degradation of cyber–physical systems under stealthy integrity attacks," *IEEE Transactions on Automatic Control*, vol. 61, no. 9, pp. 2618–2624, 2016.

[95] A. Chattopadhyay and U. Mitra, "Security against false data–injection attack in cyber–physical systems," *IEEE Transactions on Control of Network Systems*, vol. 7, no. 2, pp. 1015–1027, 2020.

[96] Z. Zhao, Y. Huang, Z. Zhen, and Y. Li, "Data–driven false data–injection attack design and detection in cyber–physical systems," *IEEE Transactions on Cybernetics*, vol. 51, no. 12, pp. 6179–6187, 2021.

[97] A. Kanellopoulos and K. G. Vamvoudakis, "A moving target defense control framework for cyber–physical systems," *IEEE Transactions on Automatic Control*, vol. 65, no. 3, pp. 1029–1043, 2020.

[98] Y. Cui, Y. Liu, W. Zhang, and F. E. Alsaadi, "Sampled–based consensus for nonlinear multiagent systems with deception attacks: The decoupled method," *IEEE Transactions on Systems, Man, and Cybernetics: Systems*, vol. 51, no. 1, pp. 561–573, 2021.

[99] J. Qin, M. Li, L. Shi, and X. Yu, "Optimal denial–of–service attack scheduling with energy constraint over packet–dropping networks," *IEEE Transactions on Automatic Control*, vol. 63, no. 6, pp. 1648–1663, 2018.

[100] A.-Y. Lu and G.-H. Yang, "Malicious adversaries against secure state estimation: Sparse sensor attack design," *Automatica*, vol. 136, p. 110037, 2022.

[101] J. Zhang, X. Wang, R. S. Blum, and L. Kaplan, "Attack detection in sensor network target localization systems with quantized data," *IEEE Transactions on Signal Processing*, vol. 66, no. 8, pp. 2070–2085, 2018.

[102] Y. Joo, Z. Qu, and T. Namerikawa, "Resilient control of cyber–physical system using nonlinear encoding signal against system integrity attacks," *IEEE Transactions on Automatic Control*, vol. 66, no. 9, pp. 4334–4341, 2021.

[103] A. Liu, Y. Ren, L. Yao, B. Niu, and P. Zhao, "Secure state estimation for cyber–physical systems by unknown input observer with adaptive switching mechanism," *Information Sciences*, vol. 647, p. 119452, 2023.

[104] C. Lee, H. Shim, and Y. Eun, "Secure and robust state estimation under sensor attacks, measurement noises, and process disturbances: Observer–based combinatorial approach," in *2015 European Control Conference (ECC)*. IEEE, 2015, pp. 1872–1877.

[105] C. Lee, H. Shim, and Y. Eun, "On redundant observability: From security index to attack detection and resilient state estimation," *IEEE Transactions on Automatic Control*, vol. 64, no. 2, pp. 775–782, 2019.

[106] Y. Nakahira and Y. Mo, "Attack–resilient $\mathcal{H}_2$, $\mathcal{H}_\infty$, and ℓ_1 state estimator," *IEEE Transactions on Automatic Control*, vol. 63, no. 12, pp. 4353–4360, 2018.

[107] H. Bai, R. A. Freeman, and K. M. Lynch, "Robust dynamic average consensus of time–varying inputs," in *49th IEEE Conference on Decision and Control (CDC)*. 2010, pp. 3104–3109.

[108] F. Pasqualetti, F. Dorfler, and F. Bullo, "Control–theoretic methods for cyber–physical security: Geometric principles for optimal cross–layer resilient control systems," *IEEE Control Systems Magazine*, vol. 35, no. 1, pp. 110–127, 2015.

[109] L. An and G.-H. Yang, "Secure state estimation against sparse sensor attacks with adaptive switching mechanism," *IEEE Transactions on Automatic Control*, vol. 63, no. 8, pp. 2596–2603, 2018.

[110] G. Wu, J. Sun, and J. Chen, "Optimal data injection attacks in cyber–physical systems," *IEEE Transactions on Systems, Man, and Cybernetics*, vol. 48, no. 12, pp. 3302–3312, 2018.

[111] W. Yang, W. Luo, and X. Zhang, "Distributed secure state estimation under stochastic linear attacks," *IEEE Transactions on Network Science and Engineering*, vol. 8, no. 3, pp. 2036–2047, 2021.

[112] G. Wen, W. Yu, Y. Lv, and P. Wang, *Cooperative control of complex network systems with dynamic topologies*. Boca Raton, FL, USA: CRC Press, 2021.

[113] A. Alarifi and W. Du, "Diversify sensor nodes to improve resilience against node compromise," in *Proceedings of the Fourth ACM Workshop on Security of ad hoc and Sensor Networks*, 2006, pp. 101–112.

[114] A. Newell, D. Obenshain, T. Tantillo, C. Nita-Rotaru, and Y. Amir, "Increasing network resiliency by optimally assigning diverse variants to routing nodes," *IEEE Transactions on Dependable and Secure Computing*, vol. 12, no. 6, pp. 602–614, 2015.

[115] F. Ghawash and W. Abbas, "Leveraging diversity for achieving resilient consensus in sparse networks," *IFAC-PapersOnLine*, vol. 52, no. 20, pp. 339–344, 2019.

[116] Z. Kazemi, A. A. Safavi, M. M. Arefi, and F. Naseri, "Finite–time secure dynamic state estimation for cyber–physical systems under unknown inputs and sensor attacks," *IEEE Transactions on Systems, Man, and Cybernetics: Systems*, vol. 52, no. 8, pp. 4950–4959, 2022.

[117] H. Sedjelmaci, N. Kaaniche, A. Boudguiga, and N. Ansari, "Secure attack detection framework for hierarchical 6G-enabled internet of vehicles," *IEEE Transactions on Vehicular Technology*, vol. 73, no. 2, pp. 2633–2642, 2024.

[118] Y. Li, C. Tang, K. Li, X. He, S. Peeta, and Y. Wang, "Consensus–based cooperative control for multi–platoon under the connected vehicles environment," *IEEE Transactions on Intelligent Transportation Systems*, vol. 20, no. 6, pp. 2220–2229, 2019.

[119] G. Wen, X. Yu, Z.-W. Liu, and W. Yu, "Adaptive consensus–based robust strategy for economic dispatch of smart grids subject to communication uncertainties," *IEEE Transactions on Industrial Informatics*, vol. 14, no. 6, pp. 2484–2496, 2018.

[120] H. Yang and D. Ye, "Observer–based fixed–time secure tracking consensus for networked high–order multiagent systems against DoS attacks," *IEEE Transactions on Cybernetics*, vol. 52, no. 4, pp. 2018–2031, 2022.

[121] A. Teixeira, I. Shames, H. Sandberg, and K. H. Johansson, "A secure control framework for resource–limited adversaries," *Automatica*, vol. 51, pp. 135–148, 2015.

[122] Y. Lv, J. Zhou, G. Wen, X. Yu, and T. Huang, "Fully distributed adaptive NN-based consensus protocol for nonlinear MASs: An attack–free approach," *IEEE Transactions on Neural Networks and Learning Systems*, vol. 33, no. 4, pp. 1561–1570, 2022.

[123] Y. Yang, Y. Li, D. Yue, Y.-C. Tian, and X. Ding, "Distributed secure consensus control with event–triggering for multiagent systems under DoS attacks," *IEEE Transactions on Cybernetics*, vol. 51, no. 6, pp. 2916–2928, 2021.

[124] M. Ruan, H. Gao, and Y. Wang, "Secure and privacy–preserving consensus," *IEEE Transactions on Automatic Control*, vol. 64, no. 10, pp. 4035–4049, 2019.

[125] J.-J. Yan and G.-H. Yang, "Secure state estimation of nonlinear cyber–physical systems against DoS attacks: A multiobserver approach," *IEEE Transactions on Cybernetics*, vol. 53, no. 3, pp. 1447–1459, 2023.

[126] S. Murugesan and Y.-C. Liu, "Resilient finite–time distributed event–triggered consensus of multi–agent systems with multiple cyber–attacks," *Communications in Nonlinear Science and Numerical Simulation*, vol. 116, p. 106876, 2023.

[127] H. Wang, W. Yu, G. Wen, and G. Chen, "Fixed–time consensus of nonlinear multi–agent systems with general directed topologies," *IEEE Transactions on Circuits and Systems II: Express Briefs*, vol. 66, no. 9, pp. 1587–1591, 2019.

[128] H. Hong, W. Yu, G. Wen, and X. Yu, "Distributed robust fixed–time consensus for nonlinear and disturbed multi–agent systems," *IEEE Transactions on Systems, Man, and Cybernetics: Systems*, vol. 47, no. 7, pp. 1464–1473, 2017.

[129] Y. Jiang, B. Niu, X. Wang, X. Zhao, H. Wang, and B. Yan, "Distributed finite–time consensus tracking control for nonlinear multi–agent systems with FDI attacks and application to single–link robots," *IEEE Transactions on Circuits and Systems II: Express Briefs*, vol. 70, no. 4, pp. 1505–1509, 2023.

[130] Z. Zuo, J. Song, B. Tian, and M. Basin, "Robust fixed–time stabilization control of generic linear systems with mismatched disturbances," *IEEE Transactions on Systems, Man, and Cybernetics: Systems*, vol. 52, no. 2, pp. 759–768, 2022.

[131] H. Zhang, J. Duan, Y. Wang, and Z. Gao, "Bipartite fixed–time output consensus of heterogeneous linear multiagent systems," *IEEE Transactions on Cybernetics*, vol. 51, no. 2, pp. 548–557, 2021.

[132] X. Shi, L. Xu, T. Yang, Z. Lin, and X. Wang, "Distributed fixed–time resource allocation algorithm for the general linear multi–agent systems," *IEEE Transactions on Circuits and Systems II: Express Briefs*, vol. 69, no. 6, pp. 2867–2871, 2022.

[133] Q. Cui, K. Liu, Z. Ji, and M. Zhao, "Finite–time and fixed–time adaptive consensus of multi–agent systems with general linear dynamics," *Mathematical Methods in the Applied Sciences*, vol. 46, no. 18, pp. 18 560–18 578, 2023.

[134] S. Gao, Z. Peng, L. Liu, D. Wang, and Q.-L. Han, "Fixed–time resilient edge–triggered estimation and control of surface vehicles for cooperative target tracking under attacks," *IEEE Transactions on Intelligent Vehicles*, vol. 8, no. 1, pp. 547–556, 2023.

[135] L. Wang and A. S. Morse, "A distributed observer for a time–invariant linear system," *IEEE Transactions on Automatic Control*, vol. 63, no. 7, pp. 2123–2130, 2017.

[136] A. Mitra and S. Sundaram, "Distributed observers for LTI systems," *IEEE Transactions on Automatic Control*, vol. 63, no. 11, pp. 3689–3704, 2018.

[137] G. Wen and W. X. Zheng, "On constructing multiple Lyapunov functions for tracking control of multiple agents with switching topologies," *IEEE Transactions on Automatic Control*, vol. 64, no. 9, pp. 3796–3803, 2019.

[138] J. G. Lee, J. Kim, and H. Shim, "Fully distributed resilient state estimation based on distributed median solver," *IEEE Transactions on Automatic Control*, vol. 65, no. 9, pp. 3935–3942, 2020.

[139] X. Ren, Y. Mo, J. Chen, and K. H. Johansson, "Secure state estimation with Byzantine sensors: A probabilistic approach," *IEEE Transactions on Automatic Control*, vol. 65, no. 9, pp. 3742–3757, 2020.

[140] Y. Jeong and Y. Eun, "A robust and resilient state estimation for linear systems," *IEEE Transactions on Automatic Control*, vol. 67, no. 5, pp. 2626–2632, 2022.

[141] F. Pasqualetti, A. Bicchi, and F. Bullo, "Consensus computation in unreliable networks: A system theoretic approach," *IEEE Transactions on Automatic Control*, vol. 57, no. 1, pp. 90–104, 2011.

[142] L. Su and S. Shahrampour, "Finite–time guarantees for Byzantine–resilient distributed state estimation with noisy measurements," *IEEE Transactions on Automatic Control*, vol. 65, no. 9, pp. 3758–3771, 2020.

[143] Y. Cui, Y. Jia, Y. Li, J. Shen, T. Huang, and X. Gong, "Byzantine resilient joint localization and target tracking of multi–vehicle systems," *IEEE Transactions on Intelligent Vehicles*, vol. 8, no. 4, pp. 2899–2913, 2023.

[144] T. Kim, C. Lee, and H. Shim, "Completely decentralized design of distributed observer for linear systems," *IEEE Transactions on Automatic Control*, vol. 65, no. 11, pp. 4664–4678, 2020.

[145] Z. Sun, A. Rantzer, Z. Li, and A. Robertsson, "Distributed adaptive stabilization," *Automatica*, vol. 129, p. 109616, 2021.

[146] S. Du, H. Sheng, D. W. C. Ho, and J. Qiao, "Secure consensus of multiagent systems with DoS attacks via fully distributed dynamic event–triggered control," *IEEE Transactions on Systems, Man, and Cybernetics: Systems*, vol. 53, no. 10, pp. 6588–6597, 2023.

[147] Y. Pan, Y. Wu, and H. K. Lam, "Security–based fuzzy control for nonlinear networked control systems with DoS attacks via a resilient event–triggered scheme," *IEEE Transactions on Fuzzy Systems*, vol. 30, no. 10, pp. 4359–4368, 2022.

[148] C. Deng and C. Wen, "MAS–based distributed resilient control for a class of cyber–physical systems with communication delays under DoS attacks," *IEEE Transactions on Cybernetics*, vol. 51, no. 5, pp. 2347–2358, 2021.

[149] Z. Zuo, X. Cao, Y. Wang, and W. Zhang, "Resilient consensus of multiagent systems against denial–of–service attacks," *IEEE Transactions on Systems, Man, and Cybernetics: Systems*, vol. 52, no. 4, pp. 2664–2675, 2021.

[150] Y. Xu, M. Fang, Z.-G. Wu, Y.-J. Pan, M. Chadli, and T. Huang, "Input–based event–triggering consensus of multiagent systems under denial–of–service attacks," *IEEE Transactions on Systems, Man, and Cybernetics: Systems*, vol. 50, no. 4, pp. 1455–1464, 2020.

[151] C. Deng, X.-Z. Jin, Z.-G. Wu, and W.-W. Che, "Data–driven–based cooperative resilient learning method for nonlinear MASs under DoS attacks," *IEEE Transactions on Neural Networks and Learning Systems*, vol. 35, no. 9, pp. 12107–12116, 2024.

[152] G. Antonelli, "Interconnected dynamic systems: An overview on distributed control," *IEEE Control Systems Magazine*, vol. 33, no. 1, pp. 76–88, 2013.

[153] W. Ren and R. W.Beard, "Consensus seeking in multiagent systems under dynamically changing interaction topologies," *IEEE Transactions on Automatic Control*, vol. 50, no. 5, pp. 655–661, 2005.

[154] D. Ding, Q.-L. Han, Y. Xiang, X. Ge, and X. Zhang, "A survey on security control and attack detection for industrial cyber–physical systems," *Neurocomputing*, vol. 275, pp. 1674–1683, 2018.

[155] A. Barboni, H. Rezaee, F. Boem, and T Parisini, "Detection of covert cyber–attacks in interconnected systems: A distributed model–based approach," *IEEE Transactions on Automatic Control*, vol. 65, no. 9, pp. 3728–3741, 2020.

[156] R. M. Kieckhafer and M. H. Azadmanesh, "Low cost approximate agreement in partially connected networks," *Journal of Computing and Information Science in Engineering*, vol. 3, no. 1, pp. 53–85, 1993.

[157] R. M. Kieckhafer and M. H. Azadmanesh, "Reaching approximate agreement with mixed–mode faults," *IEEE Transactions on Parallel and Distributed Systems*, vol. 5, no. 1, pp. 53–63, 1994.

[158] H. Zhang, E. Fata, and S. Sundaram, "A notion of robustness in complex networks," *IEEE Transactions on Control of Network Systems*, vol. 2, no. 3, pp. 310–320, 2015.

[159] Y. Shang, "Resilient consensus of switched multi–agent systems," *System & Control Letters*, vol. 122, pp. 12–18, 2018.

[160] S. M. Dibaji, H. Ishii, and R. Tempo, "Resilient randomized quantized consensus," *IEEE Transactions on Automatic Control*, vol. 63, no. 8, pp. 2508–2522, 2018.

[161] G. Wen, X. Yu, W. Yu, and J. Lv, "Coordination and control of complex network systems with switching topologies: A survey," *IEEE Transactions on Systems, Man, and Cybernetics: Systems*, vol. 51, no. 10, pp. 6342–6357, 2021.

[162] Y. Shang, "Median–based resilient consensus over time–varying random networks," *IEEE Transactions on Circuits and Systems II: Express Briefs*, vol. 60, no. 3, pp. 1203–1207, 2022.

[163] Y. Shang, "Resilient tracking consensus over dynamic random graphs: A linear system approach," *European Journal of Applied Mathematics*, vol. 34, no. 2, pp. 408–423, 2023.

[164] P. Soberón, *"The Pigeonhole Principle" in problem–solving methods in combinatorics: An approach to Olympiad problems*, Basel, Switzerland: Springer, pp. 17–26, 2013.

[165] R. Engel and G. Kreisselmeier, "A continuous–time observer which converges in finite time," *IEEE Transactions on Automatic Control*, vol. 47, no. 7, pp. 1202–1204, 2002.

[166] Y. Lv, G. Wen, and T. Huang, "Adaptive protocol design for distributed tracking with relative output information: A distributed fixed–time observer approach," *IEEE Transactions on Control of Network Systems*, vol. 7, no. 1, pp. 118–128, 2019.

[167] A. Barboni, H. Rezaee, F. Boem, and P. Thomas, "Distributed detection of covert attacks for interconnected systems," in *2019 18th European Control Conference (ECC)*. IEEE, 2019, pp. 2240–2245.

[168] A. Sargolzaei, K. Yazdani, A. Abbaspour, C. D. Crane, and W. E. Dixon, "Detection and mitigation of false data injection attacks in networked control systems," *IEEE Transactions on Industrial Informatics*, vol. 16, no. 6, pp. 4281–4292, 2020.

[169] J. Lan and R. J. Patton, "A new strategy for integration of fault estimation within fault–tolerant control," *Automatica*, vol. 69, pp. 48–59, 2016.

[170] W. T. Elsayed and E. F. El-Saadany, "A fully decentralized approach for solving the economic dispatch problem," *IEEE Transactions on Power Systems*, vol. 30, no. 4, pp. 2179–2189, 2015.

[171] A. Lubotzky, "Expander graphs in pure and applied mathematics," *Bulletin of the American Mathematical Society*, vol. 49, no. 1, pp. 113–162, 2012.

[172] T. Sung, T. Ho, C. Chang, and H. Hsu, "Optimal k–fault–tolerant networks for token rings," *Journal of Information Science and Engineering*, vol. 16, no. 3, pp. 381–390, 2000.

[173] Y. Chen, M. T. Fatehi, H. J. L. Roche, J. Z. Larsen and B. L. Nelson, "Metro optical networking," *Bell Labs Technical Journal*, vol. 4, no. 1, pp. 163–186, 1999.

[174] C. Zhao, J. He, and J. Chen, "Resilient consensus with mobile detectors against malicious attacks," *IEEE Transactions on Signal and Information Processing over Networks*, vol. 4, no. 1, pp. 60–69, 2018.

[175] L. Yuan and H. Ishii, "Resilient consensus with distributed fault detection," in *IFAC-PapersOnLine*, vol. 52, no. 20, pp. 285–290, 2019.

[176] Z. Li, W. Ren, X. Liu, and M. Fu, "Consensus of multi–agent systems with general linear and Lipschitz nonlinear dynamics using distributed adaptive protocols," *IEEE Transactions on Automatic Control*, vol. 58, no. 7, pp. 1786–1791, 2013.

[177] D. D. Šiljak, *Large–scale dynamic systems: stability and structure*. New York: North Holland, 1978.

[178] I. Shames, A. M. Teixeira, H. Sandberg, and K. H. Johansson, Distributed fault detection for interconnected second–order systems, *Automatica*, vol. 47, no. 12, pp. 2757–2764, 2011.

[179] M. Davoodi, N. Meskin, and K. Khorasani, "Simultaneous fault detection and consensus control design for a network of multi–agent systems," *Automatica*, vol. 66, pp. 185–194, 2016.

[180] M. Khalili, X. Zhang, M. M. Polycarpou, T. Parisini, and Y. Cao, "Distributed adaptive fault–tolerant control of uncertain multi–agent systems," *Automatica*, vol. 87, pp. 142–151, 2018.

[181] Y. Lv, G. Wen, T. Huang, and Z. Duan, "Adaptive attack–free protocol for consensus tracking with pure relative output information," *Automatica*, vol. 117, artno. 108998, 2020.

[182] L. Xie, Y. Mo, and B. Sinopoli, "False data injection attacks in electricity markets," in *2010 First IEEE International Conference on Smart Grid Communications*, 2010, pp. 226–231.

[183] H. Fawzi, P. Tabuada, and S. Diggavi, "Secure state–estimation for dynamical systems under active adversaries," in *49th Annual Allerton Conference on Communication, Control, and Computing (allerton)*. IEEE, 2011, pp. 337–344.

[184] A. Hoehn and P. Zhang, "Detection of replay attacks in cyber–physical systems," in *2016 American Control Conference (ACC)*. IEEE, 2016, pp. 290–295.

[185] R. S. Smith, "Covert misappropriation of networked control systems: Presenting a feedback structure," *IEEE Control Systems Magazine*, vol. 35, no. 1, pp. 82–92, 2015.

[186] P. Griffioen, S. Weerakkody, and B. Sinopoli, "A moving target defense for securing cyber–physical systems," *IEEE Transactions on Automatic Control*, vol. 66, no. 5, pp. 2016–2031, 2021.

[187] A. Höhn and P. Zhang, "Detection of covert attacks and zero dynamics attacks in cyber–physical systems," in *2016 American Control Conference (ACC)*. IEEE, 2016, pp. 302–307.

[188] M. Ghaderi, K. Gheitasi, and W. Lucia, "A novel control architecture for the detection of false data injection attacks in networked control systems," in *2019 American Control Conference. (ACC)*. IEEE, 2019, pp. 139–144.

[189] Y. Mo and B. Sinopoli, "Secure control against replay attacks," in *2009 47th Annual Allerton Conference on Communication, Control, and Computing (Allerton)*. IEEE, 2009, pp. 911–918.

[190] F. Boem, A. J. Gallo, D. M. Raimondo, and T. Parisini, "Distributed fault–tolerant control of large–scale systems: An active fault diagnosis approach," *IEEE Transactions on Control of Network Systems*, vol. 7, no. 1, pp. 288–301, 2020.

[191] Z. Li, M. Shahidehpour, A. Alabdulwahab, and A. Abusorrah, "Analyzing locally coordinated cyber–physical attacks for undetectable line outages," *IEEE Transactions on Smart Grid*, vol. 9, no. 1, pp. 35–47, 2018.

[192] A. Itai and M. Rodeh, "Finding a minimum circuit in a graph," in *Proceedings of the Ninth Annual ACM Symposium on Theory of Computing*, 1977, pp. 1–10.

[193] B. M. Nejad, S. A. Attia, and J. Raisch, "Max–consensus in a max–plus algebraic setting: The case of fixed communication topologies," in *2009 XXII International Symposium on Information, Communication and Automation Technologies*, 2009, pp. 1–7.

[194] Z. Bouzid, M. G. Potop-Butucaru, and S. Tixeuil, "Optimal Byzantine–resilient convergence in unidimensional robot networks," *Theoretical Computer Science*, vol. 411, nos. 34–36, pp. 3154–3168, 2010.

[195] J. Yan, X. Li, Y. Mo, and C. Wen, "Resilient multi–dimensional consensus in adversarial environment.," *Automatica*, vol. 145, artno. 110530, 2022.

[196] S. Kar and J. M. F. Moura, "Distributed consensus algorithms in sensor networks with imperfect communication: Link failures and channel noise," *IEEE Transactions on Signal Processing*, vol. 57, no. 1, pp. 355–369, 2008.

[197] Y. W. Wang, J. W. Xiao, and H. O. Wang, "Global synchronization of complex dynamical networks with network failures," *International Journal of Robust and Nonlinear Control*, vol. 20, no. 15, pp. 1667–1677, 2010.

[198] Y. W. Wang, H. O. Wang, J. W. Xiao, and Z. H. Guan, "Synchronization of complex dynamical networks under recoverable attacks," *Automatica*, vol. 46, no. 1, pp. 197–203, 2010.

[199] Z. Feng, G. Hu, and G. Wen, "Distributed consensus tracking for multi–agent systems under two types of attacks," *International Journal of Robust and Nonlinear Control*, vol. 26, pp. 896–918, 2016.

[200] L. Ma, Z. Wang, and Y. Yuan, "Consensus control for nonlinear multi–agent systems subject to deception attacks," in *2016 22nd International Conference on Automation and Computing (ICAC)*, 2016, pp. 21–26.

[201] C. De Persis and P. Tesi, "Input–to–state stabilizing control under denial–of–service," *IEEE Transactions on Automatic Control*, vol. 60, no. 11, pp. 2930–2944, 2015.

[202] A. Y. Lu and G. H. Yang, "Distributed consensus control for multi–agent systems under denial–of–service," *Information Sciences*, vol. 439, pp. 95–107, 2018.

[203] A. J. Gallo, M. S. Turan, P. Nahata, F. Boem, T. Parisini, and G. Ferrari-Trecate, "Distributed cyber–attack detection in the secondary control of DC microgrids," in *2018 European Control Conference (ECC)*. IEEE, 2018, pp. 344–349.

[204] A. J. Gallo, M. S. Turan, P. Nahata, F. Boem, T. Parisini, and G. Ferrari-Trecate, "A distributed cyber–attack detection scheme with application to DC microgrids," *IEEE Transactions on Automatic Control*, vol. 65, no. 9, pp. 3800–3815, 2020.

[205] G. Wen, Y. Lv, J. Zhou, and J. Fu. "Sufficient and necessary condition for resilient consensus under time–varying topologies," in *7th International Conference on Informative and Cybernetics for Computational Social Systems*, 2020, pp. 84–89.

[206] M. Taheri, K. Khorasani, I. Shames, and N. Meskin, "Undetectable cyber attacks on communication links in multi–agent cyber–physical systems," in *59th IEEE Conference on Decision and Control (CDC)*, 2020, pp. 3764–3771.

Index

For Product Safety Concerns and Information please contact our EU
representative GPSR@taylorandfrancis.com
Taylor & Francis Verlag GmbH, Kaufingerstraße 24, 80331 München, Germany

www.ingramcontent.com/pod-product-compliance
Lightning Source LLC
Chambersburg PA
CBHW082004190326
41458CB00010B/3068